THE ART OF THINGS

THE ART OF THINGS

PRODUCT DESIGN SINCE 1945

EDITED BY
DOMINIQUE FOREST

WITH
JEREMY AYNSLEY, ÁSDÍS ÓLAFSDÓTTIR
ANTY PANSERA, CONSTANCE RUBINI
PENNY SPARKE, AND MIENKE SIMON THOMAS

ABBEVILLE PRESS PUBLISHERS
New York London

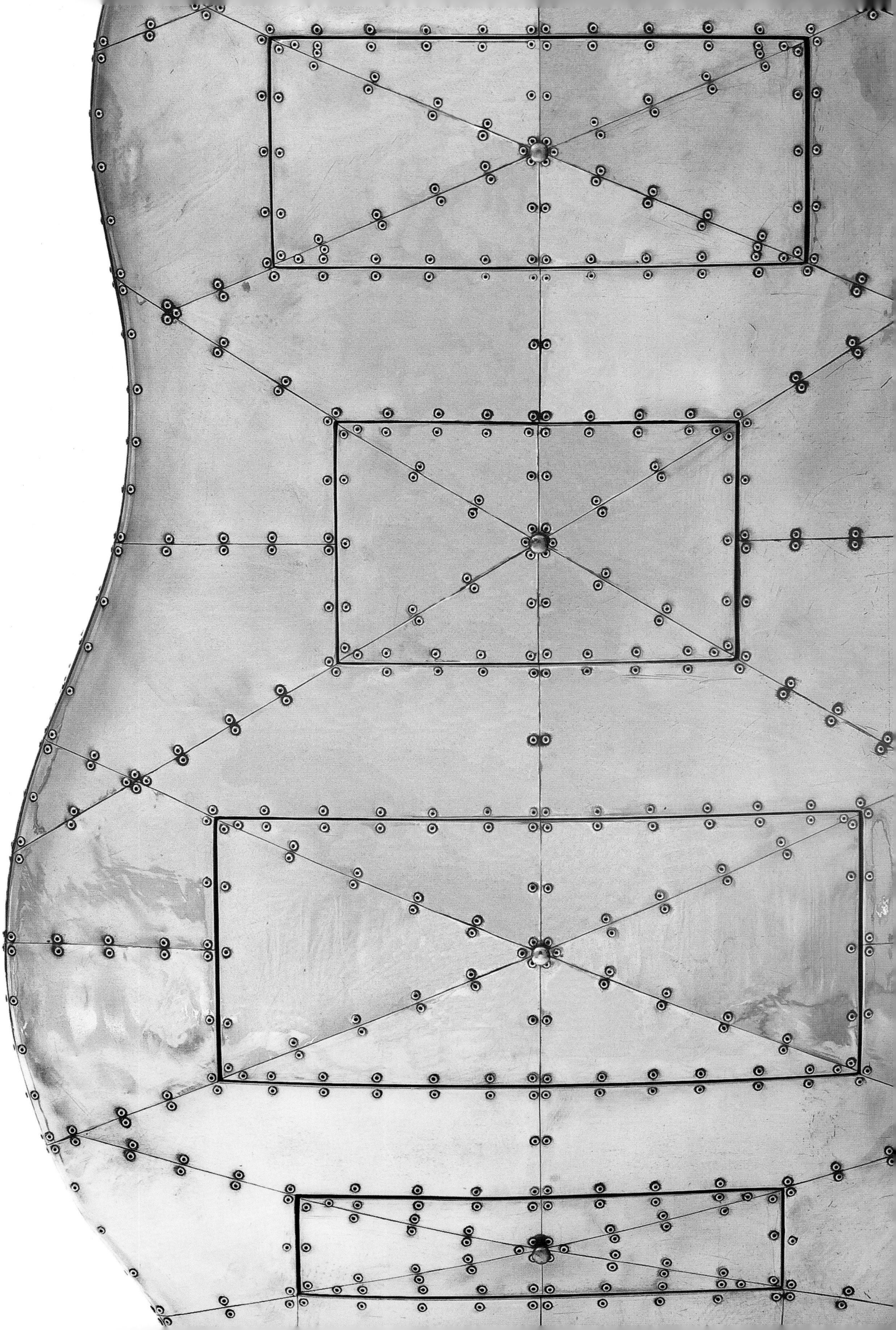

PAGE 2
1. **Charles and Ray Eames in their house in Pacific Palisades, Los Angeles** (detail), 1958
Photograph by Julius Schulman
See plate 93.

PAGE 4
2. Isamu Noguchi (1904–1988)
Akari 1AG **lantern** (detail), 1960
See plate 608.

PAGE 5
3. Marc Newson (b. 1963)
Pod of Drawers (detail),
designed 1987, made 1999
See plate 475.

PAGE 6
4. Mario Bellini (b. 1935)
Teneride **armchair** (detail), 1970
Made by Cassina (Italy)
See plate 398.

PAGE 7
5. Kaare Klint (1888–1954)
Klint 101 **pendant lamp**
(detail), 1943
Made by Le Klint (Denmark)
See plate 127.

PAGE 8
6. Jonathan De Pas (1932–1991),
Donato D'Urbino (b. 1935),
and Paolo Lomazzi (b. 1936)
Blow **armchair** (detail), 1967
Made by Zanotta (Italy)
See plate 359.

PAGE 9
7. Olivier Mourgue (b. 1939)
Djinn **chair** (detail), 1964–65
Made by Airborne (France)
See plate 546.

PAGE 10
8. Timo Sarpaneva (1926–2006)
Lancette **vase** (detail), 1952
Made by Iittala (Finland)
See plate 172.

PAGE 11
9. Roberto Pezzetta (b. 1946)
OZ **refrigerator** (detail), 1998
Made by Zanussi (Italy)
See plate 413.

PAGE 12
10. Harry Bertoia (1915–1978)
Diamond **chair** (detail), 1952
Made by Knoll International
(U.S.)
See plate 98.

PAGE 13
11. Cini Boeri (b. 1924) and
Tomu Katayanagi (b. 1950)
Ghost **chair** (detail), 1987
Made by Fiam (Italy)
See plate 336.

PAGE 14
12. Verner Panton (1926–1998)
Panton **chair** (detail), 1967
Made by Herman Miller (U.S.)
See plate 152.

PAGE 15
13. Cédric Ragot (b. 1973)
Hyper Fast **vases** (detail), 2003
Made by Ymer&Malta (France)
See plate 566.

PAGE 16
14. Wilhelm Wagenfeld
(1900–1990)
Kubus **stackable glass containers** (detail), 1938
Made by Vereinigte Lausitzer
Glaswerke (Germany)
See plate 242.

PAGE 17
15. Harley Earl (1893–1969)
Cadillac *Eldorado*
(detail of tail fins), 1959
Made by General Motors (U.S.)
See plate 85.

CONTENTS

INTRODUCTION

DOMINIQUE FOREST

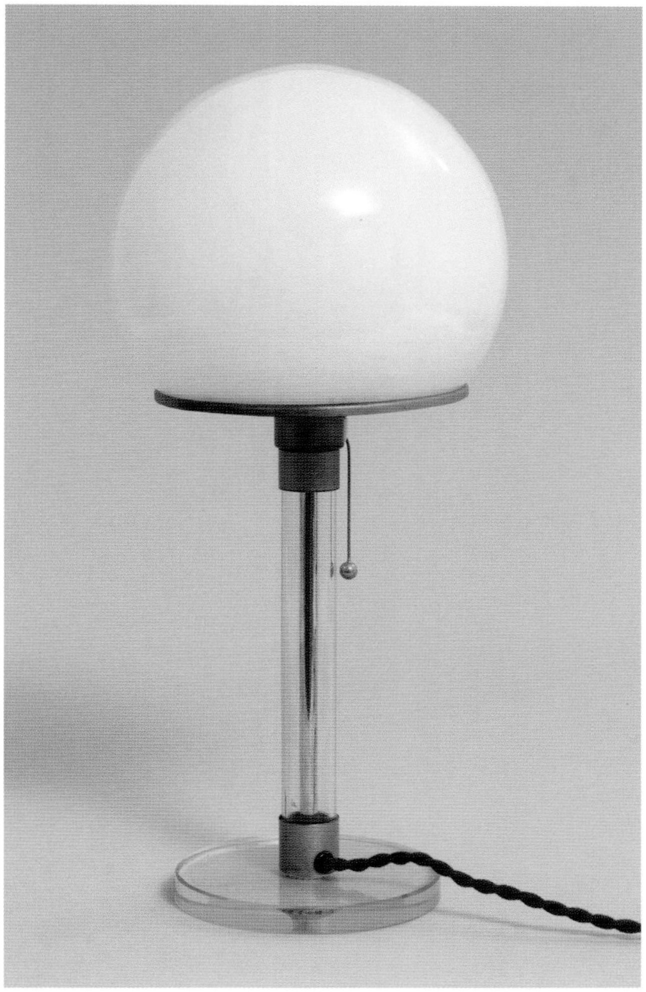

16. Wilhelm Wagenfeld (1900–1990) and Jakob Jucker (1902–1997)
Table lamp, 1923–24
Made in the metal workshops of the Weimar Bauhaus
Chrome-plated metal and glass
Museum of Modern Art, New York

The first use of the word "design" to mean a plan or pattern for a manufactured object appears in nineteenth-century England, in the writings of Henry Cole, the organizer of the Great Exhibition in London in 1851—the original World's Fair. Two years earlier, Cole had launched the *Journal of Design and Manufactures* in collaboration with Richard Redgrave.

The definition of "design" has evolved over time and now embraces a broad range of associations. It was in the United States, in the years before World War II, that it came to describe a creative discipline whose purpose was to shape products so as to enhance their appearance, and possibly their performance. A professional in this field was now referred to as a "designer." Not until the 1950s, though, did the word "design" take its place beside the term "industrial aesthetics" in continental Europe. Soon the expressions became interchangeable, and "design" ultimately triumphed, becoming commonplace in the parlance of architectural and home furnishing circles. However, many people still did not understand exactly what the word encompassed, and additional clarification was needed. The Centre de Création Industrielle (CCI), established in 1969 by the Musée des Arts Décoratifs in Paris, was thus inaugurated with an exhibition titled *Qu'est-ce que le design?* (What is design?). Charles Eames, Joe Colombo, and Roger Tallon, all legendary figures, were invited to give their own definitions and provide illustrative examples.

The history of design is inseparable from the vast changes that swept through the Western world between 1850 and 1950, particularly the extraordinary acceleration of the production of manufactured goods. Remarkable efficiencies arose from mechanization and the mastery of new materials. In the 1860s, the English Arts and Crafts movement started a debate about the respective qualities and production conditions of artisanal crafts and industrial goods. This controversy was taken up again in Germany in the early twentieth century, and figured into the new ideas and experiments that, in the 1920s, were synthesized in that laboratory of forms for the modern age, the Bauhaus (plate 16). From then on, a powerful tide of modernism would progressively eradicate the forms and styles of the past, ultimately severing ties with tradition. Because modernism established itself on the rejection of ornament, one strand of the history of design concerns the movements back and forth with respect to one enduring question: whether form and function should have primacy, or decoration.

The first study of the conditions of modernity that led to the emergence of design was *Pioneers of the Modern Movement* (1936), by the art historian Nikolaus Pevsner. The objectives of a pragmatic approach to design, serving the needs of industry and end users, were explored in the writings of American designers, including *Horizons* (1932) by Norman Bel Geddes and *Designing for People* (1955) by Henry Dreyfuss. Theorists, critics, and architects soon made their own contributions to the debate, including the Italian Gillo Dorfles, in *Il disegno industriale in Italia* (1957), and the Briton Reyner Banham, in *Theory and Design in the First*

OPPOSITE
17. Vico Magistretti (1920–2006)
***Atollo* table lamp**, 1977
Made by Oluce (Italy)
Aluminum and polyurethane
Museum of Modern Art, New York

BELOW
18. Jonathan Ive (b. 1967)
***iMac* computer**, 1997
Made by Apple Inc. (U.S.)
Musée National d'Art Moderne, Centre Georges Pompidou, Paris

Machine Age (1960). In the 1970s, the history of design began to arouse particular interest in many countries. The French scholar Jocelyn de Noblet produced one of the first syntheses of the subject in *Design, introduction à l'histoire de l'évolution des formes industrielles de 1820 à aujourd'hui* (1974).

Along with countless publications on design, a number of exhibitions have, since the early 1980s, responded to the growing public interest in the decor and products of a changing world. In 1983–84, the Philadelphia Museum of Art presented a show entitled *Design Since 1945*. Its very instructive catalog was written primarily by the leading figures of the time, designers and manufacturers both European and American: Ettore Sottsass, George Nelson, Max Bill, Timo Sarpaneva, Olivier Mourgue, Bruno Danese, Hans Wegner, and Marco Zanuso. Ten years later, *Design miroir du siècle* opened at the Grand Palais in Paris. It had the ambitious goal of demonstrating and explaining the principles of design, by means of more than 1,600 "designed" objects, ranging from coffee cups to cars.

Design has won its place slowly in the permanent collections of museums. The Museum of Modern Art in New York was among the first to add the everyday objects that are now associated with "design" to its collections, in the 1930s. Gradually, other museums of modern and decorative art, in cities ranging from London and Paris to Rotterdam, Montreal, and Munich, began to follow suit. Specialized museums devoted exclusively to design appeared considerably later. The Vitra Design Museum, designed by Frank Gehry, was established by the furniture company Vitra in Weil am Rhein in 1989, the same year that London's Design Museum opened its doors. While Italy boasted a number of company-sponsored museums (those of Kartell and Alessi, for example), it had to wait a long time for the Triennale Design Museum to open in Milan in 2007. In Tokyo, the Japanese fashion designer Issey Miyake inaugurated the center known as 21_21 Design Sight in 2008, to host temporary exhibitions on design.

While museums have certainly played an important role in raising awareness of design among the public at large, designers themselves have come together to advocate for recognition of their work by manufacturers and governments. Professional organizations have likewise contributed to the recognition and understanding of design. Over the course of the twentieth century, then, design has defined and encompassed a vast undertaking: that of constantly renewing products and adapting them to their times. Moreover, design has continued to expand its scope, which now extends from vehicles to furnishings, and from high tech to the routine objects of daily life. Because this process has manifested itself in different forms, and at different paces, in Europe, the U.S., and Asia, the contributors to this volume were given the latitude to present the history of design in each region as they saw best. Each writer has picked out the salient features and personalities—indeed, the highlights—of design in these places over the long period covered in this volume, which extends from the end of the World War II to the present day. We have called upon acknowledged experts in the field, scholars with many important works already to their credit, to contribute to this comprehensive study.

OLD CONTINENT AND NEW WORLD: THE EMERGENCE OF DESIGN

DOMINIQUE FOREST

PRECEDING PAGE
19. Gerrit Rietveld (1888–1964)
***Zig-Zag* chair**, 1934
See plate 33.

BELOW
20. William Morris
(1834–1896)
Green Dining Room, 1866
Victoria and Albert Museum, London

It was in mid-nineteenth-century England that a debate first began about the aesthetic and moral qualities of manmade goods. From the end of the eighteenth century, fundamental economic transformations had propelled the rapid development of the country's industrial base, which depended on coal as its source of energy and the steam engine as its source of power. The expertise and impressive creativity of the era's engineers led to a new mastery of metalworking and machinery. Somewhat later, this surge in industrial development reached Germany and France, and then the United States. An improvement in the quality of life—particularly in the towns and cities, where populations grew rapidly—led to a sharp increase in the demand for basic manufactured products, including both traditional goods, like textiles, and an array of new objects produced for household or office use. Mechanized or semimechanized methods of production made these items accessible at competitive prices, assuring their commercial success. In the final decades of the nineteenth century, manufacturers armed with new processes and capabilities came to rival skilled artisans, making high-quality decorative objects available to a wider clientele; the development of large-scale ceramic production in England is just one example. Many industrial products now competed directly with handcrafted works, raising deep concerns among the traditional artisanal class, and prompting British thinkers to embark upon a serious consideration of how such objects were made.

A BRITISH UTOPIA

The vigorous growth of British mills and factories alarmed many aesthetes and artists, leading them to ponder the relationship between art and industry. Their debates were in fact a continuation of ongoing conversations about the decorative arts and how they were taught and valued compared to the fine arts. Around 1860, the phenomenon known as the Arts and Crafts movement emerged at the instigation of the art critic John Ruskin (1819–1900). This movement, whose influence peaked between 1880 and 1910, aimed not to separate the artist and the artisan, but to reconcile their work in a new synthesis, echoing an idealized notion of the Middle Ages. The construction of a cathedral—in which architects, masons, and artisans skilled in stone carving, stained glass, and goldsmithing collaborated harmoniously—was cited as an example.

The proponents of Arts and Crafts were dismayed by the unequal and ultimately doomed struggle that was taking place between the artisan's workshop and the modern factory. Their concerns had less to do with the economic consequences of the conflict than with

LEFT
21. Owen Jones (1809–1874)
Decorative motifs from houses in Pompeii, from *The Grammar of Ornament*, 1868, plate 23
Bibliothèque des Arts Décoratifs, Paris

BELOW
22. **Electric radiator,** 1913
Made by Belling (U.K.)

the "quality" of the object produced and the negative impact of modern manufacturing methods on the worker. In an artisanal setting, there is a unity between, and control over, an object's conception and execution; this commonality necessarily produces an "authentic" and "virtuous" object, an article that is genuine and brings satisfaction to its creator. In contrast, the factory worker is deprived of any initiative in the creation of the product, which is thus deprived of its "authenticity"; it has been churned out with purely mercenary motivations, very far removed from the ideal of a harmonious beauty.

Ruskin and the architect Augustus Pugin (1812–1852) were among the first to articulate these issues and, in the framework of the Arts and Crafts movement, to propose solutions, which took the form of models for decorative objects, furniture, and household decor, as well as architecture. In 1859, the architect Philip Webb built the Red House near London for William Morris (1834–1896), one of the movement's most prominent figures. This residence broke with contemporary architectural conventions in its layout and design, as well as in the materials used. Morris, a designer, artisanal manufacturer, and critic, had committed himself to a militant crusade against the evils of industrialization and the resulting loss of traditional expertise and "good taste": "But it is waste of time to try to express in words due contempt of the productions of the much-praised cheapness of our epoch. It must be enough to say that this cheapness is necessary to the system of exploiting on which modern manufacture rests."

The English thinkers of the Arts and Crafts movement contested the facile acquiescence to contemporary practices. They advocated a tranquil, harmonious accord between people and the objects they make, arguing in favor of simple, functional products, respect for natural materials, and a certain restrained "rightness" in interior decor (plate 20). In so doing, they foreshadowed a debate that would arise several decades later in the other advanced industrial power, Germany, and contribute to the emergence of a new means of reconciling an era and the objects it produced: design.

DECORATION EVERYWHERE, AND FOR EVERYONE

The nineteenth century was gripped by a veritable frenzy for ornament. Everything—or practically everything—had to be decorated. New industrial processes allowed the creation of accessibly priced objects mimicking the appearance of the costly goods that had been manually produced a few decades before. Ornament invaded houses, living rooms, dining rooms, and kitchens. It was liberally applied to all industrial output, from metal boxes to bowls, plates, glasses, household fabrics, linoleum floor tiles, furniture, lighting fixtures, and wallpaper. Nothing escaped the onslaught of flowers, ribbons, neo-Renaissance grotesques, tartans, and countless juxtapositions of geometric, floral, classical, or medievalist motifs (plates 21, 22).

The working and newly minted middle classes were playing a game of catch-up: ornament had long been inaccessible to the common person. This decorative

BELOW
23. Peter Behrens (1868–1940)
Electric kettle, 1909–10
Made by AEG (Germany)
Musée National d'Art Moderne, Centre Georges Pompidou, Paris

RIGHT
24. **Poster for a Deutscher Werkbund exhibition**, 1914
Private collection

OPPOSITE
25. Peter Behrens (1868–1940)
Entry hall of the headquarters of Hoechst AG, Frankfurt, 1920–24

overload was a reaction to the forced abstinence of earlier generations, a reaction made possible by the mechanization of techniques for reproducing decorative motifs on every kind of consumer good.

The modernization of the punch card–operated Jacquard loom gave a strong impetus to the production of patterned fabrics, while, in the first half of the nineteenth century, the technique of transfer printing, or decalcomania—in which a support is used to transfer the ink to a surface such as metal, ceramic, or glass—opened the floodgates to a deluge of decorated items that were scarcely more expensive than those that were completely unadorned.

The decoration of an object had slipped from the artisan's hand. Now that the designs of the past could be reproduced en masse, they lost all value in the eyes of those who were once their creators. In the opinion of its detractors, such ornamentation could only have a pernicious impact on the purchaser and user.

ORNAMENT AND CRIME

The rapid and pervasive industrialization that Germany underwent in the late nineteenth and early twentieth centuries revolutionized perceptions and called into question the fundamental tenets of both architecture and design.

A number of the ideas originally developed in England were embraced by the German architect Hermann Muthesius (1861–1927), whose initial attraction was to Arts and Crafts architecture. He revived the arguments about form that had preoccupied the British movement, and once again posed questions concerning the unity of the arts and the relationship between artisanal crafts, artistic creation, and industrial production. Inspired by these issues, as well as by the ongoing debate about the revitalization of German art, he established the Deutscher Werkbund at Munich in October 1907. The Werkbund engaged in various forms of discussion, action, education, and propagandizing (plate 24); originally its goal was to ennoble industrial production and align it more closely with artisanal standards. Its membership included industrialists, artisans, artists, and architects, among them Peter Behrens (plate 25). Following years of wide-ranging debate, their ideas coalesced around the use of industrialized—meaning simplified and standardized—methods to make high-quality German products.

Muthesius roundly criticized the industrial output of his era, most of which perpetuated outdated forms and motifs. He and others so vigorously challenged the propensity for applying ornamentation always and everywhere that it came to be seen as an aberration. But in Germany, these debates about the meaning of ornament and the relationship between artisanal and

BELOW
26. Walter Gropius (1883–1969) and Adolf Meyer (1881–1929) **The Fagus factory in Alfeld an der Leine, Germany**, 1911–13

OPPOSITE
27. Joost Schmidt (1893–1948) **Poster for a Bauhaus exhibition in Weimar**, 1923

industrial products, or between art and craft, were underpinned by other preoccupations—namely, the economic performance and competitiveness of German products. These concerns did, however, remain somewhat veiled behind less pragmatic ones. As Muthesius commented in 1915, "It's not a question of dominating the world, nor of financing it, nor of teaching it lessons. It's a deeper question of giving the world a face."

In the same period, the Viennese architect Adolf Loos (1870–1933) published his manifesto *Ornament and Crime* (1908), which called for a radical break with the idea of ornamentation, raising fundamental questions about the purpose of artistic expression. In a 1924 interview, Loos was asked whether modern man had any need of ornament. He responded, "Modern man, a man with modern sensibilities, has no need of ornamentation; on the contrary, he detests it."[1] Loos's vision, resolutely set on the future, also had a clear economic objective: ornament was a waste of time, effort, and money for producers, and encouraged the creation of objects whose form and decoration would inevitably become outdated.

Thus were laid, in the first years of the twentieth century, the foundations of a powerfully innovative movement that would renew the realm of architecture and the decorative arts. The bursting forth of the modern world—technologically advanced and boldly inventive—was destined to make ornament obsolete.

THE BAUHAUS

Born in Berlin in 1883, Walter Gropius began his career in 1907, in the office of Peter Behrens, who proclaimed his architectural manifesto with the turbine factory he built for AEG, and also designed a broad array of electrical devices for that company (plate 23). Gropius established his own architectural firm in 1910. Like Behrens, he was affiliated with the Deutscher Werkbund and involved in the debate about how best to implement modern ideas in both objects and architecture. An architect first and foremost, Gropius focused on the potential for standardization in construction projects. He was interested in creating prefabricated components and also investigated novel construction materials and techniques.

In 1913, Gropius completed a factory commissioned by Fagus, a shoe last maker, in Alfeld an der Leine in Lower Saxony (plate 26). In this important building, he demonstrated a number of fundamental features of the architecture of the future, including lighter load-bearing elements, flat roofs, and the very innovative use of glass curtain walls. He envisioned the factory as a rational architectural ensemble, meticulously adapted to its function. The carefully conceived interior design showed his humanistic concern for the working conditions of Fagus's employees. This building was a spectacular and highly visible showcase for the ideas that would to be embraced by an entire generation of European and North American architects.

In 1919, Gropius assumed the direction of two art schools in Weimar, which he merged under the name of Staatliches Bauhaus (plate 27)—the literal meaning of *Bauhaus* being "construction house" or "building

STAATLICHES BAUHAUS
AUSSTELLUNG
JULI SEPT
1923
WEIMAR

RIGHT
28. Herber Bayer (1900–1985)
Cover of a brochure for the Bauhaus in Dessau, showing the building by Walter Gropius, 1926

OPPOSITE, TOP
29. Herbert Bayer (1900–1985)
Cover of the first Standard-Möbel catalog to feature Marcel Breuer's metal furniture, 1927

OPPOSITE, BOTTOM
30. Josef Albers (1888–1976)
Stackable tables, c. 1927
Varnished ash, painted glass
The Josef & Anni Albers Foundation, Bethany, Connecticut

house." The Weimar Bauhaus closed in 1924, and a second one was built in Dessau in Saxony and inaugurated in December 1926 (plate 28). There Gropius planned to develop and implement a new language of forms with universal application. Instruction was provided in theory, as well as in the applied arts (textiles, ceramics, goldsmithing, and metalworking), with some classes being taught by artists. Some of the greatest figures of twentieth-century modernism served on the faculty, including the architects Gropius, Ludwig Mies van der Rohe, and Marcel Breuer; the artists Wassily Kandinsky, Josef Albers, Paul Klee, and Oskar Schlemmer; and the designers László Moholy-Nagy, Wilhelm Wagenfeld, and Marianne Brandt.

In its teaching and its artisanal workshops, the Bauhaus harmonized theory and practice. The *Wassily* armchair (1926), made from chrome-plated steel tubes and leather, was designed by Breuer and produced in the workshop he supervised (plate 29). And it was in the school's metal workshop that, in 1924, Wagenfeld designed and executed the one of the masterpieces of twentieth-century design, the lamp that bears the Bauhaus name (plate 16). The cabinetmaking atelier carried out extensive research into the use of plywood and also explored modular furniture production (plate 30), while, beginning in 1929, the wallpaper studio designed and produced collections that eschewed realistic motifs and focused on the properties of the material itself. Their output was a great commercial success.

The Dessau school thus played a seminal role, providing the home for a new aesthetic that was destined to leave a lasting legacy in Europe, and even more so in the United States. A number of important Bauhaus figures moved to North America after the Nazis closed the school in 1933, including Gropius, Mies van der Rohe, and Breuer. The principles of the Bauhaus—the extreme purity of form, the priority assigned to function ("form follows function"), and the elimination of anything extraneous—were integrally associated with the artistic ferment that stirred certain European circles at the beginning of the twentieth century. A number of avant-garde painters rejected the figural representation of reality with a fervor that had spiritual overtones; these artists included Kandinsky, Kazimir Malevich, František Kupka, and Piet Mondrian, who began working with Theo van Doesburg in 1917 to establish the neoplasticism movement, De Stijl, in the Netherlands. As Van Doesburg explained, "After 1913, we all felt a need for abstraction and simplification. A mathematical approach seemed necessary to counter impressionism, which we rejected." Gropius's theories embraced this fresh perspective. He was convinced that the force of the mind would prevail in the creation of the modern world.

DESIGN FOR ARCHITECTURE

Marcel Breuer created his *Wassily* armchair in 1926, the year the Bauhaus transferred its classes and workshops to the new building in Dessau. In fact, the chair was designed to harmonize with its architectural setting: new architecture called for new materials and new furnishings. Like Gropius and Mies van der Rohe, Breuer was an architect. He was particularly interested in furniture design and advocated an absence of style in his

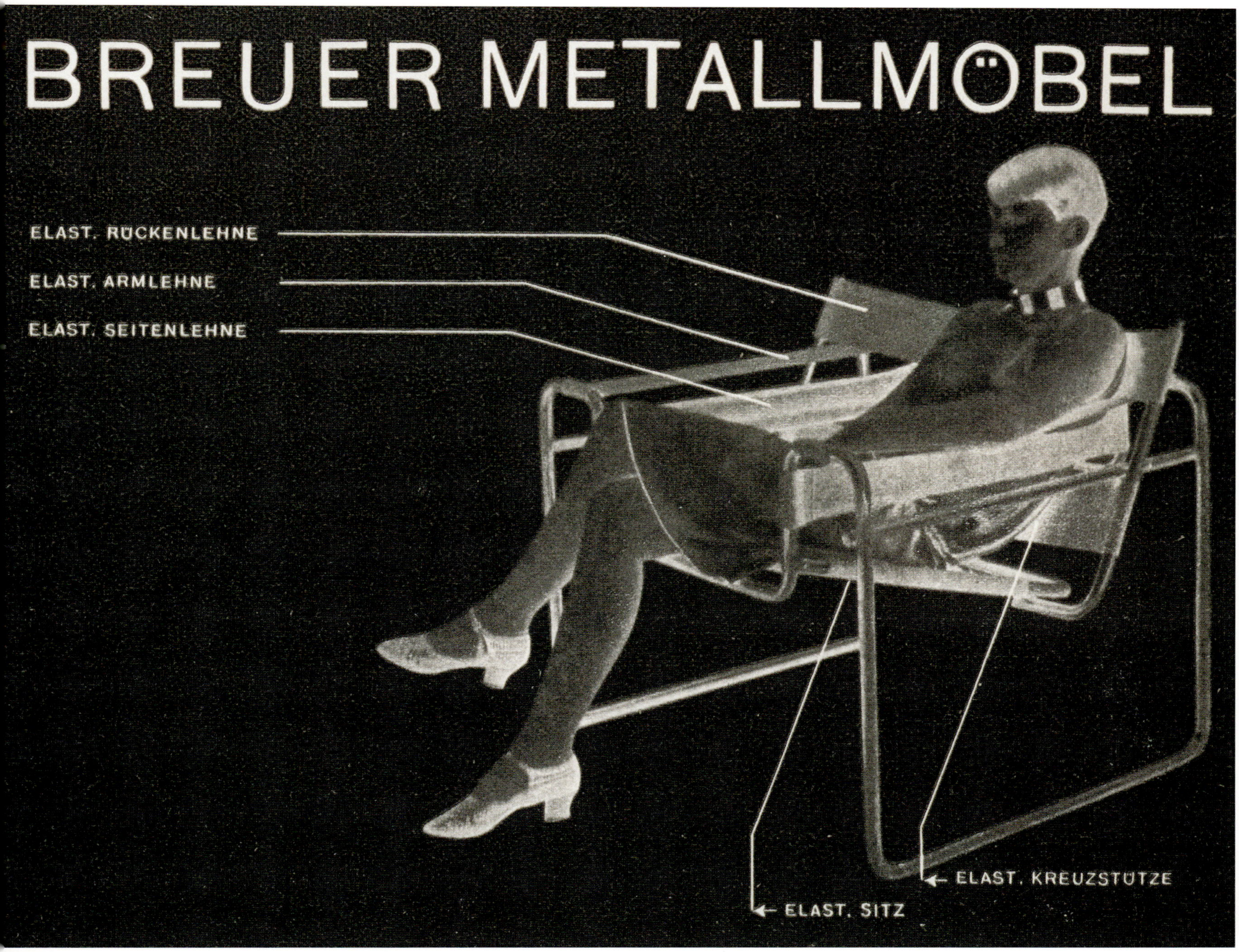
BREUER METALLMÖBEL
ELAST. RÜCKENLEHNE
ELAST. ARMLEHNE
ELAST. SEITENLEHNE
ELAST. KREUZSTÜTZE
ELAST. SITZ

BELOW
31. Le Corbusier (1887–1965), Charlotte Perriand (1903–1999), and Pierre Jeanneret (1896–1967)
Reclining chaise longue, 1928
Made by Thonet Frères (France)
Chrome-plated and painted steel, leather, fabric
Museum of Modern Art, New York

OPPOSITE
32. **Charlotte Perriand on the reclining chaise longue she designed with Le Corbusier and Pierre Jeanneret**, 1928
Charlotte Perriand Archives

creations, focusing exclusively on function. In Breuer's opinion, "We don't expect things to express anything other than their function and the structure necessary for that function." Mies van der Rohe expressed a similar view: "Less is more."

Breuer left the Bauhaus in 1928 because he disagreed with the school's emphasis on teaching and practicing the decorative arts, which he felt came at the expense of architectural instruction. He pursued his architectural career in Berlin, and subsequently worked in England and the U.S., where he moved in 1937, before later returning to Europe. Breuer was particularly attentive to the interiors of his buildings, striving for coherence between the architecture and furnishings. From the outset, he paid close attention to the selection of materials (solid wood, chrome-plated steel tubing, laminates, and aluminum) and created several pieces of furniture that set the standard for the European avant-garde. In 1932, he designed his *B32* chair with its distinctive cantilevered seat. This chair, which is still in production, is a cornerstone of the reputation that Breuer enjoys not only as an architect but also as a designer.

Other architects attuned to Bauhaus ideas also designed furnishings: Mies van der Rohe (the *Barcelona* chair, 1929), Le Corbusier (the *Grand Confort* chair and *Reclining chaise longue* with Charlotte Perriand and Pierre Jeanneret, 1928; plates 31, 32), Gerrit Rietveld (the *Zig-Zag* chair, 1934; plate 33), and Alvar Aalto, who designed the *Paimio* chair (1930–31; plate 34), made from glue-laminated wood, for the patients of a sanatorium he built in the Finnish town of that name between 1929 and 1933.

For many modern architects, designing furniture for public buildings (such as hospitals, religious structures, and schools) was an essential part of their work, allowing them to link the interior to the exterior and thereby ensure the coherence and unity of the structure. Equally they wanted to prevent the installation of unrelated furnishings that would obscure the forms and volumes of their interiors, or be incompatible with the new architectural materials they employed, such as concrete, glass curtain walls, and metal. In principle, the architect was supposed to create a complete environment, following Aalto's example in his sanatorium in Paimio. Aalto asserted, with logic and simplicity, that furnishings "are accessories to architecture."

While few of the "new" architects were as involved with furniture design as Aalto and Breuer, a number of them did, in the early decades of the twentieth century, create furnishings and accessories that remain iconic in the history of design. Some of these pieces are, indeed, still being produced and distributed today, and remain sought after for their comfort, their classic beauty, and the quality of their materials.

OPPOSITE
33. Gerrit Rietveld (1888–1964)
***Zig-Zag* chair**, 1934
Oak
Museum of Modern Art,
New York

ABOVE
34. Alvar Aalto (1898–1976)
***Paimio* armchair** (first series),
1930–31
Made by Oy Huonekalu-ja
Rakennustyötehdas (Finland)
Bent plywood, beech
Musée National d'Art Moderne,
Centre Georges Pompidou,
Paris

35. **Cover of a catalog for Thonet's furniture in tubular steel**, 1933
Musée National d'Art Moderne, Centre Georges Pompidou, Paris

TRADITION AND THE AVANT-GARDE

The adaptation of everyday objects to the contemporary world that occurred between 1920 and 1950 did not result from a popular embrace of Bauhaus principles. Gropius may have declared that objects "must be perfectly adapted to their purpose, that is to say fulfill their practical function, and be durable, economical and attractive,"[2] but the reality of the situation was otherwise. The infinite variety of geometric and floral motifs had mounted its assault well before the Exposition Internationale des Arts Décoratifs et Industriels Modernes of 1925 gave art deco its name. Ornamentation invaded all kinds of everyday objects (furniture, glassware, fabrics, wallpaper), as well as household fixtures (heating, lighting, and so forth).

In France, this craze for ornament aroused great dismay among the ranks of the avant-garde, particularly the members of the UAM (Union des Artistes Modernes), who had hoped that the industrial recovery of the interwar years would serve as a means of disseminating their modernist ideas. Established in 1929, the UAM included many architects, decorators, poster artists, and designers who had broken ties with the far more conservative Salon des Artistes Décorateurs. Whether they were founding members of the UAM or not, all the supporters of the avant-garde were to be found at its shows—from Francis Jourdain to René Herbst and Robert Mallet-Stevens, as well as Pierre Chareau, Charlotte Perriand, Eileen Gray, and Le Corbusier. However, their avowed objective of destroying the "parasites of the aesthetic" came up against the powerful opposing forces of tradition, and their visionary new directions were ignored or rejected by the single-minded logic of commerce.

The concept of modernity was evidently perceived very differently by a receptive elite than by a mass audience! It would take decades for new ideas and forms, sometimes poorly executed, to make their way into the households of the future. Moreover, the emergence of a new way of living would depend, in its early stages, not only on a strong aesthetic program, but also on the mastery of new technologies and materials (structural steel, reinforced concrete, and architectural glass) and the use of industrial materials (such as nickel-plated metal tubing and laminates) in new contexts (plates 35–38).

NEW THINGS IN OLD GUISES

The end of World War I, which the U.S. entered in 1917, marked for that country a fresh surge in the economic growth it had experienced in the preceding decades. This spurred an increase in the urban population, as well as in the purchasing power of the city dwellers, who had a strong desire for modern consumer goods. Manufacturers grew and diversified at a rapid rate, becoming archetypal representatives of a newly powerful and dynamic America. The country was now undeniably an industrial powerhouse, confident of its technological skills and productivity. Cutting-edge industries included the transportation sector (airplanes, automobiles), chemistry, communications, and domestic appliances. In the two decades between 1910 and 1930, car ownership in the U.S. soared from one in

BELOW LEFT
36. René Herbst (1891–1982)
Chair, 1930
Made by Établissements René Herbst (France)
Chrome-plated steel, elastic cords
Musée National d'Art Moderne, Centre Georges Pompidou, Paris

BELOW RIGHT
37. René Coulon (1908–1997)
Armchair, 1938
Made by Saint-Gobain (France)
Tempered curved glass, leather seat
Musée des Arts Décoratifs, Paris

BOTTOM
38. Eileen Gray (1878–1976)
Cabinet with pivoting drawers from the E-1027 house in Roquebrune-Cap-Martin, 1926–29
Painted wood
Musée National d'Art Moderne, Centre Georges Pompidou, Paris

BELOW
39. Henry Dreyfuss (1904–1972)
Vacuum model 150, 1935
Made by Hoover (U.S.)
Musée des Beaux-Arts, Montreal

OPPOSITE
40. **The Singer Building under construction in New York**, 1905–8

184 people to one in five. Nevertheless, as in Europe, many manufacturers of the 1920s continued to produce outdated designs little suited to the enthusiastic consumerism of a vigorous, forward-looking America.

This was the era when electromechanical devices (refrigerators, fans, space heaters, kitchen utensils, vacuum cleaners, phonographs, telephones, light fixtures) were rapidly developed for a vast new domestic market, bringing about revolutionary changes in lifestyles and daily activities. However, the manufacturers of these objects had no frame of reference for their design besides existing knowledge and habits, and thus the new products of the early twentieth century were still laden with antiquated motifs and ornamentation. Between 1925 and 1930, though, the art deco style was imported from Europe and established a firm hold on American design.

THE LITTLE SEAMSTRESS

The sewing machine was the first precision mechanical device to make its way into a significant fraction of European and American households, in the last two decades of the nineteenth century. Considerable quantities were produced in Europe, and then even greater numbers in the U.S. When it was completed in 1908, the skyscraper constructed by the Singer sewing machine company in New York was the tallest building in the world, a spectacular demonstration of the firm's success and prosperity (plate 40).

Since 1850, all sewing machine manufacturers had faced the same problem: how to make a cast-iron and steel machine aesthetically presentable. It was a fairly complex device, with wheels, gears, and other elements suggestive of workshops and factories. Manufacturers had to find a way to ensure user safety, while masking the technical and utilitarian aspect of a machine that was increasingly making its way into well-appointed domestic interiors. Understandably, producers turned to ornament to obscure any allusion to the world of hard labor and engine grease, and to transform the little sewing machine into a decorative object suitable for a sitting room. Its somber black surface was lavishly transfer-printed with gilding and elaborate motifs like sphinxes, garlands, and floral sprays. As freestanding pieces of furniture, too large to be concealed in a closet, treadle-powered models were often encased in the sort of cabinetry considered suitable to a civilized domestic interior.

Quite a number of domestic novelties fell victim to the chameleon syndrome, including the phonographs and radios manufactured by the Edison Company, which were concealed in small, elaborately decorated chests, adorned with inlays of exotic woods. Vacuum cleaners were sometimes covered with colored artificial leather. In 1910, "technological" objects like these were still considered overly reminiscent of the industrial workplace, and were hidden beneath decorative overlays. However, a completely different approach was to be adopted in the coming decades: instead of being camouflaged, technology and materials gradually began to be emphasized. The influence of the designer was making itself felt (plate 39).

THE IMAGE OF SPEED

The imperatives of a production-driven economy, the modernizing impulse of skilled designers, and the resolutely optimistic attitude of American consumers enjoying rapid progress all led to the aesthetic renewal of the ordinary objects used in the home and the workplace. Components made of new materials like aluminum were not concealed beneath paint but enthusiastically displayed, as symbols of a product's "industrial" efficiency. Steel, now protected from corrosion by chrome plating, lent itself to a wide range of uses, while new metal-stamping techniques facilitated the production of small parts, from the housing of a toaster to a pen cap. Indeed, stamping gave designers an unprecedented freedom in determining the outward shape of manufactured goods.

In the 1920s, the popular imagination began to associate modernity with transatlantic crossings by great

THE
EVENING MA
One Cent Buys the
ROEBUCK'S
HANAN & SON

NEW YORK
CENTRAL
SYSTEM

OPPOSITE
41. Henry Dreyfuss (1904–1972)
***20th Century Limited* locomotive**, 1938

BELOW
42. Raymond Loewy (1893–1986)
Drawing of the *K4S* locomotive for the Pennsylvania Railroad Company, New York, 1936–37
Cooper-Hewitt, National Design Museum, New York

RIGHT
43. **Poster for the Hamburg-American Line**, c. 1930

ocean liners whose hulls were designed to cut through the waves (plate 43), and even more with air travel, which experienced rapid growth in the U.S.: in the decade between 1920 and 1930, the country boasted some forty airlines. Manufacturers coated the fuselages of their planes with Duralumin, a hardened aluminum alloy, and researched the best designs to facilitate airflow over aircraft wings and bodies. These flying machines, which allowed Americans to traverse their country's vast expanses, became symbols of a modern, rapid, and efficient new world.

The airplane and the train engaged in a fierce competition. The railroads, which had made such a vital contribution to the development of the U.S., confronted the rounded, gleaming fuselages of the new airplanes with old-fashioned locomotives that belched smoke and noisily spurted water and steam. These behemoths were burdened with a complex of rods, gears, and pistons, all dominated by an immense horizontal boiler painted a somber black. Technology without adornment, indeed, but not in this case much admired: trains had begun to seem rather old-fashioned by the 1930s. Thus, in 1932, the Union Pacific Railroad called in William Bushnell Stout, a legendary designer and engineer in the automotive and aeronautic fields. He designed for them the Pullman *M10000*, featuring a Duralumin body with a smooth, simplified profile, the first "streamlined" train. In 1934, the Chicago, Burlington and Quincy railroad introduced the *Zephyr*, also the work of an aeronautical engineer, who covered the entire body in stainless steel. Other railroads emulated these examples, retrofitting their old locomotives and giving new ones a sleek look adapted to contemporary tastes.

Three of America's most important avant-garde designers developed locomotives, virtually building their careers on these spectacular machines: Norman Bel Geddes, Raymond Loewy, and Henry Dreyfuss (plates 41, 42). Their locomotives were smooth, sleek, and rounded, creating an aerodynamic look that gave at least the illusion of enhanced performance. What mattered was image, communication, style—the idea of power and speed. Now trains were a match for planes!

In the 1930s, the major automakers began to alter the look of their new models as well. Cars had generally retained a tall, upright profile, but this was now replaced by sleek, rounded forms (plate 44). Manufacturers also developed concept cars to evoke consumer fantasies. One example was Buick's beautiful *Y-Job* of 1938, designed by Harley J. Earl (plate 45).

BELOW
44. **Advertisement for Buick automobiles**, 1945
Private collection

RIGHT
45. Harley Earl (1893–1969)
Buick *Y-Job*, 1938
General Motors Heritage Center, Sterling Heights, Michigan

Buick 'Y'

MODERN TIMES

As exemplars of American modernity, of a society on the move, manufacturers of all kinds eagerly adopted the new visual idiom. As early as 1925, the major industrial groups were recruiting designers, sometimes as outside consultants and sometimes as employees, but always with the same objective: to increase sales. The electrical appliance maker Westinghouse called in Donald H. Donner in 1926, while Ray Patten collaborated with General Electric to design an alarm clock in the purest streamlined style, like a railroad locomotive. From 1933–35, as the country began its long emergence from the Depression and industry recovered, the design of many mass-market products was refreshed. In 1936, the industrial designer Egmont Arens created the definitive model of the KitchenAid food mixer, which became a classic of household design and is still being produced today (plate 48). For the same company, Hobart Manufacturing, Arens designed a remarkable kitchen slicer in 1940; it was fashioned from cast aluminum and steel with fluid, aerodynamic lines (plate 47).

In the years preceding World War II, American design was dominated by four men: Walter Dorwin Teague (1883–1960), Norman Bel Geddes (1893–1958), the French-born Raymond Loewy (1893–1986), and Henry Dreyfuss (1904–1972). All four worked to define the field of industrial design in theory and practice, and to clarify its role in the economy. Moreover, they all established studios in New York that welcomed rising young American and European designers. Teague proved to be an astute and talented businessman, collaborating with corporations including Kodak and Texaco (plate 49). A theorist as well as a practicing designer, Bel Geddes was very conscious of the economic impact of his profession. In 1932, he published *Horizons*, a book that that would long be influential among thinkers and designers. This work endorsed the streamlined style as an apt expression of contemporary America (plate 51), and predicted that industrial design would be an important factor in the nation's growth and prosperity. Indeed, the book was an enthusiastic attempt to anticipate the future: for Bel Geddes, speed was "the cry of our era, and greater speed one of the goals of tomorrow." A designer-architect (like Teague) and a friend of Erich Mendelsohn and Frank Lloyd Wright, Bel Geddes designed the General Motors pavilion at the 1939 World's Fair in New York (plate 50). Within this spectacular structure, called Futurama, visitors rode a conveyor belt over an enormous diorama of a possible

OPPOSITE
46. **Film still from Charlie Chaplin's *Modern Times***, 1936

BELOW
47. Egmont Arens (1889–1966)
Slicing machine model *410*,
c. 1940
Made by Hobart (U.S.)
Cast aluminum, steel, rubber
Museum of Modern Art,
New York

OVERLEAF
48. Egmont Arens (1889–1966)
Kitchen mixer model *K 45*,
1962
Made by KitchenAid (U.S.)
Musée National d'Art Moderne,
Centre Georges Pompidou,
Paris

PAGE 49
49. Walter Dorwin Teague
(1883–1960)
Camera model *800*, c. 1957
Made by Polaroid Corporation
(U.S.)
Cooper-Hewitt, National
Design Museum, New York

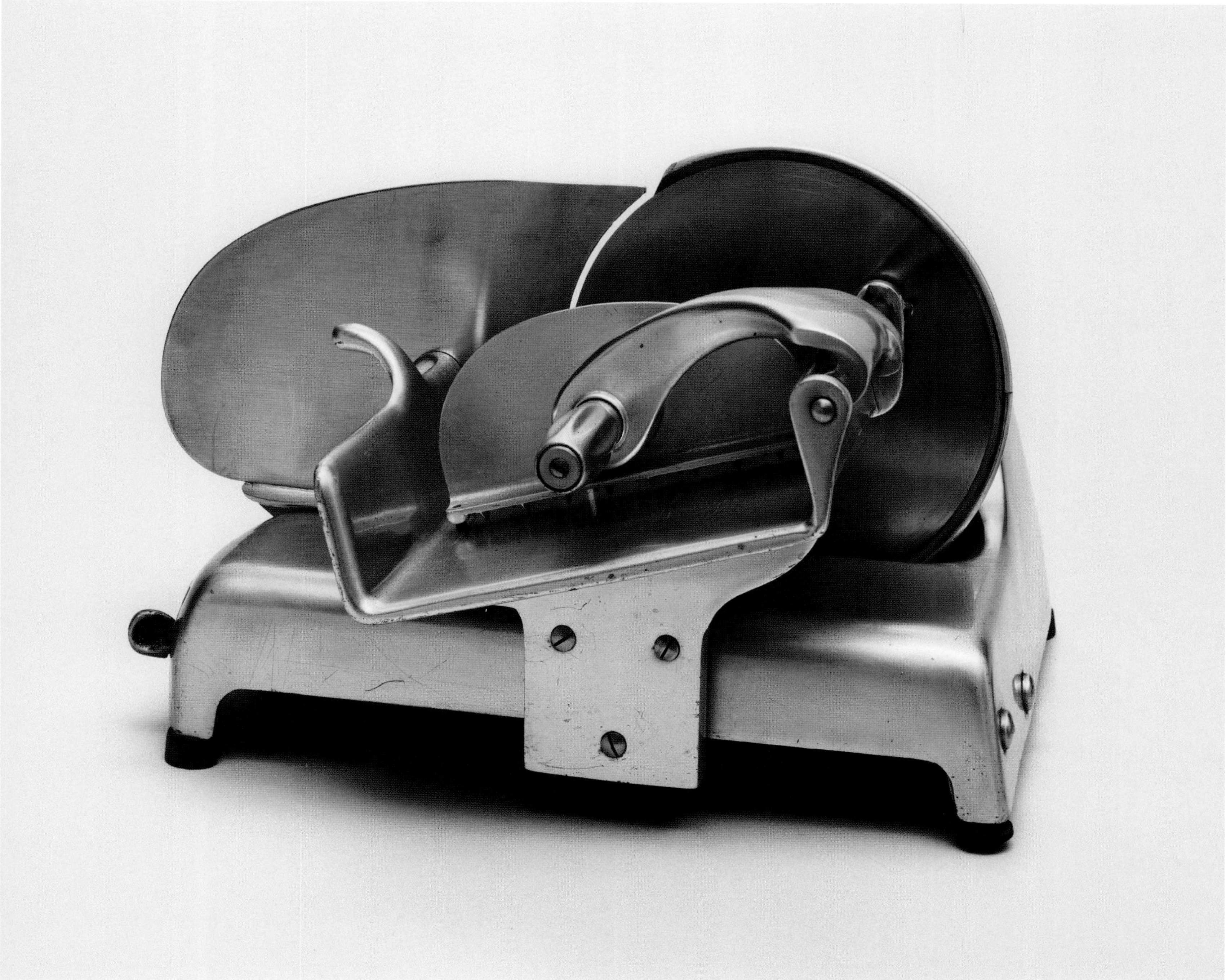

EV
10
POLAROID
Land Camera
The 800
B
I

future America, a tantalizing vision of gleaming cities, suburbs, and freeways.

During the 1930s, the aerodynamic look influenced a vast array of manufactured products aimed at the American mass market. Toasters, refrigerators, radios, photographic equipment, lamps, and office supplies all assumed rounded forms with a strong emphasis on horizontality. At times they resembled something quite other than themselves: Loewy's pencil sharpener of 1933 is a literal quotation of aeronautic design (plate 52).

In the U.S., streamlined design overlapped with "nautical moderne" in architecture. Between 1935 and 1940, the geometric forms of art deco buildings gave way to stylistic elements borrowed from transatlantic liners, including flat roofs, an emphasis on horizontality, rounded corners, and decorative schemes with clearly marked parallel bands. This taste also carried over into the design of certain products—radios, for example. Nautical moderne had an influence in Europe, as well, which is visible in the architecture and furniture of the years immediately preceding World War II.

Between the two world wars, at a time when the country was shaken by a major economic crisis, American manufacturers and designers were able to develop an original and effective formal language, popular in its appeal, and structured around the theme of speed and power. This streamlined look, the American style of 1930 to 1940, would endure in Europe in the years immediately following the war, as a symbol of the enviable American way of life. In the flourishing U.S. of the 1950s, the streamlined look transformed itself to evoke anew the spirit of success, progress, and dynamism: it became the "rocket style," which made abundant use of colored plastics and chrome trim. The earlier

"American style," reduced to a generalized fluidity and roundness of form, still persisted for a time in a postwar Europe in search of models, notably in the British automobile industry and among manufacturers of "modern" consumer conveniences. But now, in the 1950s, design, so long shaped by the dialogue between Europe and the United States, went in search of new territories.

OPPOSITE, TOP
50. Norman Bel Geddes
(1893–1958)
Illustration of Futurama, the General Motors pavilion at the New York World's Fair, 1939
General Motors Heritage Center, Sterling Heights, Michigan

OPPOSITE, BOTTOM
51. Norman Bel Geddes
(1893–1958)
***Soda King* siphon bottle**, 1938
Made by Walter Kidde (U.S.)
Cooper-Hewitt, National Design Museum, New York

ABOVE
52. Raymond Loewy
(1893–1986)
Office pencil sharpener, 1933
Prototype
Private collection

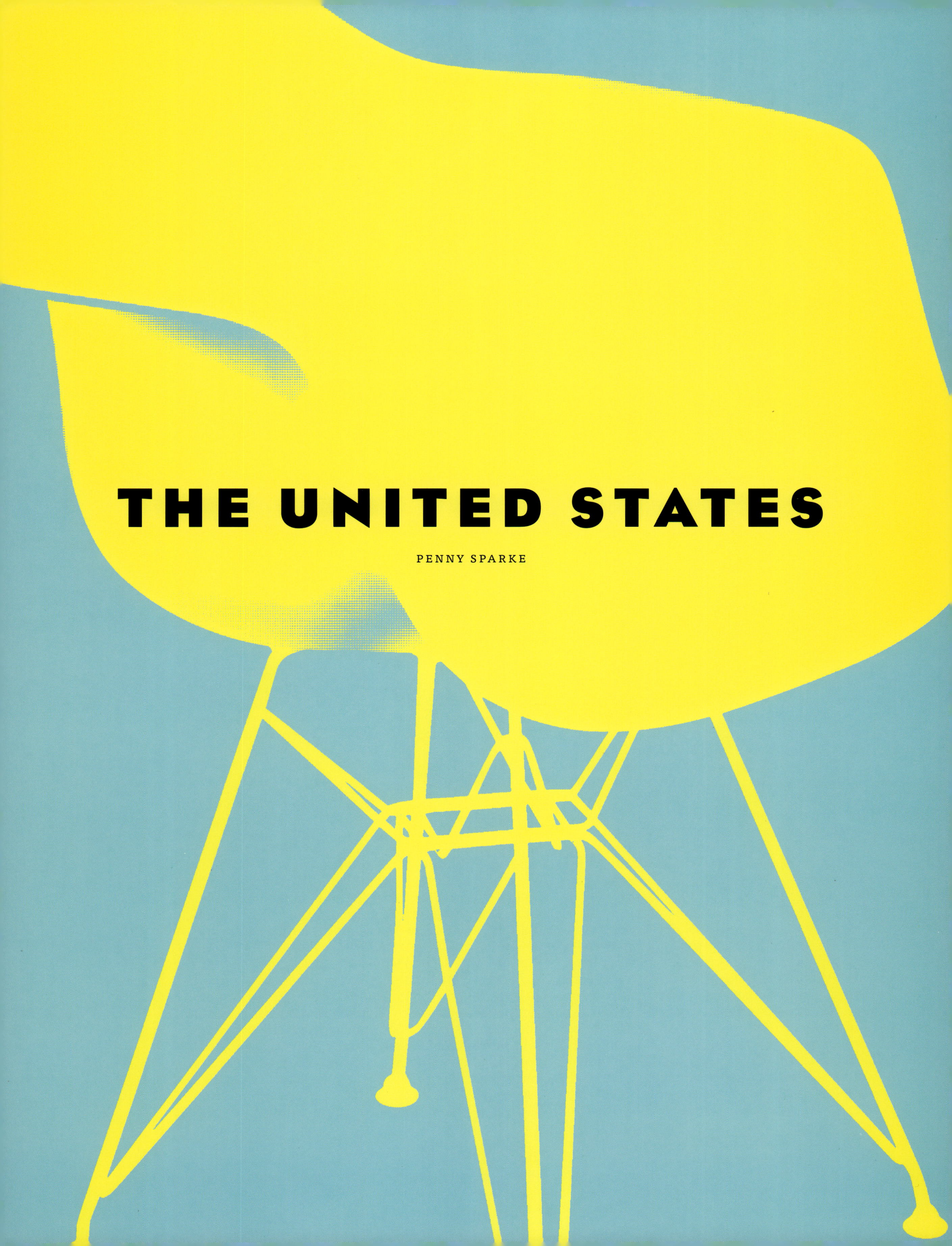

THE UNITED STATES

PENNY SPARKE

PRECEDING PAGE
53. Charles Eames (1907–1978) and Ray Eames (1912–1988)
***DAR* (*Dining Armchair Rod*) chair**, 1948–50
See plate 87.

OPPOSITE
54. **Electric toaster**, c. 1910
Made by General Electric (U.S.)

Design must play a central role in any discussion of the United States in the period after 1945. In the nineteenth century, the U.S. had been a leader in mass production and the development of advanced technology, and in the interwar years it had helped to create the modern profession of industrial design, which was emulated around the globe. Building on these accomplishments, the country went on, after the war, to define the modern consumer society, in which design played an important role, and to export this lifestyle to the rest of the world. The U.S. led the way in designing vehicles and kitchen appliances, and in making the "dream home" the key site of consumer aspiration and desire (plate 54). Its leading designers of the postwar era—Charles and Ray Eames, Eero Saarinen, George Nelson, and later Robert Venturi, Michael Graves, and others—acquired international reputations and became household names.

Postwar design in the U.S. built on two parallel traditions. One derived from the interwar focus on the mass production of technological objects, such as automobiles and refrigerators. It that context, the role of design was to make products more attractive in order to increase consumer desire. The evocative aesthetic of streamlining developed out of this objective in the 1930s, and after 1945, even more evocative forms were developed in the areas of automotive design and kitchen gadgets. The other strand of American design borrowed the simple, geometric forms of European modernism. Manifesting itself first in the traditional decorative arts—furniture in particular—this approach led to an upmarket, sophisticated domestic style epitomized by the work of Charles and Ray Eames. While streamlined objects found their way onto the streets and into the office and the kitchen, European-inspired designs were much more at home in the living room.

In the years after 1945, the U.S. exported a model of modern living that combined affluence with technological utopianism and had design at its core. Leading the way in the space race of the 1950s and '60s, the U.S. persuaded the world that the future lay in the application of technology to our everyday lives. In the twenty-first century, the U.S. has once more taken the lead, this time in the development of advanced electronics and in the application of "design thinking" to contexts beyond the designed object. Most significantly, the U.S. has reinvented the very notion of design several times over the last century and a half. More than any other country, perhaps, it has established design as a flexible, responsive concept that can meet the needs of the day.

DESIGN BEFORE 1945

In the international history of industrial design, the 1920s and '30s in the U.S. represent a crucial period. It was then that the role of the consulting industrial designer came into being, as it became clear that design constituted a form of added value, with the power to dictate commercial success or failure. The background to this thinking lay in the nation's earlier success in mass production, which has been charted by the historian of technology David A. Hounshell in his book *From the American System to Mass Production, 1800–1932: The Development of Manufacturing Technology in the United States*. Hounshell starts with the changes that took place in U.S. arms production in the nineteenth century and continues through automobile manufacturing in the 1920s, taking in clocks, sewing machines, bicycles, and agricultural machinery along the way. He argues that Henry Ford's *Model T* represented a system of pure mass production (plate 55), which, however, lasted only twelve years, until Alfred P. Sloan at General Motors introduced flexible mass production in response to the need for product differentiation.

Henry Ford's system of mass production depended, above all, on the concept of standardization. His famous dictum that people could have any color "as long

RIGHT
55. ***Model T* Ford**, 1908

BELOW
56. ***Model A* Ford**, 1920

as it's black" led to the idea of a widely available basic car, made from standardized components that could be easily manufactured, assembled, and repaired. The status that the *Model T* conferred on its purchasers was linked solely to the fact that they had exchanged their horse for an automobile. It did not matter in the context of that first purchase what the object looked like, merely that it was affordable. When it came to the acquisition of a second car, however, the principle of outdoing one's neighbor came into play. Then it became necessary to provide some form of added value through design. At that point, General Motors' principle of flexible mass production came to the fore. In response, Ford placed his philosophy of manufacturing under review, closing his plant for a year in order to readjust his production processes and add design to the mix. The result was the *Model A* Ford (plate 56).

Hounshell's text concludes just at the point when the industrial designer (better known as a stylist at that time), in the form of Harley Earl, was parachuted into General Motors to create a level of differentiation for its automobiles. Hounshell's study emphasizes an important moment in the story of modern design, one in which technology and culture came into direct conflict with each other. In the event, culture won the day, and design provided the means by which the two forces could achieve a workable compromise. From that point onward, the new mass-production industries were sensitive to the vagaries of the fashion system. Understanding that they had to accommodate art within their rational operations, manufacturers came to value the role of the designer.

THE EMERGENCE OF THE DESIGN PROFESSION

The competition between Ford and General Motors in the mid-1920s marked an important moment in the emergence of the industrial designer. Alfred P. Sloan, vice president of General Motors, hired the coach builder Harley Earl to make his cars look more appealing. This strategy was rapidly emulated by the manufacturers of other new technological goods, including domestic appliances like refrigerators (plate 57), office machines, telephones, and radio sets. Challenged by the Great Depression, these manufacturers decided to inject artistic styling into their products in an attempt to fight off their competitors. Raymond Loewy's 1929 design for the Gestetner duplicator is a case in point: he

57. **Advertisement for the *Shelvador* refrigerator, manufactured by Crosley**, 1936

LEFT
58. Raymond Loewy (1893–1986)
Duplicator, 1929
Made by Gestetner (U.K.)
Victoria and Albert Museum, London

BELOW
59. Walter Dorwin Teague (1883–1960)
***Bantam Special* camera,** c. 1936
Made by Eastman Kodak (U.S.)

OPPOSITE
60. Henry Dreyfuss (1904–1972)
Model *302* telephone for Bell Telephone Laboratories, 1937
Made by Western Electric (U.S.)
Yale University Art Gallery, New Haven, Connecticut

modernized that rather crude-looking piece of office machinery by adding a smooth body shell (plate 58).

The emergence of American industrial design in the late 1920s and '30s has been documented by a number of historians. Jeffrey Meikle's 1979 book *Twentieth Century Limited: Industrial Design in America, 1925–1939* gives one account of the development of the design profession. His analysis focuses on the fact that "technological innovation and mass production brought former luxury items to people at lower income levels" and on the aspiration of industrial designers to inject a level of modern luxury into new goods. It was no coincidence that several of the leading industrial designers had begun their careers in advertising graphics and department stores, which, in emulation of French models, had been quick to embrace the modern style. Loewy, for example, had worked at Saks Fifth Avenue in New York, designing the elevator attendants' uniforms, among other things.

In his book *All Consuming Images: The Politics of Style in Contemporary Culture,* Stuart Ewen approaches the emergence of American industrial design from a cultural perspective. He claims that industrialization "displaced the customary fabric of culture" and that the marriage between art and commerce offset that displacement by bridging the gap between production and consumption.

The best-known American industrial designers of the day—among them Loewy, Walter Dorwin Teague (plate 59), Henry Dreyfuss (plate 60), and Norman Bel Geddes (plate 61)—penned their own accounts of their careers, in which they stressed their perception of themselves as idealistic modernists. Linking themselves to the modern European architectural movement allowed them to separate themselves from the commercial context that had created them. In his 1940 text *Design This Day: The Technique of Order in the Machine Age*, Teague, for example, makes repeated references to Le Corbusier, Walter Gropius, and Mies van der Rohe, suggesting, perhaps, that his name should be added to the list.

Most accounts of the work of the American consultant designers of the 1930s concentrate on their advocacy of dramatic streamlined forms depicting the future and play down their market-oriented pragmatism. However, the majority of their time was, in reality, devoted to redesigning consumer goods. If the consultant designers were heralded by the press of the day—and *Time* and *Life* magazines did showcase their work

VALENTINE
2424
RM. 210

BELOW
61. Norman Bel Geddes (1893–1958)
***Patriot* radio**, c. 1940
Made by Emerson Radio Corporation (U.S.)
Victoria and Albert Museum, London

OPPOSITE
62. **Postcard showing the Trylon and the Perisphere at the 1939 New York World's Fair**

and report on the details of their daily lives as if they were Hollywood stars—it must be realized that their celebrity status was crucial to the manufacturers who employed them, bestowing as it did an instant added value on their products. The designers themselves became brands, and their names were used as a form of product endorsement. In turn, the products became advertisements for the designers.

Industrial designers quickly became an intrinsic component of the American commercial system that had engendered them. Their arrival marked the moment when "designer culture"—that is, the attribution of added value to an object, image, or environment because it has a well-known designer's name attached to it—came into being. Such was the power of the first generation of American consultant designers that the application of their names to the design of a train, or of a biscuit, bestowed an instant aura on those artifacts. A number of explanations have been offered for this phenomenon. The most persuasive suggests that it represents consumers' need to construct their identities through the acquisition of commodities associated with the name of an individual known to possess a high level of taste. This explanation builds on the fact that, in the years before industrialization and the rupture between production and consumption, the upper classes had been guaranteed social status through the acquisition of custom-made, handcrafted artifacts. The association of a name with otherwise anonymous, mass-produced goods could restore to them a certain individualism. And as the numbers of standardized products increased, the need to individualize them became even more pressing. The consultant designers were acutely aware of the diversity of consumer tastes in the marketplace. Bel Geddes, for example, acquired his knowledge of the market by undertaking extensive

consumer surveys, and he proposed four different radio designs to the Philco Radio Company—the *Highboy*, the *Lowboy*, the *Lazy-boy*, and a radio-phonograph combination—each styled with a different market in mind.

While the automotive industry usually preferred to keep its stylists in-house and maintain their anonymity, many other new industries engaged in producing consumer goods, including domestic appliances and office machinery, benefited from bringing in generalist consultant designers who had a special overview of the sector as a whole. The model of consultant design that emerged in the U.S. starting in the late 1920s was rapidly emulated by other countries that wished to embed design in their emerging industries and thereby increase consumer desire. And in the years after 1945, the dissemination of this model of design was one of the strategies of American cultural imperialism.

THE 1939 NEW YORK WORLD'S FAIR

The U.S. entered World War II in 1941, following the Japanese attack on Pearl Harbor. Only two years earlier, however, it had been celebrating its achievements in technology and design, and looking forward to an optimistic future in which those two forces would take the American people to a new level of material affluence. The 1939 New York World's Fair, held in the borough of Queens, across the Hudson River from Manhattan, was conceived as a celebration of industrial achievement. Its main theme was "Building the World of Tomorrow," and the key question addressed was where the products and processes of modern civilization were leading. The industrial designer, rather than the architect, was the hero of the event, and the obvious figure to visualize America's future. As Walter Dorwin Teague, who was on the fair's board, explained, "Because the industrial designers are supposed to understand public taste and be able to speak in a popular tongue as a profession and they are bound to disregard traditional forms and solutions and to think in terms of today and tomorrow, it was natural that the Board of Design should turn to them for the planning of the major exhibits in which the theme of the Fair is to be expressed." Teague was personally responsible for the design of the Ford, U.S. Steel, Eastman Kodak, and National Cash Register pavilions.

The style of the temporary pavilions was influenced by product streamlining, and smooth, rounded surfaces were widely employed. The 610 foot tall (186 m) Trylon and 180 foot tall (55 m) Perisphere visually dominated the event (plate 62). The avenues that radiated out from the white center of the site were surfaced with vivid colors that grew darker as they moved toward the periphery. Dreyfuss designed the Perisphere's interior exhibit, *The City of Tomorrow*. Transport featured strongly at the fair, with an emphasis on the future role of new road and rail systems. Loewy's *Transport of Tomorrow* exhibit included a streamlined taxi, an ocean liner, cars, and trucks, as well as a rocket ship that was predicted to travel between New York and London in 1960. Bel Geddes showcased the results of his transportation experiments in the General Motors pavilion, Futurama. His *Highways and Horizons* exhibit envisioned a network of motorways covering the U.S. Together, the industrial designers created a popular, futuristic fantasy, projected forward from the technological and design advances already achieved, that captured the public's imagination in an unprecedented way. Little did they know that, before their dream could be realized, the world was to experience a hiatus during which all progress would be refocused in a new direction, namely victory in war.

BELOW
63. Charles Eames (1907–1978)
Leg splint, 1942
Made by Evans Products Company, Venice, California (U.S.)
Molded plywood
Museum of Modern Art, New York

OPPOSITE
64. ***Stratoliner* passenger plane**, 1940
Made by Boeing (U.S.)

WARTIME DEVELOPMENTS

The war years were crucial in preparing the U.S. for the future, particularly for what came to be seen in Europe as its dominance of postwar design. Most significant was the leap forward that occurred in the invention of new materials, and of new applications for old ones (plate 63). Much has been written about nylon (mostly in the context of the sheer leg coverings that became cheap substitutes for silk stockings), but little has been said about its use in parachutes. Aluminum was widely employed in the war effort, and women were asked to part with their saucepans so that the metal could be melted down and repurposed. Products that had hitherto depended on aluminum had to make do with substitute materials. For example, Parker's *Vacumatic* pen, which had previously featured an aluminum plunger, was redesigned with a celluloid one.

The streamlined forms of prewar automobiles were applied to aircraft fuselages. Take for instance Boeing's *Stratoliner*, introduced as a commercial craft sixteen months before the U.S. entered the war but given over to military use (plate 64). This plane was larger than any that had come before. Its cabin, nearly twelve feet (3.7 m) wide, was big enough to provide berths for overnight travelers. The *Stratoliner* represented the new world of air travel and was admired internationally.

The consultant designers of the interwar years played an important, albeit low-key, role during the conflict, mostly waiting for an era of peace in which they could once again display their talents. One young designer—Eero Saarinen, the son of the Finnish architect Eliel Saarinen who headed the Cranbrook Academy of Art—is known to have had a significant part in the war effort. He was one of a handful of designers employed by the U.S. Office of Strategic Services, formed in 1942 and run by William Donovan. As head of the Special Exhibitions Section, Saarinen used his skills to make models for new weapons and devices, as well as props for the White House war room.

Wartime design focused, therefore, on the inventive reuse of materials and artifacts and the development of new ones, as well as on the strategic application of design techniques. While it did not necessarily produce any new icons, it helped in the war effort, and in preparing society and designers for the dramatic shift that was to occur in the years after 1945.

NX 19901
NX
19901

LEFT
65. **Russel Wright holding a clay model of a chair designed and manufactured for a national furnishings program**, c. 1940

BELOW
66. **George Nelson posing with the *Action Office System* manufactured by Herman Miller**, c. 1964
Vitra Design Museum, Weil am Rhein, Germany

OPPOSITE
67. **Advertisement for Orlon acrylic fiber, manufactured by DuPont**, c. 1950

THE MODERN CONSUMER SOCIETY

The postwar U.S. experienced social and economic changes unmatched elsewhere. A sophisticated consumer society emerged as a significant sector of the population sought modern homes, cars, and household goods with which to express its new affluence. The "American dream" represented a new world of plenty in which consumer goods played a key role. Inevitably, designers were needed to create them. Viewed from abroad, the new homes that occupied the suburban areas around American cities stood for the country's perceived cultural, political, and economic superiority, and in the context of the Cold War, the U.S. was quick to exploit that fact by exporting its vision of domestic life as widely as it could.

In 1953, for example, Edgar Kaufmann Jr., a curator at the Museum of Modern Art (MoMA) in New York, developed an exhibition entitled *American Design for Home and Decorative Use* for display in Finland. The names of well-known American designers appeared in the catalog—among them Eero Saarinen, Charles Eames, Don Wallance, Eva Zeisel, Harry Bertoia, Russel Wright (plate 65), and George Nelson (plate 66). This exhibition was one instance of the U.S. using a cultural program to further its economic agenda in Europe. Indeed, since 1951 MoMA had played a key role in promoting the U.S. as a style leader in Europe. The underlying message was that Europe should embrace the American model, especially its emphasis on the modern home and its accessories as *the* markers of the modern consumer society (plates 68–72). The U.S. saw consumerism as something to be celebrated, a sign of affluence and part of a modernizing strategy that depended strongly on the power of design. The emphasis was firmly placed on the twin themes of innovation and U.S. superiority. On one level, the strategy was successful, as American consumerism and the mass material culture that emanated from it were widely embraced in Europe. On another, it provoked a strong anti-Americanism that, by the mid-1950s, manifested itself in a number of different places across Europe.

Initiatives such as Kaufmann's exhibition rode on the back of the Marshall Plan of 1948–52, under which the U.S. poured large amounts of financial aid into Western Europe in order to strengthen its industrial sector and create a barrier against the Eastern bloc. This program played a key role in the reconstruction of postwar Europe.

While consumer demand continued to set the pace for design innovation in the U.S. in the years after 1945, it was the speed of technological change that made such innovation possible. New production technologies dramatically influenced the way in which manufacturing was organized. Mass production became the common means of ensuring that consumers had an abundance of goods available to them at prices they could afford.

ORLON—new blueprint in fashions with a care-free future!
The young idea in clothes—as in homes—is to look smart with the least upkeep. And Du Pont "Orlon" acrylic fiber makes it possible with women's suits and dresses that have a warm, soft feel . . . an extravagant look—but new and better behavior. Their press is not just for today—but for tomorrow and tomorrow. They're neat-keeping, pleat-keeping . . . and keep their first-day look with just easy care.
In fabrics for your home, in clothes for all the family, you'll find lasting good looks and easy upkeep in fabrics made from "Orlon" acrylic fiber. Whether it's a man's rugged tweed or a curtain marquisette, life will be more fun and less fuss when you look for fabrics of "Orlon."
ORLON is Du Pont's trade-mark for its acrylic fiber.
DU PONT
BETTER THINGS FOR BETTER LIVING
. . . THROUGH CHEMISTRY
E. I. du Pont de Nemours & Co. (Inc.)
Textile Fibers Department
Wilmington 98, Delaware
ORLON®
ACRYLIC FIBER
one of Du Pont's modern-living fibers

BELOW
68. **Advertisement for DuPont paints**, c. 1950

OPPOSITE
69. Greta von Nessen (1900–1978)
***Anywhere* lamp**, 1951
Made by Nessen Studio Inc. (U.S.)
Aluminum, enameled steel
Museum of Modern Art, New York

In addition, technology joined with design in a hungry pursuit of new materials with which to meet consumers' growing desire for novelty and industry's demands for ever cheaper means of manufacturing goods. It continued to be the designer's task to ensure that these new materials acquired forms and meanings both appropriate and desirable.

The period after 1945 saw the beginning of a new phase of industrial expansion in the Western world, and an intensification of the corporate globalism that was to become such a strong feature of late twentieth- and early twenty-first-century culture. It was characterized by the consolidation of the large American corporations formed in the early part of the century—General Motors, DuPont, and General Electric, among many others—and their increasing presence in Europe and beyond. Many of the technological breakthroughs that had occurred during the war years were now exploited commercially, and initiatives put on hold by the war were resumed. As an in-house DuPont publication explained, "postwar America had new needs and made new demands. It was to satisfy these that the company now turned its attention. The program included modernizing plants and adding manufacturing facilities to fill expanded markets for pre-war product lines. Also it encompassed facilities to develop new processes and products which had been delayed by the war." After the war, DuPont moved into the development and manufacture of a range of new synthetic fabrics such as Orlon and Dacron, in addition to the already familiar nylon (plate 67). DuPont's experience echoed that of corporations in many other industries, including the automakers General Motors, Ford, and Chrysler, as well as the large-scale manufacturers of household goods, such as the Tupperware Corporation and General Electric. Although they were all committed to standardized mass production, their growing orientation toward the consumer required them to become increasingly design-led and to offer the market a wide variety of goods. In 1957, for example, one American furniture manufacturer offered "twenty different styles of standard sofas . . . eighteen styles of love seats and thirty-nine upholstered chairs." Inevitably they ranged from traditional to modern in style. Some of the companies that produced consumer goods—including General Electric, Westinghouse, and Chrysler—also derived an extra level of economic stability from military contracts.

BELOW
70. **Advertisement for a General Electric refrigerator with a freezer compartment**, 1953

OPPOSITE, TOP
71. Russel Wright (1904–1976)
American Modern **dinnerware**, 1937
Made by Steubenville Pottery Company (U.S.)
Glazed earthenware
Metropolitan Museum of Art, New York

OPPOSITE, BOTTOM
72. Don Wallance (1909–1990)
One **flatware**, 1952
Made by H. E. Lauffer (Germany)
Stainless steel
Museum of Modern Art, New York

Smart Riding Habit —

"RIDE THE ROCKET!"

Like to meet the most *glamorous* of all the "Rockets"? Just pay a visit to your Oldsmobile dealer and ask for an introduction . . . *to the "98" Holiday Coupé.* Once you've admired those dream-car lines . . . once you've relaxed in that luxury interior . . . you'll lose your heart for sure. But wait! The real thrill comes when you take the wheel. When you feel the surging power of the "Rocket" Engine . . . the carefree smoothness of Hydra-Matic Drive* . . . the gliding ease of Oldsmobile's "Rocket Ride." Here indeed is a "smart riding habit" . . . a very good habit for *you!*

"98" *Below, glamorous "Rocket 98" *Hydra-Matic Drive optional at extra cost. Equipment, accessories, and trim illustrated subject to change without notice.*

A General Motors Value

"ROCKET" OLDSMOBILE

OPPOSITE
73. **Advertisement for the Oldsmobile *Rocket 98***, 1951

BELOW
74. **Students at Black Mountain College, North Carolina**, 1945

THE DESIGN INFRASTRUCTURE

By the late 1940s, design was not just the means by which new consumer goods reached new audiences; it had also developed its own infrastructure and was becoming a force to be reckoned with. Professional designers had begun to organize themselves in the late 1920s in order to protect their creative efforts. In 1944, fifteen well-known designers came together to create the Society of Industrial Designers, whose first president was Walter Dorwin Teague.

Design education was also well established by the early postwar years. One important school, Black Mountain College, had been established in North Carolina in 1933 (plate 74). Although its teaching had a fine and performing arts emphasis, the important American designer and inventor Buckminster Fuller was an early alumnus, while the Bauhaus's preeminent textile designer, Anni Albers, was a member of its teaching staff. In fact, a number of the Bauhaus designers who had fled Germany in the 1930s became influential in the U.S. The "New Bauhaus," founded in Chicago in 1937, clearly drew on the work of its German predecessor; former Bauhaus teacher László Moholy-Nagy was its founding director. In 1944, it changed its name to the Institute of Design. Gradually the European staff was supplemented by Americans, and in the 1950s the school became part of the Illinois Institute of Technology. The Cranbrook Academy of Art, established in 1932, also played an important role in this context. Set up by the Finnish architect Eliel Saarinen, its alumni included his son Eero, Charles and Ray Eames, and Florence Knoll.

The imported ideas of European modernism remained an important component of postwar American design. Their influence was felt most strongly in the areas of architecture, furniture, and the decorative arts. This contrasted significantly with the distinctively American sphere of product and automobile design, which was more populist and focused on the idea of "dreams that money can buy" (plate 73). Not surprisingly, these two approaches—one oriented toward high culture, and the other toward a mass audience—frequently came into conflict. Where the consuming public was concerned, however, it was merely a matter of stylistic choice. While the Bauhaus look suited the new open-plan lifestyle of the domestic living and dining areas, the kitchen and the garage housed goods that were less high-minded and more stylistically evocative. Automobiles and refrigerators boasted bulbous metal forms, while dining chairs featured slim, tapered shapes of Scandinavian origin.

75. Conrad Buff III (1926–1989), Calvin Straub (1920–1998), and Donald Hensman (1924–2002)
***Case Study House No. 20,* Altadena, California**, 1958
Photograph by Julius Schulman
J. Paul Getty Trust Research Institute, Malibu

THE AMERICAN KITCHEN

By the 1950s, the American dream was unequivocally evoked by the suburban home (plate 75). Popular magazines printed in strong colors featured loving married couples and whole families living the ideal life in their horseshoe-shaped kitchens. Gone was the small, all-white laboratory kitchen of the interwar years. Instead, 1950s open-plan kitchens, managed by glamorous housewife-hostesses, featured bulbous refrigerators in a range of colors. In 1955, Frigidaire, for example, offered its refrigerators in "Sherwood green," "Stratford yellow," and "Snowy white."

These could be accompanied by floral wallpaper and an array of new, equally colorful streamlined gadgets, from mixers to toasters. Postwar mixers were among the most stylized products in the world. By the 1950s, a wide range of manufacturers, including Sunbeam, Dormeyer, Hamilton Beach, Westinghouse, KitchenAid, Montgomery Ward, and Universal, had entered into the highly competitive marketplace for mixers. The iconic Sunbeam *Mixmaster* owed a huge debt to contemporary automobile styling, especially door-handle designs (plates 76, 77).

One product, Tupperware, may serve to represent a number of characteristics of the American consumer society of the 1950s. While the polyethylene food containers made by the Tupperware company were simple and functional, rather than overtly stylish (plate 78), they played an important role in the public acceptance of plastic goods.

Tupperware was invented by Earl Silas Tupper, a former tree surgeon who moved into the plastics industry and set out to create an airtight food container. His aim was to ensure that leftover food could be stored in refrigerators, thereby avoiding waste. The simple products he created were not intended for display, therefore, but rather to be tucked away out of sight. As such, they could not be marketed on the basis of their appearance. Their most important selling point was, in fact, the seal joining the lids to the containers to make them airtight, a design feature that could be recognized only by the noise emitted when they were opened, dubbed the Tupperware "burp."

The containers' near invisibility provided a serious selling challenge that was addressed in the early 1950s by one of the company's door-to-door saleswomen, a single mother named Brownie Wise. Her strategy was to completely sidestep the retail sector and develop a method of direct selling that involved working face-to-face with female consumers in their own homes.

By the middle of the decade, the Tupperware party had taken shape, and countless suburban hostesses were selling Tupperware items at gatherings in their living rooms (plate 79). This subtle sales strategy built on the shift in emphasis, initiated in the prewar years, from the housewife's identity as homemaker and consumer to her new one as hostess. Wise "Tupperized" suburban women by providing them with what they needed most—a social life, rather than food containers—and by building on their desires. At first Tupperware items were bought as gifts, but gradually women felt able to purchase them for themselves. To offset the inherent rationality (waste-saving) of the Tupperware product and transform them into objects of desire, Wise introduced a high level of glamour into the party culture she promoted, setting the pace herself by wearing fashionable clothes, upholstering her furniture in flamingo pink, and driving a pink Cadillac. Wise, in effect, designed an image and a lifestyle for Tupperware, which women bought along with these simple plastic goods.

In 1956, like the Jeep before it, the neutral Tupperware container was selected as an item of "good design" by the curators at MoMA. By that time, it had come to be seen, in the words of Alison J. Clarke, as a "shining beacon of hope in a period defined by an unprecedented rise in mass consumption and a perceived decline in mass consumer taste." This unobtrusive, practical object had apparently reversed that trend and repositioned the designer-engineer, rather than the popular stylist, as the hero of the modern world. However, MoMA's validation of the Tupperware container ignored the fact that its success was due to its novel method of marketing rather than the inherent effectiveness of its simple design.

BELOW
76. ***Mixmaster* stand mixer**, 1953
Made by Sunbeam Corporation (U.S.)
Science Museum, London

RIGHT
77. **Advertisement for the Sunbeam *Mixmaster Junior***, 1956

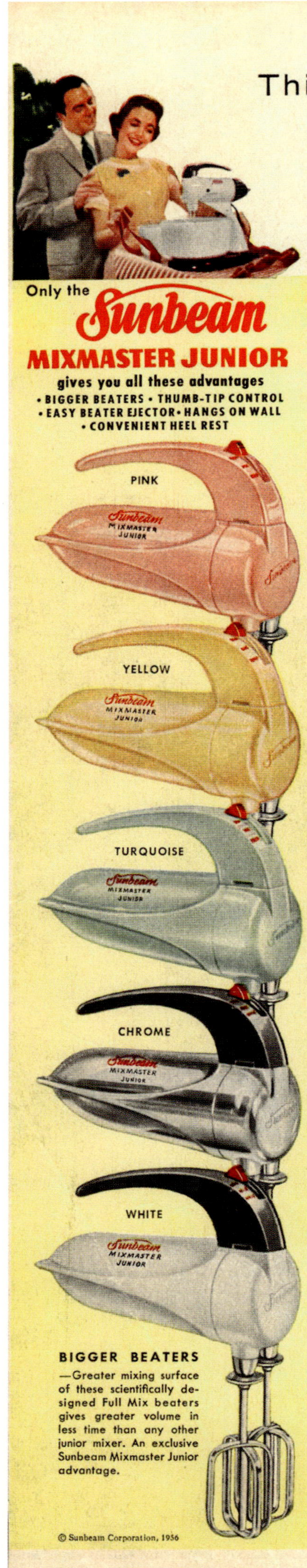

LEFT
78. Earl S. Tupper (1907–1983)
Milk jug and water jug, 1946
Made by Tupper Corporation (U.S.)
Polyethylene
Museum of Modern Art, New York

BELOW
79. **Tupperware party, Sarasota, Florida**, 1958
Museum of Modern Art, New York

BELOW
80. **Bill Mitchell (left) and Paul Gillian with a model of the *Firebird IV* in the General Motors pavilion at the 1964 New York World's Fair**
General Motors Heritage Center, Sterling Heights, Michigan

OPPOSITE, TOP
81. **Advertisement for General Motors' Pontiac brand**, 1958

OPPOSITE, BOTTOM
82. **Advertisement for the Studebaker *Commander***, c. 1950

THE AUTOMOBILE INDUSTRY

The automobile played a major role in the expansion of the suburbs in the postwar years. Every housewife had to have a car to do the shopping, while her husband drove to work on the new highways. Increased affluence made the purchase of a car possible for more people than ever before

Even more than the creators of household gadgets, the American automobile designers of the 1950s took the idea of styling to its limits and were responsible for some of the most visually extreme products of the era. While some independent companies—Kaiser Frazer, Tucker, and Studebaker—produced striking cars in the early postwar years (plate 82), the leading forces in automotive design were General Motors, Chrysler, and Ford. Competing neck and neck from the mid-1950s onward, these companies incrementally added extra features to their cars in the battle to attract customers. The add-ons included more chrome trim and increasingly stylized tail fins, taillights, and radiator grilles, inspired by jet and rocket imagery; technical novelties, like power steering and electric windows; and luxurious interior amenities such as beauty accessories, drink cabinets, and cigarette lighters (plate 81). The designers created a technological utopia that was enormously attractive to a generation living in the early days of space exploration and believing in the redemptive power of modern technology. Many of them had worked in aeronautics during the war, and some had served apprenticeships in Harley Earl's styling section at General Motors. Typically (with the exception of Raymond Loewy for Studebaker) automobile designers worked on an in-house, rather than a consultancy, basis. Virgil Exner, Bob Bourke, Bill Mitchell, George Walker, Eugene Boudinat, Howard Darrin, and Alex Tremulis (most of them unknown to the general public, as it was the name of the car, not its stylist, that was advertised) are the among the designer-heroes of the era (plates 80 and 83).

Earl maintained his position as the leading automobile designer of the day. He was heavily influenced by jet fighters and space travel, and believed that getting into a car should be like going on vacation. It was he who introduced the tail fin, in the 1951 Buick *LeSabre* concept car, named after the F-86 Sabre fighter jet (plate 84). Earl also developed the curved windshield, which was inspired by airplane cockpits. His pièce de résistance was the 1959 Cadillac *Eldorado* (plate 85), whose huge fins sported "rocket" taillights.

You get the solid quality of Body by Fisher.
Who in the World Built this Beauty?
Big and wonderful and beautiful things have happened to Pontiac—America's Number One Road Car! And they start with an engineering advance no other car has.
Engineering's hottest team has now changed the very foundation of automotive design to bring you Wide-Track Wheels and a ride that never before existed! This is the spark that ignited a chain reaction of new ideas. You'll discover fresh new lines . . . vista-lounge interiors with full 360-degree visibility . . . seats wider than a sofa . . . True-Contour Air-Cooled Brakes for smooth, easy, unvarying stops. And the industry's most advanced V-8 reaches new peaks of perfection in Pontiac's spectacular new Tempest 420 V-8—with sound and vibration reduced almost beyond belief.
For extra gas mileage there's a new companion engine, optional at no extra cost, that gives you full V-8 power and pep plus the mileage of smaller cars with so-called "economy engines"—and it uses regular gas!
If all that sounds like more than enough for one car—that's the new Pontiac! Come in to see this new wide-track Pontiac. Bring all the doubts you can muster—you won't leave with a single one!
PONTIAC!
America's Number 1 Road Car
3 Totally New Series • Catalina • Star Chief • Bonneville
EXCLUSIVELY YOURS—WIDE-TRACK WHEELS
Wheels moved out a full 5 inches for the widest, steadiest stance in America—better cooling for engine and brakes—lower center of gravity for better grip on the road, safer cornering, smoother ride. You get the most beautiful roadability in the whole wide world!
PONTIAC MOTOR DIVISION . GENERAL MOTORS CORPORATION

BELOW, TOP
83. Alex Tremulis (1914–1991)
Tucker *Sedan*, 1948
Made by the Tucker Corporation (U.S.)

BELOW, BOTTOM
84. Harley Earl (1893–1969)
Buick *LeSabre* convertible, 1951
Made by General Motors (U.S.)

OPPOSITE
85. Harley Earl (1893–1969)
Cadillac *Eldorado* (detail of tail fins), 1959
Made by General Motors (U.S.)

THE MUSEUM OF MODERN ART AND "GOOD DESIGN"

Sharply opposing the image of excess and luxury conveyed by American automobiles in this era, MoMA set out its shingle as an arbiter of good design, which it equated with the concept of modernist design. Edgar Kaufmann Jr.'s rhetorical description of the concept of good design stressed the qualities of "integrity, clarity and harmony." Significantly, the examples he cited were all chairs created by early modernists such as Marcel Breuer and Le Corbusier, as well as by a number of later adherents of the same school of design, including the Scandinavians Finn Juhl and Bruno Mathsson and the Americans Charles Eames and George Nelson. Just as Nikolaus Pevsner had identified a stable of pioneer modernists in the mid-1930s, so Kaufmann's work at MoMA established an elite cadre of innovative second-generation modernists who became known internationally. MoMA's message was clear: Modernism was alive and well in postwar Scandinavia and the U.S., epitomized by the work of a certain group of designers. As for work that catered to other tastes, in the resounding words of Kaufmann, "Streamlining is not good design." American modernists like him felt that the country's postwar design strategy should be to emulate Europe, rather than to acknowledge the indigenous industrial design movement of the interwar years, which they considered vulgar by comparison. The specter of the superiority of European taste lingered on.

Exhibitions continued to place design before the American public in the 1940s and '50s, although none of the postwar events held at MoMA were conceived on the same scale, or communicated the same level of popular excitement, as the 1939 World's Fair. Nevertheless, the public was presented with a concept of good modern design that, they were told, would improve their lives significantly. The good-design campaign was managed not by the federal government but by the culture industry, which took on the task of helping manufacturers and retailers persuade consumers to shop with discretion.

Culture and commerce collaborated closely; one example is MoMA's work with the Merchandise Mart, a wholesale marketplace in Chicago. In 1950, as director of the Department of Industrial Design at MoMA, Kaufmann responded to a request from the Merchandise Mart to help it publicize good, modern design to manufacturers and consumers. The result was a series of six annual *Good Design* shows (plate 86), which

LEFT
86. **View of the *Good Design* exhibition at the Museum of Modern Art, New York**, 1953
Museum of Modern Art, New York

OPPOSITE
87. Charles Eames (1907–1978) and Ray Eames (1912–1988)
***DAR* (*Dining Armchair Rod*) chair**, 1948–50
Made by Herman Miller (U.S.)
Chrome-plated tubular steel, fiberglass-reinforced polyester
Museum of Modern Art, New York

proved very popular with the American public and provided an opportunity for the concept of good design to be debated in the press. These exhibitions were dominated by furniture and domestic items, a large fraction of which was either Scandinavian in origin or heavily influenced by the Scandinavian craft-based aesthetic.

Two cultures of design were clearly emerging in the postwar U.S.: one, destined for the living room, was strongly influenced by Europe and had a clear social cachet; the other, an indigenous movement with more popular origins and appeal, was visible in the street and in the kitchen, in the streamlined forms of automobiles and refrigerators. MoMA did try to bridge the cultural gap between these two schools of design by holding two exhibitions dedicated to the automobile, in 1951 and 1953. Its selection process, however, reflected its taste in furniture, as the cars on view all had a strong European flavor.

THE FURNITURE DESIGNERS

In 1941, the team of Charles Eames and Eero Saarinen was awarded first prize in a competition called Organic Furnishings in the Home, organized by MoMA. Eliot Noyes, the impetus behind the competition, had offered the following challenge: "A new way of living is developing, [which] requires a fresh approach to the design problems and a new expression." Eames valued natural materials, but, in seeking to give them new forms, he also pushed them to the edges of their naturalness. The organically shaped, molded plywood chair shells he developed with Saarinen for the 1941 competition represented one of the most significant technological breakthroughs in twentieth-century design, and the chairs, screens, and coffee tables he created with his wife Ray through the 1940s went on to become design classics.

A later MoMA project, the Low Cost Furniture Competition of 1948, saw the Eameses present their *DAR* chair to the world (plate 87). This molded fiberglass shell mounted on a metal frame was an important step toward the solution of a problem that occupied designers and manufacturers from the late 1940s through the 1960s, that of creating a one-piece plastic chair.

Working in the middle years of the century, Charles and Ray Eames were inspired by the innovations of the modernist architect-designers of the prewar era, in both Europe and the U.S. At the same time, they had the distinct advantage of being able to see where the ideals that drove modernist architecture and design had led and, more important, where they had failed. There can be no doubt that modernism underpinned the Eameses' oeuvre and determined its core values, however. Always their vision was inspired by the challenges of new technologies and forms and implemented to meet the needs of a lifestyle defined by modernity. Nowhere was this more evident than in their designs for furniture (plates 88, 89). From a background in architecture, Charles Eames had ventured into furniture design as means of testing new materials and production techniques, and of discovering new forms that would enable new ways of living.

The Eameses continued to pursue the formal challenges of new technologies in their furniture designs of the late 1940s and the 1950s. For example, their molded plywood chairs of 1946 with metal-rod legs exploited

BELOW
88. Charles Eames (1907–1978) and Ray Eames (1912–1988)
Three-legged side chair,
c. 1944
Made by Evans Products Company (U.S.)
Molded plywood, lacquered metal, rubber
Museum of Modern Art, New York

OPPOSITE
89. Charles Eames (1907–1978) and Ray Eames (1912–1988)
***RAR* rocking armchair**,
1948–50
Made by Herman Miller (U.S.)
Fiberglass-reinforced polyester, metal, birch, rubber
Museum of Modern Art, New York

an innovative method of gluing metal to wood with the addition of rubber pads, while their fiberglass *DAR* chair of 1948 was well ahead of its time. It was, as Edgar Kaufmann Jr. explained in his *Introductions to Modern Design*, the "first one-piece plastic chair to feature the natural surface of its material, variegated and satiny." The modular storage cabinets of 1950 were, Kaufmann wrote, "the first . . . to forsake traditional furniture construction." Perhaps no single designed artifact expressed the idea of the modern lifestyle better than the Eameses' famous rosewood lounge chair of 1956, a compromise between contemporary style and comfort (plates 91, 92).

Through the late 1950s and the 1960s, the Eameses went on to create a series of designs in aluminum, from lounge chairs to secretaries' chairs and multiple seating (plate 90). They quickly became iconic objects in a range of modern public interiors, from offices to airports, transforming those spaces with the clean, rational lines of modernism but adding a significant level of lightness and comfort at the same time. The aluminum pieces were both aesthetically and technologically sophisticated. The 1956 lounge chair, for example, reclined and swiveled. It had a thin, flat profile, and the structural ribs of the continuous seat and back echoed those of the body that sat in it. These elegant, modern objects were at home in an executive office or a reception area but could, like the lounge chair, equally find a place in the domestic arena.

The Eameses also designed their own home in Pacific Palisades between 1946 and 1949. Throughout his career, Charles saw himself as an architect first and foremost. "I think of myself officially as an architect," he explained. "I can't help but look at the problems around us as problems of structure—and structure is architecture." Charles's love affair with new materials and production techniques underpinned the design of the Eameses' house, which was made entirely from prefabricated elements.

Throughout their working lives, the Eameses' commitment to modernism was absolute and uncompromising. Ray Eames had a background in avant-garde art—painting and sculpture—and before studying at Cranbrook in the late 1930s (where she met Charles, then head of the design department, and married him in 1941), she had worked with Hans Hofmann in New York. Ray had gravitated to graphic design, and her contribution to the couple's work lay in the application of her tasteful eye and her ability to "decorate." Thus, while Charles's interest in structure led him to create the shell of the Pacific Palisades house, Ray's eye for detail turned the house into a home and, in this way, made modernism acceptable.

The numerous photographs of the interior of the Eameses' house, in addition to the short film they made about it in 1955, entitled *House after Five Years of Living*, testify to the way in which they changed a static, technologically determined modernism into an organic, living process that embraced its participants (plate 93).

90. **Chairs by Charles and Ray Eames presented in a Herman Miller catalog**, c. 1955
Vitra Design Museum, Weil am Rhein, Germany

INCORPORATING REVOLUTIONARY PRINCIPLES IN BOTH DESIGN AND CONSTRUCTION, THESE CHAIRS AND TABLES DESIGNED BY CHARLES EAMES HAVE WON WORLD-WIDE ACCLAIM AS FOREMOST EXPRESSIONS OF GOOD MODERN DESIGN. THEIR EXCEPTIONAL COMFORT, PRACTICALITY AND CRISP SILHOUETTES, COUPLED WITH THEIR EXTREMELY MODEST PRICES, HAVE GAINED THEM A CONSTANTLY EXPANDING ACCEPTANCE FROM COAST TO COAST.

THESE CHAIRS AND TABLES CAN BE PURCHASED THROUGH FURNITURE DEALERS EVERYWHERE. ON VIEW AT OUR SHOWROOMS IN NEW YORK, CHICAGO, GRAND RAPIDS, KANSAS CITY AND LOS ANGELES.

herman miller furniture company

TABLES PLYWOOD AND PLASTIC TABLES FOR MANY USES ARE AN IMPORTANT PART OF THE EAMES GROUP. DINING, CARD, COFFEE AND INCIDENTAL TABLES ARE INCLUDED, SEVERAL WITH FOLDING OR DETACHABLE LEGS.

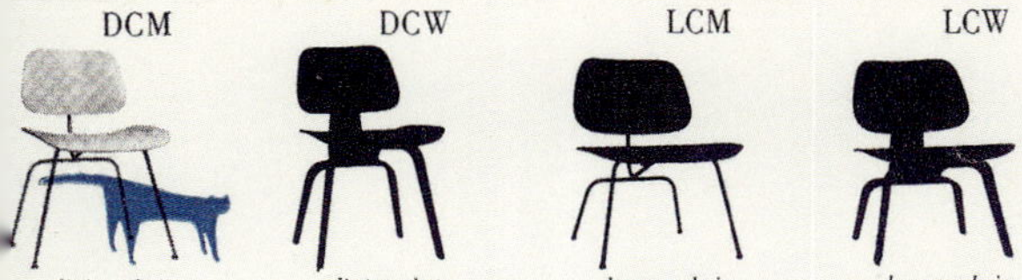

DCM	DCW	LCM	LCW
dining chair rod base seat ht. – 17¾"	*dining chair wood base seat ht. – 17½"*	*lounge chair rod base seat ht. – 15½"*	*lounge chair wood base seat ht. – 15½"*

WOOD:

MOLDED PLYWOOD SEATS AND BACKS, ATTACHED BY RUBBER SHOCK MOUNTS TO FRAMES OF HEAVILY PLATED METAL OR WOOD, YIELD A RESILIENCE AND COMFORT NORMALLY FOUND ONLY IN UPHOLSTERED CHAIRS. AVAILABLE IN WALNUT, BIRCH, CALICO ASH, RED OR BLACK, IN LOUNGE AND DINING HEIGHTS.

FINISHES

DAR	PAW	DAW	DAX	SAX	RAR	LAX	LAR
ing or desk chair wire base seat ht. – 17⅞"	*swivel chair wood base seat ht. – 17⅞"*	*dining or desk chair wood base seat ht. – 17⅞"*	*dining or desk chair rod base seat ht. – 17⅞"*	*standard height chair rod base seat ht. – 16¾"*	*rocking chair rocker base seat ht. – 16"*	*lounge chair rod base seat ht. – 14⅞"*	*low lounge chair wire base seat ht. – 12¼"*

PLASTIC:

A PLASTIC REINFORCED WITH FIBERGLAS IS SKILFULLY MOLDED TO FORM AN EXCEPTIONALLY COMFORTABLE AND VIRTUALLY INDESTRUCTIBLE ARMCHAIR. THE PLASTIC SHELL COMES IN EIGHT ATTRACTIVE COLORS; CHOICE OF SIX DIFFERENT BASES, OF BOTH WOOD AND METAL, INCLUDING A ROCKER AND SWIVEL DESK CHAIR.

FINISHES

DKR	PKW	DKW	RKR	LKX	LKR
ing or desk chair wire base seat ht. – 18½"	*swivel chair wood base seat ht. – 17⅞"*	*dining or desk chair wood base seat ht. – 17⅞"*	*rocking chair rocker base seat ht. – 17"*	*lounge chair rod base seat ht. – 16"*	*low lounge chair wire base seat ht. – 13"*

WIRE:

A MOLDED WIRE SHELL, UPHOLSTERED WITH A ONE OR TWO PIECE CUSHION, EASILY REMOVABLE, OFFERS EXCEPTIONAL SEATING COMFORT AT LOW COST. CUSHIONS CAN BE HAD IN A GOOD-LOOKING TEXTURED FABRIC OR GENUINE LEATHER. THERE ARE SIX BASES, EMPLOYING BOTH METAL AND WOOD, IN STYLES SUITABLE FOR DINING, READING AND LOUNGING.

FINISHES

CUSHION IS EASILY REMOVED FROM WIRE SHELL. CUSHIONS ARE INTERCHANGEABLE.

BELOW
91. **Exploded view of the *Lounge Chair 670* by Charles and Ray Eames**, 1956
Eames Office LLC, Santa Monica, California

BOTTOM
92. Charles Eames (1907–1978) and Ray Eames (1912–1988)
***Lounge Chair 670* with ottoman**, 1956
Made by Herman Miller (U.S.)
Polished aluminum, molded rosewood veneer, latex, down fill, leather, plastic
Museum of Modern Art, New York

OPPOSITE
93. **Charles and Ray Eames in their house in Pacific Palisades, Los Angeles**, 1958
Photograph by Julius Schulman
J. Paul Getty Trust Research Institute, Malibu

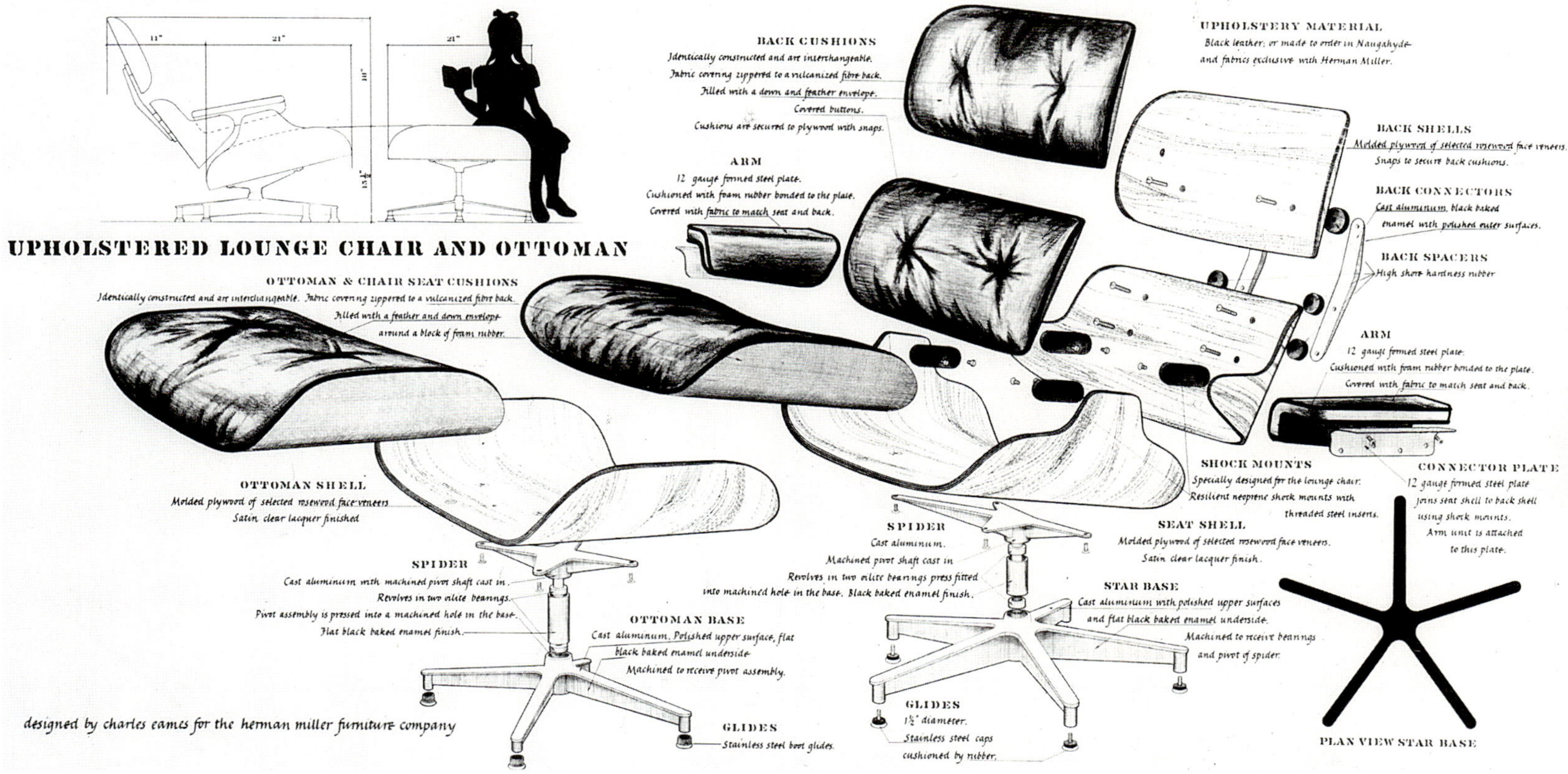

The countless "primitive" objects collected on trips abroad, the toys they loved so much, and the decorated textiles and historical artifacts that caught their eyes increasingly filled the interior spaces of the house and transformed a lifeless structure into an animated container filled by the lives of its inhabitants.

A move into exhibition design, multiple-screen communication, and film came in the 1950s, '60s, and '70s. Eames Demetrios, Charles Eames's grandson, has explained that the pair "applied the structure and discipline of architecture to filmmaking." Their film *Blacktop* of 1952, for example, focused on patterns made by soap bubbles. *A Communications Primer* of 1953 was one of the Eameses' first attempts to use film as an educational tool, in this instance to explain communications theory to architects.

Much of the Eameses' best work was created in collaboration with IBM. In 1961, they designed an exhibition called *Mathematica* for the company at the California Museum of Science and Industry. This project increased their interest in educating the public, and the interactive installations they created remained on display for thirty-seven years—a testament to their ability to communicate ideas to a mass audience. Demetrios has commented that "Ray . . . and Charles saw no distinction between education and fun and play." This approach lay at the very heart of their endeavors.

Charles and Ray Eames dominated American design in the middle years of the twentieth century and beyond. Their interiors became *the* idealized vision with which modern society defined and expressed

itself, both at home and at work. The links the couple made with American corporate culture through their close association with companies such as IBM positioned them at the global center of things.

The Eameses' colleague Eero Saarinen, the son of the Finnish architect Eliel Saarinen, who had founded the Cranbrook Academy of Art, also took up the challenge of exploiting the properties of plastic, in his *Tulip*

BELOW
94. Eero Saarinen (1910–1961)
***Tulip* table and chairs**, 1956
Made by Knoll International (U.S.)

OPPOSITE
95. Eero Saarinen (1910–1961)
***Tulip* chair**, 1956
Made by Knoll International (U.S.)
Cast aluminum, polyester, latex foam, fiberglass, fabric
Musée National d'Art Moderne, Centre Georges Pompidou, Paris

chair of 1956. While the *Tulip* appeared to be a one-piece plastic chair, it actually consisted of a molded plastic shell mounted on a metal pedestal (plate 95). The practicalities of removing plastic objects from molds, combined with the limited structural strength of the plastics then available, made it impossible to produce a true all-plastic chair at that time.

Whereas the Eameses rose to fame in collaboration with the furniture manufacturer Herman Miller, Saarinen worked with his fellow Cranbrook graduate Florence (Schust) Knoll, who, along with her husband Hans, built the Knoll furniture company into one of the leading producers of progressive modern furniture in the U.S. Together Saarinen and Knoll oversaw the production of a range of chairs: the *Grasshopper* lounge chair and ottoman of 1946, the *Womb* chair and ottoman of 1948 (plate 97), the *Womb* settee of 1950, and several side chairs and armchairs between 1948 and 1950. The *Tulip* group, introduced in 1956, also included side chairs and armchairs, as well as dining, coffee, and side tables and a stool—all characterized by a modern simplicity and elegance and evoking a sophisticated modern lifestyle (plate 94).

Another Cranbrook alumnus, Harry Bertoia, also made a mark for himself in these years by collaborating with Knoll. Although a sculptor and a jewelry designer, among other things, he is best known for his series of wire chairs. Constructed from a lattice of welded steel,

RIGHT
96. Harry Bertoia (1915–1978)
420A chair, 1952
Made by Knoll International (U.S.)
Steel, metal tubing, Rilsan, latex foam, polyvinyl chloride
Musée des Arts Décoratifs, Paris

BELOW
97. Eero Saarinen (1910–1961)
***Womb* chair and ottoman**, 1947–48
Made by Knoll International (U.S.)
Steel, polyester reinforced with fiberglass and wood inserts, lacquered tubular steel, foam, fabric
Musée National d'Art Moderne, Centre Georges Pompidou, Paris

these skeletal structures allowed the passage of light, and thus did not interrupt the flow of space in the rooms they occupied (plate 96). The most successful of the group was dubbed the *Diamond* chair because of the pointed form of its back (plate 98). These pieces were as near to being made of air as any seat could be and appealed greatly to the modern desire for a new simplicity in interiors. Its affinity with sculpture also gave the *Bertoia Collection* an appealing distinctiveness in that era of mass culture and mass consumption. In essence, Bertoia and his fellow Cranbrook graduates brought a sense of European style and sophistication to an American market that was, at the same time, enthralled by Harley Earl's monster automobiles.

Another American designer of the 1950s who understood the world of "good design" imported from across the Atlantic was George Nelson. Partly following the model of the interwar consultant designer and partly adhering to the new modern taste, Nelson made a name for himself with a range of products that included a clock with rays emanating from it (plate 100) and some innovative items of furniture (plates 99, 102). Nelson is perhaps best known for his book *Tomorrow's House*, coauthored with Henry Wright, in which he introduced a number of new concepts, among them the "family room" and the "storage wall." This book brought him to the notice of the chairman of Herman Miller, and Nelson served as the company's director of design from 1945 to 1972.

The decorative arts were also given the modern treatment in this period. Russel Wright had begun his career as an industrial designer in the interwar years, but in 1939 he created a set of colorful ceramic dinnerware, eventually known as *American Modern*, that became hugely popular and remained in production until 1959. Along with his aluminum pieces and furniture

items, Wright's ceramic designs brought modernism within the reach of the ordinary consumer. He also worked in plastics—melamine in particular—and injected a new modern aesthetic into items made from that material. Eva Zeisel, a Hungarian immigrant, was responsible for a number of modern ceramic designs that transformed the domestic landscape, beginning in the interwar years. Her style, which combined an organic aesthetic with a strong sense of color, continued to be hugely popular in the postwar era. In 1946, she was given the first one-woman show at MoMA.

98. Harry Bertoia (1915–1978)
***Diamond* chair**, 1952
Made by Knoll International (U.S.)
Soldered chrome-plated steel wire and tubing, latex, fabric
Musée des Arts Décoratifs, Paris

OPPOSITE
99. George Nelson (1908–1986)
***Coconut* armchairs**, 1955
Made by Herman Miller (U.S.)
Vitra Design Museum,
Weil am Rhein, Germany

BELOW
100. George Nelson
(1908–1986)
***Atomic* clock**, 1949
Made by Howard Miller
Clock Co. (U.S.)
Lacquered wood, brass
Musée National d'Art Moderne,
Centre Georges Pompidou,
Paris

BELOW
101. **Illustration for the AA armchair, or *Butterfly* chair, created in 1938 by Jorge Ferrari Hardoy, Juan Kurchan, and Antonio Bonet, and sold in the U.S. by Knoll,** 1951
Private collection

OPPOSITE
102. George Nelson (1908–1986)
***Marshmallow* sofa,** 1956
Made by Herman Miller (U.S.)
Lacquered steel, vinyl-upholstered cushions
Vitra Design Museum, Weil am Rhein, Germany

INTERIOR DESIGN

Founded in 1905 and based in Zeeland, Michigan, Herman Miller was one of the first American furniture manufacturers to enter the design field, beginning its collaboration with the modernist Gilbert Rohde in 1932. As noted, the company hired George Nelson as its director of design in 1945, and he enlisted the services of such figures as Eames, Saarinen, Bertoia, and Noguchi. Nelson continued to play a key role at Herman Miller into the 1970s; important projects initiated by him included *Action Office I* and *Action Office II*, thoroughgoing reconsiderations of the modern workspace that were designed by Robert Probst in the 1960s, on the basis of extensive research.

In the 1970s, Don Chadwick and Bill Stumpf became two of Herman Miller's key designers. Together they created the *Equa* office chair of 1984 and the highly successful *Aeron* chair of 1994, which became one of the most iconic office chairs of that decade and beyond, able to adapt naturally to most bodies and 94 percent recyclable. Following the earlier efforts of Henry Dreyfuss and others, the emphasis in these designs was on ergonomics, or the human-object interface.

From its ongoing commitment to rigorous research, which resulted in new approaches to the design of offices and office furniture, Herman Miller went on to address the highly topical issue of sustainability by eliminating harmful chemicals from its products, generating no hazardous waste, developing a zero air emissions policy, and using 100 percent green electrical energy. It also initiated a recycling program and ensured that its products could be easily disassembled at the end of their useful life. The company's present factory, designed by William McDonough + Partners and opened in 1995, incorporates green design principles and is known as the *Greenhouse*.

The other furniture and interior design company with a long history that has continued to be active and influential in the early twenty-first century is Knoll. Founded in 1938 by Hans and Florence Knoll and based in Greenville, Pennsylvania, this firm has always seen its task as that of selling innovative pieces by progressive designers, both by commissioning new designs (plates 103, 105) and by acquiring the rights to a number of seminal international designs, such as, in 1953, the furniture designs of Mies van der Rohe and, in 1968, Marcel Breuer's *Wassily* chair from the Bauhaus (plates 101, 104). Today Knoll's products—both its classic modern and its contemporary pieces—continue to be

103. Florence Knoll (b. 1917)
Credenza, 1950
Made by Knoll International (U.S.)
Solid varnished teak, chrome-plated metal legs
Musée d'Art Moderne, Saint-Étienne Métropole

popular with architects specifying furnishings for both residential and public interiors.

In the twenty-first century, the U.S. continues to distinguish itself in the field of furniture and interior design, especially office design. It still, for example, supports strong manufacturers of designed goods, such as Steelcase, a century-old producer of office furniture and systems, based in Grand Rapids, Michigan, that operates globally. While Steelcase does not engage with design in a particularly innovative way, it recognizes, nonetheless, that it needs to bring together marketing and design to maintain its existing markets and develop new ones.

As product design has become increasingly strategic—often abandoning the physical object to pursue the type of business consultancy known as "design thinking," discussed below—interior design has expanded to become big business and continues to fulfill one of the most important roles of American design since World War II, that of bridging the gap between style and social status. In 2009, for instance, the Californian Michael S. Smith was selected to design the private living areas of the Obama White House. Like many other interior designers of his generation, Smith has a talent for combining tradition with style, and thus was seen to be the right choice for this prestigious commission.

While the U.S. has continued to produce its own tastemakers in furnishings and interior design, it has continued to recognize the leadership of Europe in this context. In the 1980s and '90s, for example, the hotelier and entrepreneur Ian Schrager brought in a number of notable European designers—Philippe Starck and Andreé Putman among them—to create strikingly avant-garde interiors that would appeal to style-conscious customers. The first Schrager hotel, the Morgans in New York City, was designed by Putman and opened in 1984. It was followed by the Royalton and the Paramount, both Starck creations. Starting in 2005, Schrager worked with Julian Schnabel, the Swiss architects Herzog & de Meuron, and the British architect John Pawson on several other hotel and residential projects. Another acknowledgment of European dominance in furnishings and interiors came in the form of a large exhibition of contemporary European design at the Indianapolis Museum of Art in 2009.

THE END OF CONSUMERISM

In the wake of the flamboyant consumerism and exuberant design of the 1950s, the 1960s in the U.S. looked a little more staid. Automobiles became more boxlike in shape, and consumers accepted a more sober aesthetic in their domestic lives. Refrigerators acquired sharper corners, and the color palette for the home became more restrained. This pendulum swing was largely in response to a number of wake-up calls about the waste that was being created through overconsumption and about the irresponsibility inherent in the materialism of the earlier postwar decades.

Two key writers need to be mentioned in this context. First, there is Vance Packard, who published a series of books that focused attention on the downside of the consumer society. *The Hidden Persuaders* of 1957 sought to expose the workings of American advertising and its subliminal methods. Two years later *The Status Seekers* appeared, while in 1960 *The Waste Makers* explained how the American economy depended on the concept of planned obsolescence. Nowhere was this

RIGHT
104. Herbert Matter
(1907–1984)
Advertisement for chairs by Harry Bertoia sold by Knoll International
L'oeil, No. 17, May 1956

BELOW
105. Warren Platner
(1919–2006)
***Platner* chairs, stool, and table**, 1966
Made by Knoll International (U.S.)
Steel wire, cloth upholstery

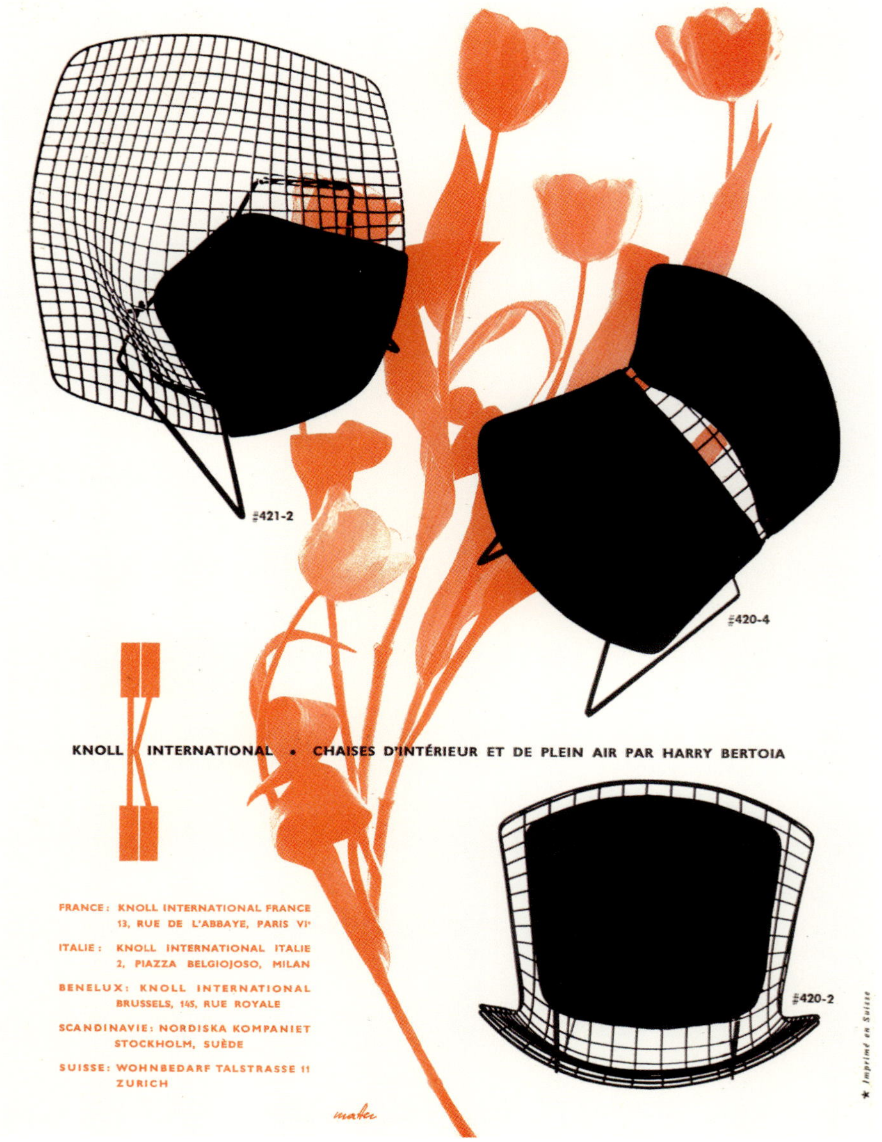

106. Henry Dreyfuss
(1904–1972)
"Human Measurements," from *Designing for People* (New York: Simon & Schuster, 1955)
Bibliothèque Forney, Paris

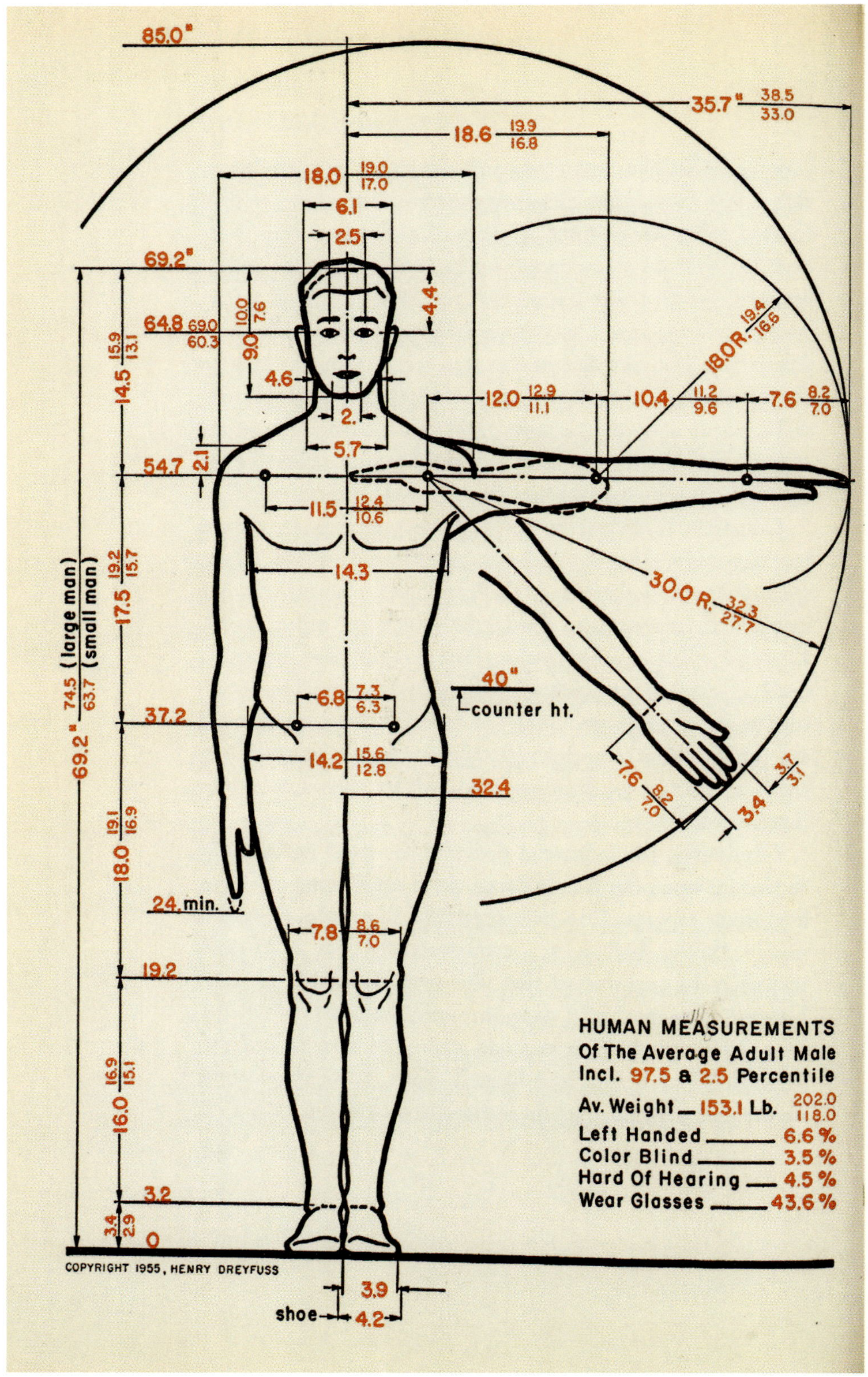

more apparent than in the automotive industry, whose products had a life span built into them. Packard's books were hugely influential in awakening the popular conscience, and the positive image of the consumer society went into a rapid decline. This trend was only reinforced by the writings of the political activist Ralph Nader, whose best-selling *Unsafe at Any Speed* (1965) pointed out the safety hazards of American cars in general and of the Chevrolet Corvair in particular. Through the course of the 1960s, it became increasingly obvious that the era of excess was over and that design had to reorient itself to a new context.

The careers of the pioneering industrial designers were also coming to an end, and a new generation was emerging that understood design in a new way. Henry Dreyfuss led the way in this regard: his 1955 book *Designing for People* explained that designed objects needed to fit into the lives of their users rather than serve as advertisements for manufacturers or symbols of consumer society. He invented two characters, Joe and Josephine, an everyman and everywoman who needed to be taken into account in the design process, especially in terms of their measurements and those of the objects with which they surrounded themselves (plate 106). His next book, *The Measure of Man* (1960), took Dreyfuss's interest in anthropometrics and ergonomics one step further.

CORPORATE IDENTITY

One response to the shift away from consumer culture and the private individual was a new emphasis on the corporate world, for which design was seen as a means of creating a brand image and identity. The link between design and branding had long been part of American culture, but it took on a new life in the years after 1945.

At that time, a number of corporations began to define themselves as transnational entities. The most obvious case was that of Coca-Cola, which had continually expanded to embrace world markets since its founding in 1886. In fact, the term "coca-colonization" was frequently used to describe the impact of American culture in general on the rest of the world after World War II. Described as a megabrand, the Coca-Cola product—a sweet liquid, in essence—was highly dependent upon design for its identity, not only through packaging but also, more subtly, through the company's corporate image. Much effort had gone into advertising and promoting the product and the brand both before and during the war, but the pace quickened after 1945, with a new focus on the product's position within a lifestyle. As an advertisement explained, "Creative entertaining is part of today's good life . . . and you can count on Coca-Cola to make its own contribution to the good taste of your arrangements." Here was an example of design doing its work without objects, an evocation

BELOW
107. Paul Rand (1914–1996)
***Eye-Bee-M* (IBM) poster**, 1982
Museum of Modern Art,
New York

RIGHT
108. Eliot Noyes
(1910–1977)
***Selectric I* typewriter**, 1961
Made by IBM (U.S.)
Musée National d'Art
Moderne, Centre Georges
Pompidou, Paris

of the concept of good taste in the abstract, linked to a cold drink. As Stephen Bayley points out, "By 1969 Coca-Cola was very much more than a drink. It was a talisman." Although Coca-Cola was international, it was also, paradoxically, American (plate 109).

The idea of corporate identity also took hold among the manufacturers of new electrical and electronic machines. IBM, for instance, recruited the architect and product designer Eliot Noyes to breathe new life into its machines and its global image. Coming from a career that had involved working for Walter Gropius and Marcel Breuer, designing military gliders during the war, and serving as the first director of industrial design at MoMA, he was well prepared to set up one of the first corporate design departments at IBM (which he would also do at Mobil Oil and Westinghouse). Hired by IBM's Thomas J. Watson Jr. in 1956, Noyes also enlisted the services of the Eameses and Paul Rand (plate 107). Noyes personally designed several products and buildings for the company, including the 1961 *Selectric* typewriter, which proved a great success internationally (plate 108), and the 1964 IBM Aerospace Building in Los Angeles. His work at Mobil included a corporate rebranding project in 1964; his brief was for an instantly recognizable and aesthetically pleasing design, and the results encompassed color schemes, forms, and logos. The first rebranded Mobil station opened in 1966.

SPACE RACE SPIN-OFFS

One of the ways in which the U.S. achieved global technological superiority in the postwar years was through its space program, which was at its peak in the 1960s. Design, in tandem with technology, inevitably played an important role in the space program, in terms of both the equipment (from rockets to space suits) needed to support it directly and the products spun off from it for the commercial marketplace and, in some cases, even for the home. The Apollo program, initiated by President John F. Kennedy with the goal of putting a man on the moon, was particularly productive in this respect. Its spin-offs included scratch-resistant lenses, freeze-dried food, and cordless power tools, as well as athletic shoes, or "sneakers," which were associated with shock absorption, stability, and motion control. This use of spin-off technology to create consumer goods resulted in a new set of industries that capitalized on the novelty of these products. The sneakers produced by companies like Nike—originally established as Blue Ribbon Sports in 1964—filled a huge gap at a time when people were becoming increasingly involved in sports. Nike, which began opening its own retail stores in 1967, remains one of the country's most successful brands to this day. It created a new kind of product that commanded high prices from technologically conscious consumers.

BELOW
109. **Coca-Cola advertisement**, 1947

OPPOSITE
110. **Victor Papanek and James Hennessey, illustration from *Nomadic Furniture* (London: Studio Vista, 1974)** Bibliothèque des Arts Décoratifs, Paris

USEFUL DESIGN

The 1970s saw an even greater move away from the consumer society of the 1950s and its objects of mass luxury. As the economy declined and the student movements of the late 1960s called materialism even more into question, a new mood was established that required a new approach to design. By this time, the automobile industry had tamed its approach to styling, and a new generation of consultant designers was abandoning the old manufacturing sector for the software and electronics companies that were emerging on the West Coast.

In the early 1970s, Victor Papanek's book *Design for the Real World* established the idea that designers should develop a sense of social responsibility and distinguish between people's wants and needs. Papanek also proposed that they turn their attention to the needs of the developing world and consider the problem of dwindling natural resources (plate 110). In many ways, Papanek's arguments followed directly from Vance Packard's, but he voiced them at a moment when there was a receptive audience, and his book was hugely influential. It belonged to the new West Coast milieu that rejected advanced consumerism and sought to develop a lifestyle based on closeness to nature and making things by hand. The *Whole Earth Catalog*, also published at this time, showed that it was possible to avoid being part of the consumer society and to meet one's material needs by alternative means.

POSTMODERNISM

While Papanek's words were widely listened to, they were not to have a significant impact for more than three decades. Design became increasingly interchangeable with taste and lifestyle in the 1970s and '80s, and the boundary between design as a professional and an amateur practice was increasingly blurred. Recognizing the dramatic cultural shifts that were taking place, many architects and designers set out to challenge the assumptions that had guided their profession throughout the twentieth century. As a result, modernist design was replaced by a new movement that avoided judgments about good and bad, and which acknowledged the importance of consumption, market values, and popular taste.

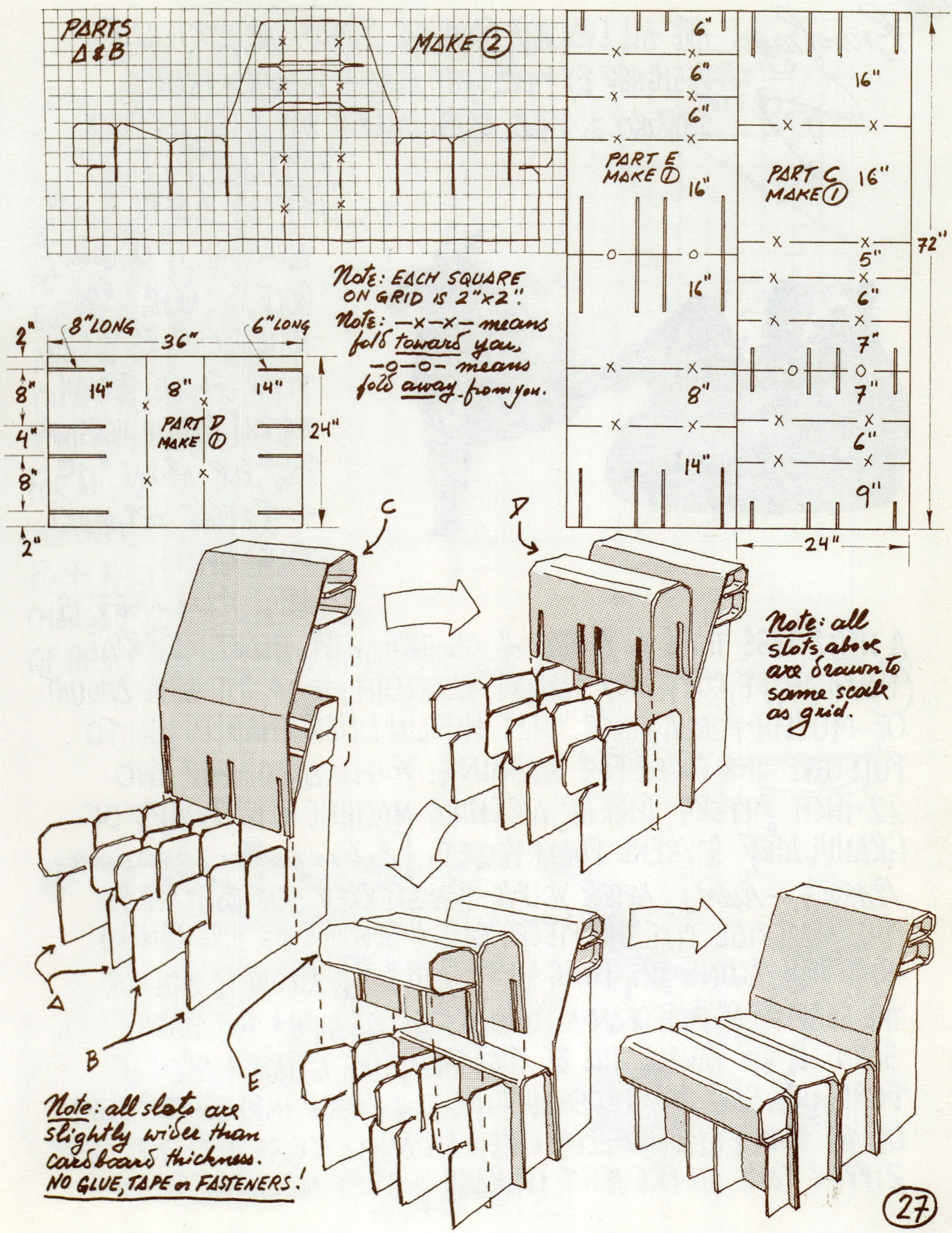

PARTS A & B
MAKE ②
PART E
MAKE ①
6"
6"
6"
16"
16"
8"
14"
PART C
MAKE ①
16"
16"
5"
6"
7"
7"
6"
9"
72"
24"
Note: EACH SQUARE ON GRID IS 2"×2".
Note: —x—x— means fold toward you, —o—o— means fold away from you.
8" LONG
6" LONG
36"
2"
8"
4"
8"
2"
14"
8"
14"
24"
PART D
MAKE ①
C
D
Note: all slots above are drawn to same scale as grid.
A
B
E
Note: all slots are slightly wider than cardboard thickness.
NO GLUE, TAPE or FASTENERS!

OPPOSITE
111. Frank O. Gehry (b. 1929)
Wiggle Side Chair, 1972
Made by Jack Brogan
Laminated corrugated cardboard
Museum of Modern Art, New York

BELOW
112. Robert Venturi (b. 1925)
***Gothic Revival* chair** (painted by Michael Wommack), 1984
Made by Knoll International (U.S.)
Lacquered wood, plastic
Philadelphia Museum of Art

RIGHT
113. Robert Venturi (b. 1925)
***Queen Anne* side chair**, 1984
Made by Knoll International (U.S.)
Steel, plywood, and plastic
Museum of Modern Art, New York

Back in 1966, the American architect Robert Venturi had set the tone in his book *Complexity and Contradiction in Architecture*. This influential text laid out the basic tenets of what, in the 1970s and '80s, came to be called the postmodern movement in architecture and design. "I prefer," explained Venturi, "both-and to either-or, black and white, and sometimes gray, to black or white." He called for an end to the distinction between high and popular culture, and for architecture and design to embrace popular values. His words suggested a definition of design that was rooted in consumption rather than production, that favored "form follows expression" over the modernist "form follows function," and that reflected a complex and pluralistic state of "post-modernity," in continual flux.

Venturi's approach required designed objects to be expressive as well as functional, and opened the floodgates to design that delighted in surface pattern, decoration, and references to the past (plate 111). Above all, it finally made sense of the challenges that had confronted designers in trying to apply the modernist principles of "form follows function" and "truth to materials" to technologically complex objects such as vacuum cleaners. Now they were free to inject meaning into those objects so that they looked outward to their users, rather than visually reflecting their internal components or their means of production.

Postmodernism made an enormous impact in the U.S., where it enabled a group of progressive architects to discover a new expressive language. Venturi turned his own words into practice through his collaboration with Knoll International in the early 1980s. The result was a set of plastic-laminated, bent-plywood furniture available in nine different styles, from Queen Anne to art nouveau (plates 112, 113). The coexistence of these varied styles, which were manufactured by a single production process with superficial variations, echoed the customized mass production pioneered by General Motors back in the 1920s, which had put an end to Henry Ford's "pure" mass production.

Other examples of American postmodern design in the 1980s included buildings by Richard Meier and by the former arch-modernist Philip Johnson. The latter's "Chippendale" building for AT&T in New York, which featured a cornice inspired by the back of a Chippendale chair, represented a dramatic change of direction from his earlier modernist designs, such as the ultraminimalist Glass House he built for himself in New Canaan, Connecticut. Several manufacturers besides Knoll embraced the postmodern style, including Swid Powell, which was founded in New York in 1983. The objects it produced were designed by the Italian Ettore Sottsass and the American architects Robert Venturi and Michael Graves, among others. Graves was a prolific designer in this context, responsible for many postmodern items from the mid-1970s onward, including furniture for Sunar Hauserman and Memphis (plate 115). His most famous design, however, is the *Bird-whistle* kettle he created for Alessi (plate 114). For Moller International, Graves designed two other kettles, one featuring Mickey Mouse; in fact, he also worked directly for the Walt Disney Company, thereby consolidating the inherent link between postmodernism and popular culture.

RIGHT
114. Michael Graves (b. 1934)
***Bird-whistle* kettle**, 1985
Made by Alessi (Italy)
Stainless steel, handle and bird-whistle in two-color polyamide
Centre National des Arts Plastiques/Fonds National d'Art Contemporain, Paris, on loan to the Musée des Arts Décoratifs, Lyon

OPPOSITE
115. Michael Graves (b. 1934)
***Plaza* dressing table**, 1981
Made by Memphis S.R.L. (Italy)
Painted wood, briar, mirror, lightbulbs
Centre National des Arts Plastiques/Fonds National d'Art Contemporain, Paris, on loan to the Musée des Arts Décoratifs, Paris

DESIGN THINKING

While one aspect of twentieth-century design was linked to the concept of lifestyle, another took a step back from the designed artifact and retreated into the world of the design process. This approach was pioneered by the West Coast design consultancy IDEO. Formed in 1991 through the merger of David Kelley Design and ID2, IDEO was run by British designer Bill Moggridge. In its early years, it concentrated on designing user-friendly high-tech goods, from computers to cameras, but it quickly became one of the first firms to move from designing objects to designing experiences. In this process, the idea of the named designer was much less relevant than an interdisciplinary team that included, in addition to designers, social scientists, architects, engineers, and others. A 2003 project undertaken for the health care organization Kaiser Permanente, for example, brought just such a group together to work with the client's nurses, doctors, and facility managers. The aim was to attract more patients and cut costs. The team discovered, among other things, that checking in was a nightmare for patients, waiting rooms were uncomfortable, and the doctors and medical assistants sat too far apart. The cognitive psychologists brought in by IDEO found that patients' friends and families were not allowed to talk to the doctors, and the sociologists noted that patients did not like the examination rooms, because they were left there half-naked for up to twenty minutes. The final conclusion was not that new buildings or products were required but that the patient experience needed changing. As one journalist put it, "Kaiser learned from IDEO that seeking medical care is much like shopping—it is a social experience shared with others."

The concept of "design thinking" was much discussed in the early years of the twenty-first century. IDEO's CEO Tim Brown explained that it involved the use of both consumer/user insights and rapid prototyping to get quick, innovative results that might or might not relate to product development. In his words, the aim was "to get beyond the assumptions that block effective solutions." The emphasis on prototyping showed how the design process, as it had been developed in collaboration with the manufacturing industry over two centuries, could be applied to a new context and provide a new way to reach innovative solutions to problems of all kinds.

Design thinking affected both practicing designers and design education. One of its earliest educational manifestations was at Stanford University's D. School, established in the 1980s and formally known as the Hasso Plattner Institute of Design. By the start of the twenty-first century, the D. School focused on the use of design thinking by multidisciplinary teams to solve problems in ways that did not necessarily result in new products or services, but which might produce new business, implementation, or strategic plans for companies or other stakeholders. While most projects were business-focused, some were more altruistic in nature, such as dealing with the aftermath of a disaster.

By 2006, 60 percent of IDEO's projects were strategic rather than design-focused. The firm used the technique of storytelling, employing filmmakers and writers for that purpose. It worked on the principle that designers have always had to combine their own know-how with a bit of everybody else's discipline. Moreover, its designers were encouraged to question briefs rather than merely fulfill them. Indeed, for Brown, "figuring out the right problem" was the most important part of the process. The work at both Stanford and IDEO was premised on the idea that designers should be leaders. Besides Apple Inc., one company frequently cited at that time for its innovative practices was the coffee retailer Starbucks, which had provided a completely new space for young adults that was not a bar, a school, or a home. Starbucks was also admired for the fact that it had redefined its customer base, rather than simply catering more effectively to its existing one.

In the first decade of the twenty-first century, a number of design consultancies followed the model of IDEO and engaged in multidisciplinary work. Other design groups used the D. School model, some of them adding the factor of culture (that is, the client's values) to the mix. All their activities were underpinned by the concept of innovation, which they defined as "invention with socioeconomic impact."

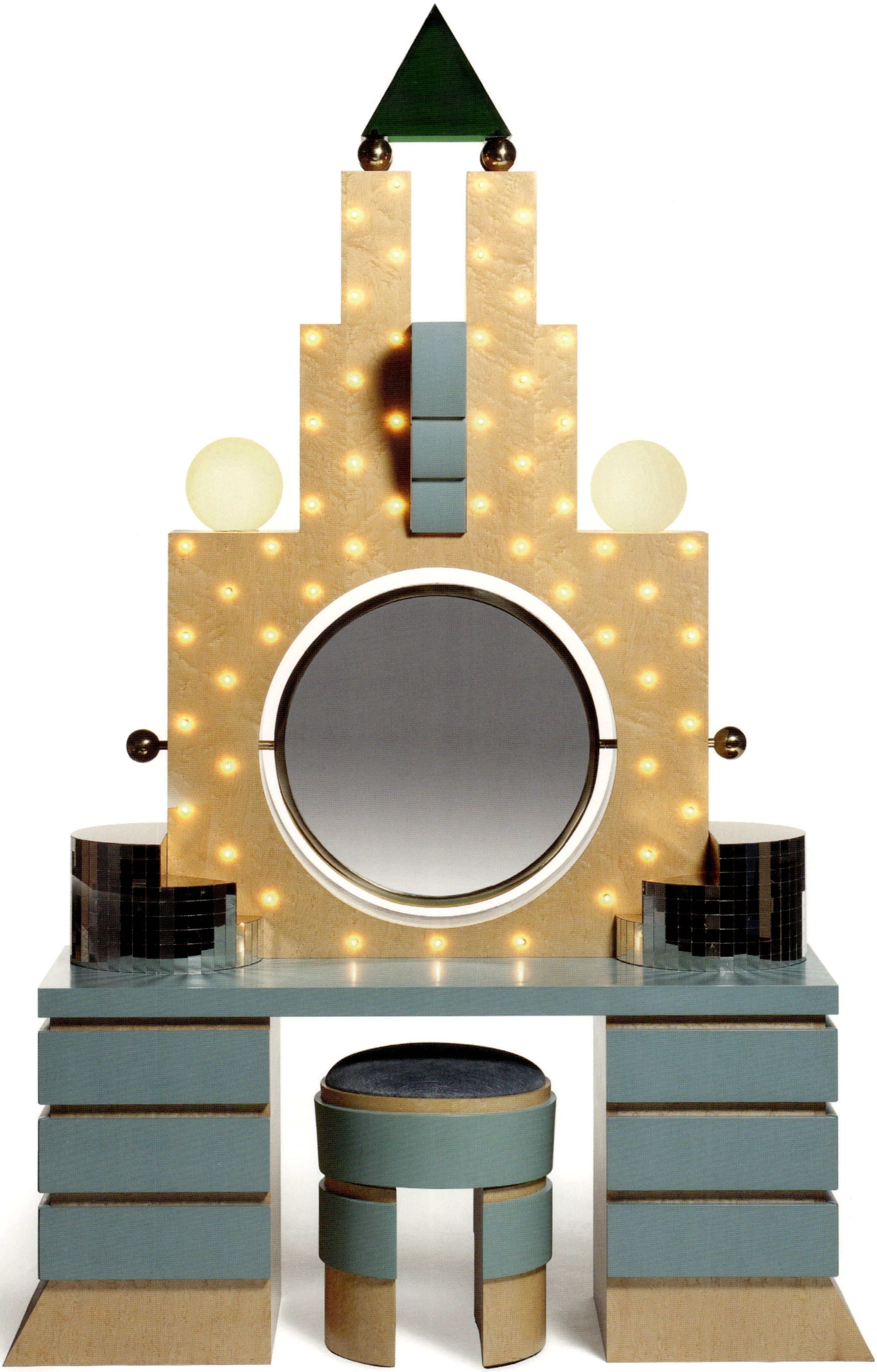

LEFT
116. ***iPhone 3G***, 2007
Made by Apple Inc. (U.S.)

BELOW
117. Jonathan Ive (b. 1967)
iPod, 2001
Made by Apple Inc. (U.S.)
Museum of Modern Art, New York

OPPOSITE
118. **Advertisement for the Apple *iMac G3* by Jonathan Ive**, 1998

SILICON VALLEY DESIGN

By far the most important developments in American design since 1980 have taken place on the West Coast, in the area south of San Francisco, around the town of Palo Alto, that became known as Silicon Valley. While the first phase of modern American design had been linked to the dominance of electrical power, this new phase was connected with important developments in electronics and information technology, and the consumer-oriented industries that grew up on their backs. Designers became important in both the hardware and software aspects of these industries, presenting advanced technologies to the consuming public in a way that made them accessible, usable, and desirable.

In this new context, an important shift occurred from the single object that performed a single function (and whose form was determined by that function, as modernism had decreed) to the complex, multifunctional object. For example, Apple's *iPad*—visually speaking, a simple neo-modern rectangular tablet—combined the functions of a television, radio, telephone, typewriter, hi-fi system, camera, photo album, and address book, among countless other things. Through its link to the Internet, this seemingly banal artifact in fact offered unprecedented possibilities for entertainment, information, and communication. The crucial fact is that its popularity was determined by what it did, not just what it looked like.

One of the first design companies to locate itself in Silicon Valley was frog design (now known simply as frog), led by Hartmut Esslinger. Founded in Germany in 1969, it moved first to Palo Alto and then to San Francisco, although it has branches all over the world. While its origins were in industrial design, it has expanded into the wider field of innovation. Its most renowned designs have been for computer and consumer electronics brands such as Sony and Apple.

Frog design's *Snow White* computer for Apple (1984) represented an important early attempt to design a neat little machine with a user-friendly appearance. Given the severe constraints of the brief—essentially, to create a box housing the computer's inner workings—close attention was paid to those areas in which a designer could make a contribution, namely, the curves, lines, and color of the shell. Apple also strove to make the computer interface itself user-friendly, conscious that the feel of the object was just as important as its performance in helping consumers make a purchasing decision. Later, in the 1990s, Apple pursued this principle even further, through its collaboration with the English designer Jonathan Ive, which resulted in the transformation of a series of electronic devices into lifestyle objects (plate 118). By 2010, the *iPhone*, the *iPod*, and the *iPad* had all joined the stable of Apple goods designed by Ive (plate 116, 117). By that time, the functionality of Apple's products had come to overshadow their user-friendliness. While Apple's sophisticated touch screens and the visual language of its operating system were designed to minimize user discomfort, it was the technological virtuosity of the machines' applications, and the increasingly novel combinations of them, that most impressed consumers. While, for example, the *iPad 2* (2011) came in a variety of colors,

0870 2410212

Collect all five.

The new iMac. Now in five flavours. Think different.

BELOW
119. Yves Béhar (b. 1967)/ Fuseproject and Issey Miyake (b. 1938)
Vue watch, 2010
Made by Fuseproject (U.S.)

OPPOSITE
120. Yves Béhar (b. 1967)/ Fuseproject
***XO One Laptop per Child* computer**, 2005 (shown open and closed)
Made by Quanta (U.S.)
Museum of Modern Art, New York

thereby reemphasizing that such devices were still lifestyle choices, the visual and material considerations of traditional product design were far outweighed by the tablet's sophisticated systems and interface design.

Apple led, and continues to lead, the field in the design of consumer electronics. Ive's creation of a new, highly appealing minimalist language for Apple's products, as well as the fact that the company included him on its board and considered design to be a fundamental feature of its products, singled the company out from its Silicon Valley neighbors. However, while many of the giant companies that use the Internet to sell their services—including Google, Amazon, Microsoft, Facebook, and eBay—focused on engineering in their early days, they also came to realize that the user experience was key to their success, and required the input of designers. In the wake of Apple, several of them brought designers into their midst. By 2012, Microsoft, for example, had hired six hundred designers, among them Albert Shum, who had previously worked for Nike. In 2008, he became head of the Windows Phone design team. Microsoft used designers to distance itself gradually from its reputation as an old-fashioned software company. For instance, its development of Metro, a new "design language" for its products, was strongly influenced by Swiss modern typography.

Other giants, like Facebook and Google, also realized that they needed designers to bring their Web sites and apps to life, and the profession of interface design grew apace. In June 2011, Facebook bought the Dutch software design company Sofa, just one of the many examples of a huge company buying a small startup for its design talent. Then, in August 2011, Facebook went on to acquire Push Pop Press, a digital publishing company cofounded by interface designer Mike Matas. Similarly, Google brought in Daniel Burka, a former creative director at the design firm Digg. Both Apple and Google look for talent through an invite-only site called Dribbble, on which designers show off their work. As the interest in interface and user experience design has grown, its practitioners have been able to command increasingly high salaries.

The presence of so many new, increasingly design-conscious companies in Silicon Valley has encouraged many young designers to locate there. One frog design employee who broke away to set up on his own was the Swiss-born Yves Béhar, who established the firm fuseproject in San Francisco in 1999. Fuseproject, which currently employs around forty people, began by working for American giants like Hewlett Packard. Béhar quickly became one of the West Coast's best-known and most successful designers, working in fields ranging from furniture to watches (plate 119), consumer electronics, and online design, as well as social projects for the nonprofit sector. One of his most-discussed designs, initiated in 2005, was for the world's first hundred-dollar laptop computer, the *XO* (plate 120), developed for Nicholas Negroponte's One Laptop

BELOW
121. Yves Béhar (b. 1967)/
Fuseproject
***Jambox* wireless speaker**, 2012
Made by Jawbone (U.S.)

OPPOSITE
122. Yves Béhar (b. 1967)/
Fuseproject
***Up* digital wristband**, 2013
Made by Jawbone (U.S.)

per Child organization, which aims to bring technology to the poorest areas of the world.

Béhar has also done seminal work for the wearable technology company Jawbone, whose products address the need for increased awareness of health and well-being (plates 121, 122). In addition to creating the firm's brand identity, he has also been responsible for items including *Up*, a wristband that monitors its wearer's movements and can be programmed to give reminders, with a small vibrating motor, about when to exercise, go to sleep, and so forth. Wearers can also use *Up* to monitor how they react to certain foods. Designed to be worn at all times, the band is covered in a sweat- and waterproof thermoplastic rubber and works in conjunction with an app for *iPhones* and *iPods*.

Other Béhar designs to date include an identity and strategy for Google's home page, the NYC condom for the New York City Department of Health, a reflective red-lacquered laptop for Toshiba, and a recycling project for Coca-Cola. Many of his designs are highly progressive both technologically and socially, and all are characterized by a strong humanism.

CONCLUSION

In the years since 1945, the United States has seen design make several pendulum swings, adapting itself to changing circumstances in order to continue playing a key role in both the economic and the cultural life of the country. Most significantly, design helped create a modern consumer society that led the way out of the Great Depression, made new technological breakthroughs available to everybody, and eventually spread to the rest of the world. The U.S. has continued to respect Europe as the home of good modern design, but it has also made European design its own, aligning it to American capitalism by replacing ideology with style. While American design has repeatedly shifted its allegiance between mass production and consumption, playing a transformative role in both contexts, it has always remained close to commerce, the force that has driven it forward.

More recently, design in the U.S. has, at least partly, abandoned its earlier roles within industrial production and mass consumption and redefined itself as a key player in addressing the big questions of the day—economic, social, cultural, and political. In this new approach, which links design thinking with innovation, the U.S. is leading the way and showing Europe where to go.

JAWBONE
JAWBONE

SCANDINAVIA

ÁSDÍS ÓLAFSDÓTTIR

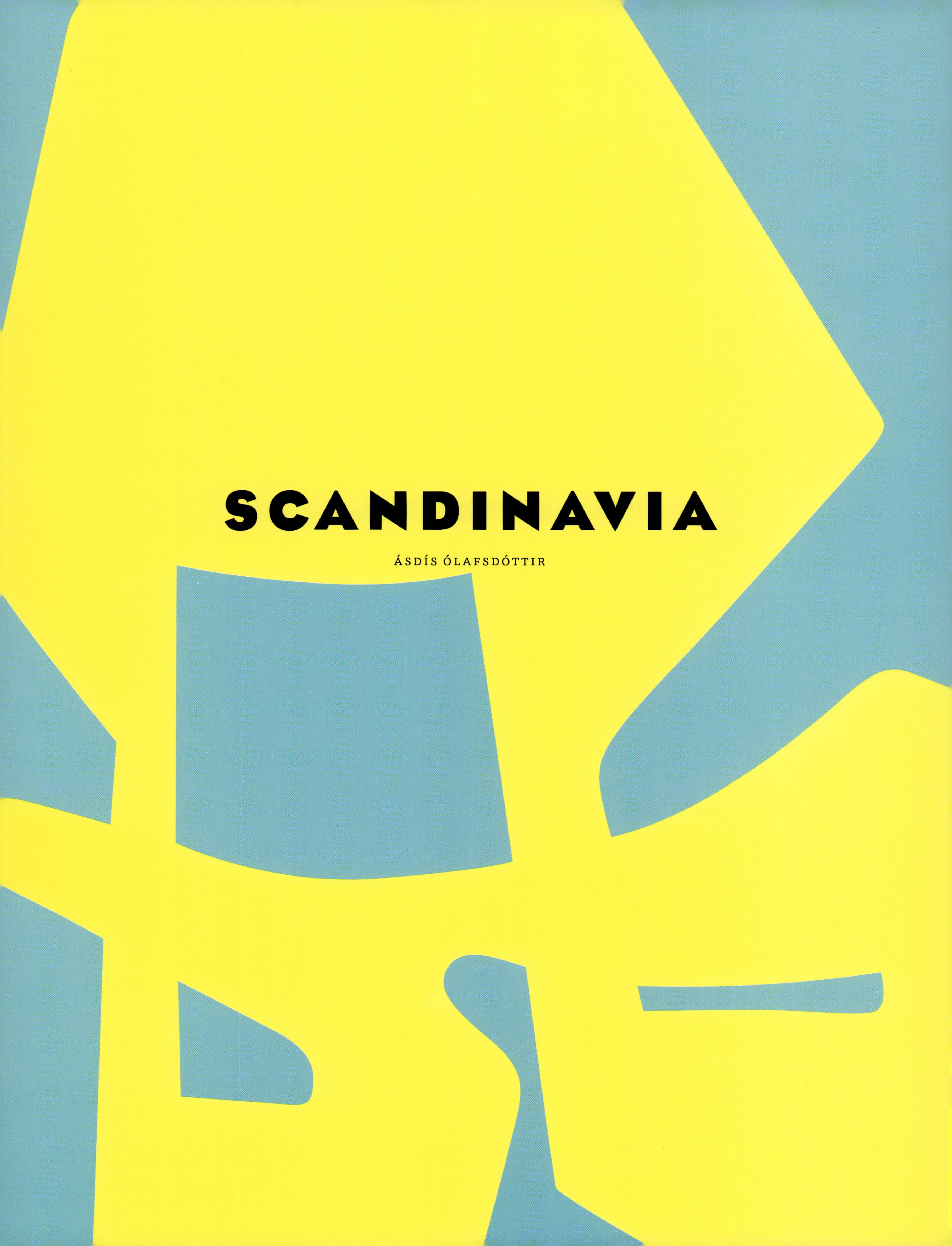

PRECEDING PAGE
123. Grete Jalk (1920–2006)
Chair, 1963
Made by Poul Jeppensen (Denmark)
See plate 139.

OPPOSITE
124. Carl Larsson (1853–1919)
The Kitchen, c. 1895
Watercolor on paper
Nationalmuseum, Stockholm

During the 1950s and '60s, Scandinavian design earned a reputation for fine, well-made objects, suited to modern lifestyles. Blond wood, Scandinavia's material of choice, corresponded to a society rebuilding itself on optimism and lightness. The sober and unadorned aesthetic of Nordic design was especially appealing to Americans, who shared with Scandinavians a somewhat puritanical vision of life. The fashion spread to Europe as well, where exhibitions and specialized boutiques offered the latest in Nordic design.

The quality of Scandinavian design relies a great deal on the proper balance between craft and well-considered industry. Quality craftsmanship had a long history in Scandinavia, where industrialization came late enough that manual skills were not lost. Nordic designers generally had a solid education in craft techniques, which led them to respect their materials and bring out their best properties. In northern regions, domestic interiors are important, and the quality of everyday objects is appreciated. Moreover, even before World War II, the social-democrat societies of Scandinavia were concerned with equality and stability, and designers intended their products to be for the masses. Everyone was to have access to beauty and quality.

BEAUTY FOR ALL

The founding texts of Nordic modernism identify beauty as a human and social objective. The Swedish philosopher, feminist, and theorist Ellen Key included a text called "Beauty in the Home" in her 1899 book *Beauty for All*. According to her, the new aesthetic sensibility began with the home, where life and creativity took place, and where women played a dominant role. Beauty within the home transformed life and, in turn, society as a whole. Key's discourse was influenced by the English Arts and Crafts movement, whose principles were adopted by the Swedish Society of Arts and Crafts, but she also thought the Nordic countries had their own part to play in the dissemination of new ideas throughout the European continent. She believed that the residents of the North had a special sensitivity to nature—to its light, silence, and solitude—and therefore a particular aptitude for communicating the experience of it to others. This philosophy is illustrated in the watercolors of her friend Carl Larsson, who depicted the simple life of a family in a light and luminous environment, close to nature and to traditional crafts (plate 124). Larsson's book illustrations were very popular in their time and contributed significantly to the spread of new ideas.

Twenty years after Key's text appeared, Gregor Paulsson published a pamphlet entitled *Better Things for Everyday Life*. As an art historian and the director of the Swedish Society of Arts and Crafts, Paulsson was one of the most influential theorists of functionalism in Scandinavian architecture and design. He did not advocate the abandonment of industrial production in favor of old-fashioned craftsmanship, but rather sought to introduce beauty into manufactured products. As a result, ugliness would be eliminated from modern society, and mankind would be liberated from the historic styles that muddled contemporary architecture and decorative art. For Paulsson, beauty was an important sociopolitical tool, contributing to the idealistic reforms he longed for.

The Stockholm Exhibition of 1930, organized by Paulsson and the Swedish Society of Arts and Crafts, marked the advent of international modernism in Scandinavia. The event drew large crowds and triggered debate about modernist precepts. A year later, Paulsson and the architects Uno Ahrén, Gunnar Asplund, Wolter

125. Gunnar Asplund
(1885–1940)
***Senna* chair**, 1925
Reissued by Cassina (Italy), 1983
Leather, cherry wood
Private collection

Gahn, Sven Markelius, and Eskil Sundahl—all deeply involved in the Stockholm Exhibition—published the manifesto *Acceptera* (To Accept). This long illustrated text, intentionally confrontational, defended the modernist point of view: It claimed that the present standard of beauty was associated with the bygone aesthetics of a dominant class and prevented the discovery of new forms suitable for the times. A new form of beauty was needed, linked to function and the industrial age.

Despite this radical text, the industrial aesthetic of the Bauhaus did not really take root in Scandinavia, even among those who had penned the manifesto. The tubular furniture of Mart Stam and Marcel Breuer was well known in Scandinavia and could be ordered through the Thonet catalogs, but this austere look was generally softened, and steel quickly replaced by wood. Nevertheless, the formula "form follows function" could easily be applied to the Nordic penchant for simplicity and practical, well-made objects.

MODERNISTS OF THE 1920S AND '30S

One of the authors of the *Acceptera* manifesto, Gunnar Asplund, was the most important Swedish architect of the interwar period. In particular, he designed the steel and glass buildings at the 1930 Stockholm Exhibition. Asplund's furniture is a good illustration of Nordic design at the time, from his prestigious *Senna* chair in walnut and leather for the 1925 Paris International Exposition (plate 125) to the 1931 *Karmstol* chair in tubular steel and leather. For his addition to the Göteborg town hall (1931–34), Asplund again designed warm, simple, and refined furniture in wood and leather.

In Denmark, the extraordinary tradition of cabinetwork left almost no room for steel. Designer Kaare Klint made furniture based on existing models, often favoring maritime or colonial pieces because they were simple, folding, and functional. As an example, his *Safari* armchair (plate 126) and *Deck* chair from 1933 are elegant reinterpretations with references to England and the East. In 1924, Klint helped found the department of furniture and interior design at the Royal Danish Academy of Fine Arts in Copenhagen. His teaching and his research in anthropometrics greatly inspired the generation of designers who worked to revive Danish design after World War II. Klint also designed lamps with simple shades of folded paper (plate 127). Since the 1920s, another Dane, Poul Henningsen, had

BELOW
126. Kaare Klint (1888–1954)
***Safari* chair**, 1933
Made by Rud. Rasmussens Snedkerier (Denmark)
Leather, maple
Private collection

OVERLEAF
127. Kaare Klint (1888–1954)
***Klint 101* pendant lamp**, 1943
Made by Le Klint (Denmark)
Polycarbonate
Musée d'Art Moderne, Saint-Étienne Métropole

PAGE 119
128. Poul Henningsen (1894–1967)
Lamp, c. 1929
Made by Louis Poulsen (Denmark)
Chrome-plated metal, reflectors in lacquered metal
Musée National d'Art Moderne, Centre Georges Pompidou, Paris

BELOW
129. Alvar Aalto (1898–1976)
Stackable stools, 1932–33
Made by Oy Huonekalu-ja Rakennustyötehdas (Finland)
Cross-laminated birch
Museum of Modern Art, New York

OPPOSITE
130. Alvar Aalto (1898–1976)
***Savoy* vase**, 1937
Made by Iittala (Finland)
Mold-blown glass
Victoria and Albert Museum, London

been developing lamps intended to reduce the blinding brightness of electric light by precisely overlapping convex and concave pieces (plate 128). His timeless and elegant *PH* series continues to be manufactured by Louis Poulsen. Through the radical positions he took as a journalist and publisher, Henningsen also contributed to the discourse on Scandinavian design in the interwar period.

The Finn Alvar Aalto was close to the Swedish modernists and a friend of Henningsen, whose lamps he used in his first modern buildings. Along with his wife Aino Marsio-Aalto, also an architect and designer, he perfected furniture in bent, glue-laminated birch that met the same demands for solidity and hygiene as tubular steel, while being warmer and more human (plate 129). In 1935, the couple founded Artek, which is still responsible for the production and distribution of their work today. In 1937, Aalto designed the Finnish pavilion for the Paris Exposition and exhibited a few of his pieces there, notably the *Savoy* vase (plate 130). In this object, which became an icon of twentieth-century design, there emerged the curved and organic forms, inspired by both nature and modern art, that were so characteristic of Aalto's work.

After being one of the pioneers of Finnish art nouveau, the architect Eliel Saarinen emigrated to the U.S. in 1923. He designed the buildings for, and later taught at and directed, the Cranbrook Academy of Art near Detroit, a school that aimed to be the American equivalent of the Bauhaus. He and his wife, the textile artist Loja Saarinen, trained an entire generation of designers working after World War II, including Ray and Charles Eames, Harry Bertoia, Florence Knoll, and their own son Eero Saarinen.

KARHULA

BELOW
131. **Finnish pavilion at the 1951 Triennale di Milano**
Triennale di Milano, Milan

OPPOSITE
132. **Poster for *Design in Scandinavia: An Exhibition of Objects for the Home from Denmark, Finland, Norway, and Sweden*,** 1954
Published by A. Remlov, Oslo
Brooklyn Museum Libraries and Archives

THE SPREAD OF SCANDINAVIAN MODERN

Since the 1930s, the Nordic countries—Sweden and Finland in particular—have participated in the Triennale di Milano, an important international exhibition of architecture and decorative art. Alvar Aalto notably exhibited his glue-laminated wood furniture abroad for the first time at the inaugural Triennale in 1933. Nevertheless, it was after World War II that Scandinavia received particular attention at these showcases of fine design. Denmark, Finland, Norway, and Sweden exhibited regularly at the postwar Triennales and received a good share of the medals awarded. With its award-winning pavilion designed by Tapio Wirkkala, Finland stood out among the exhibitors at the 1951 Triennale (plate 131).

Following that edition of the Triennale, the American magazine *House Beautiful* organized a traveling exhibition of Nordic design. *Design in Scandinavia* opened in 1954 in Richmond, Virginia, and went on to twenty-four other locations, in seventeen U.S. states and three Canadian provinces, over the next three and a half years (plate 132). The success of this exhibition was immense, and the popular notion of Scandinavian good taste took root. The 1958 exhibition *Formes Scandinaves* at the Musée des Arts Décoratifs presented Nordic design to the French public. On this occasion, Iceland joined the four other Scandinavian countries as a participant. A revamped version of *Design in Scandinavia* was presented in seven Australian cities in 1968. The triumph of the 1950s and '60s eventually faded, and the exhibition *Scandinavian Modern Design 1880–1980*, presented at three American museums in 1982 and 1983, emphasized artisanal crafts and the familiar work of the postwar period.

The Scandinavian countries efficiently organized the export of their products, and also held salons and fairs that attracted designers and manufacturers from other parts of the world. The workshops of the North became an important stop on the circuit traveled by French designers like Jean Royère, Pierre Paulin, Marc Held, and Olivier Mourgue. Scandinavian design thus served as a kind of school and was widely copied around the world.

In organizing common events and choosing to exhibit together, the Nordic countries consciously built a unified image and reputation. Effacing national differences, they deliberately promoted the idea of a sleek "Scandinavian" aesthetic, which concealed an output that was, in reality, rather more varied and at times iconoclastic. In analyzing the design of these Nordic countries, let us try to understand what unites them and also where their differences lie.

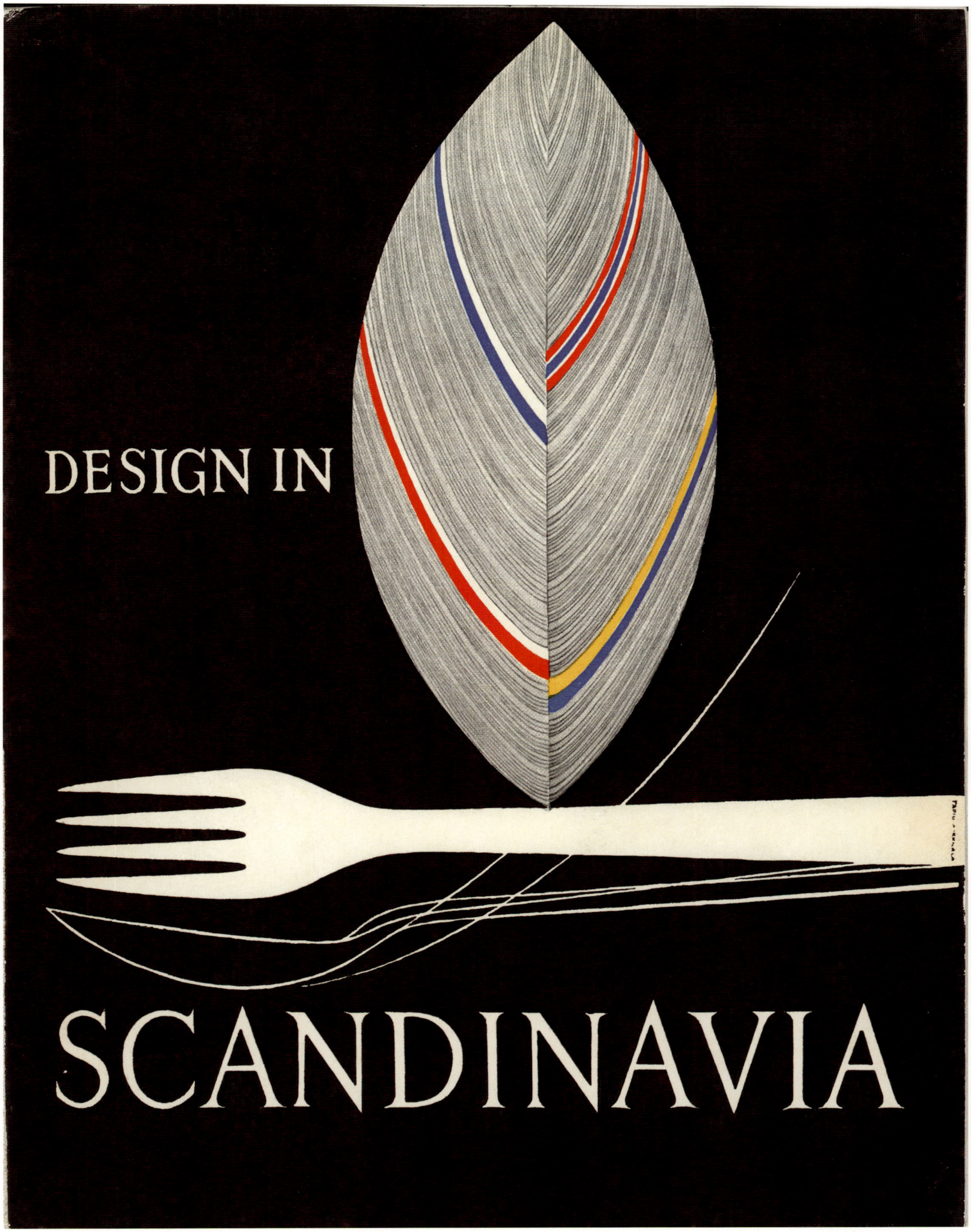
DESIGN IN
SCANDINAVIA

RIGHT
133. Hans Wegner (1914–2007)
***Peacock* chair**, 1947
Made by PP Møbler (Denmark)
Ash, teak, woven cord
Kunstmuseet Trapholt, Kolding

OPPOSITE
134. Børge Mogensen (1914–1972)
***Spanish* armchair**, 1959
Made by Fredericia (Denmark)
Oak, leather
Kunstmuseet Trapholt, Kolding

DENMARK

The smallest of the Nordic countries has always compensated for its size with navigation and trade, even colonialism. Consequently, foreign influences, from both the East and the West, have been naturally absorbed through the centuries. The late industrialization of the country meant that craft survived as one of the pillars of production. Danish design is characterized by both functionalism and a respect for tradition. In the vein of the pioneer Kaare Klint, postwar furniture designers drew their inspiration from the American Shakers, the English Arts and Crafts movement, and also Asian craftsmanship. Denmark was by far the largest manufacturer of furniture in Scandinavia: exports to Europe and especially the U.S. at one point accounted for about 80 percent of the country's total furniture production, which attests to the international influence of "Danish modern" in the 1950s and '60s.

Behind the seeming unity of Danish furniture can be distinguished two trends, which emerged as early as the 1940s. On the one hand, the Klint school worked on archetypes of traditional furniture—simplifying them, ridding them of all ornamentation, and seeking to attain their essence, while taking into consideration the user's needs and ergonomics. Børge Mogensen and Hans Wegner are the best representatives of this movement. On the other hand, designers like Finn Juhl sought to extricate themselves from this relatively conservative mode of thinking and bring innovation and fantasy to Danish design. Both movements shared the traditional bases of Danish furniture: craft knowledge and a respect for materials—wood in particular.

The powerful Copenhagen Cabinetmakers' Guild (Københavns Snedkerlaug), founded in 1554, brought together a number of small and medium-size workshops. From 1927 to 1966, the guild presented the furniture produced by its members at an annual exhibition. What made this exhibition unique is that a jury chose projects submitted by young designers to be manufactured and exhibited by a workshop of cabinetmakers. Before and after World War II, most Danish designers began their careers in this way. The exhibitions were much anticipated, and were the subject of critical discussion and debate in the media, evidence of the influence that the decorative arts—and furniture in particular—enjoyed in Danish society in this period.

THE TRADITION OF CABINETWORK: BØRGE MOGENSEN AND HANS WEGNER

The direct heir of Kaare Klint was undoubtedly Børge Mogensen. After attending the School of Arts and Crafts (Kunsthåndværkerskolen) in Copenhagen, Mogensen studied under Klint at the Academy of Fine Art between 1938 and 1941 and became his assistant a few years later. From 1942 to 1950, Mogensen was also head of the bureau of furniture design at the Danish Cooperative Union (FDB). There his task was to create simple and reasonably priced furniture for small apartments. The characteristics of his design emerged in these unadorned and functional pieces, for example the *J-39 Shaker* chair of 1944, made from beechwood and woven cord, and inspired by the simplicity and solidity of American Shaker furniture. The following year, he exhibited his famous *Tremmesofa* couch, manufactured by Fritz Hansen, which can be expanded by lowering the sides. Inspired by a traditional model, it enjoyed great popular success and was reissued in 1963.

In the 1950s, Mogensen developed the *Øresund* and *Boligens* shelving systems in collaboration with Grethe

Meyer. These designs were based on in-depth studies of the housing and organizational needs of middle-class families. Mogensen also designed textiles with Lis Ahlmann and furniture for Søborg Møbelfabrik, Fredericia, and Karl Andersson & Søner; his pieces of this era include the *2213* leather sofa, in two widths, and the *Spanish* armchair of 1959 (plate 134). Inspired by a traditional model, the latter chair, with its solid construction and virile forms, represents the simplified and unadorned aesthetic of its maker. Mogensen sought to design honest furniture that was well made but also accessible to the masses. He favored local materials like beech, oak, and birch, remaining uninterested in the fashion for exotic woods. His designs enjoyed great popularity, and many of them continue to be manufactured today.

A love of wood led Hans Wegner to become a qualified cabinetmaker at the age of seventeen. Originally from Jutland like his friend Børge Mogensen, Wegner studied design in Copenhagen at the end of the 1930s. In 1940, he worked for Arne Jacobsen and Erik Møller on the construction of the Aarhus town hall, for which he headed the furniture program. During the war, he collaborated with Mogensen on simple furniture for the FDB. Wegner also developed furniture with the cabinetmaker Johannes Hansen. These pieces were presented at the exhibition of the Cabinetmakers' Guild (of which Hansen was the president starting in 1941) and set Wegner on his way to designing hundreds of models for several manufacturers. His first notably successful piece was the *Chinese* chair that he designed for Fritz Hansen in 1943. This reinterpretation of a traditional form, simplified and modernized, already presented certain characteristics of Wegner's design: a fluid line between back and arm, solid but elegant proportions, and, needless to say, perfect execution. In 1947 followed the *Peacock*, an expressive variation on the Windsor chair (plate 133). Wegner continued to develop variations of the *Chinese* chair and, in 1949, finally succeeded in reducing its form to its purest possible expression (plate 136). Dubbed *The Chair* by the Americans, this piece was a resounding success abroad. Presented in 1950 on the cover of *Interiors* as "the most beautiful chair in the world," it launched the international export of Wegner's furniture. In the famous televised debate between John F. Kennedy and Richard Nixon in 1960, the candidates and their moderators were comfortably seated on these chairs. Another great success was the *Y* chair of 1950, one of the most elegant dining-room chairs of the twentieth century. Winning the Lunning Prize (founded to recognize eminent Nordic designers) in 1951, Wegner was celebrated in Scandinavia, and also abroad. His furniture was sold by the Danish manufacturer Fritz Hansen and by the export company Salesco, which served several manufacturers. Today, PP Møbler is the largest manufacturer of his assembled furniture, while Carl Hansen & Søn are responsible for his more industrial pieces.

Wegner preferred to work with locally available wood, which in Denmark meant beech, oak, ash, maple, and birch. His training as a cabinetmaker taught him about manufacturing processes, and his designs always take the user into consideration: "A chair is only finished when someone is seated in it," he liked to say.

OPPOSITE, LEFT
135. Finn Juhl (1912–1989)
***45* chair**, 1945
Made by Niels Vodder (Denmark)
Walnut, cloth upholstery
Kunstmuseet Trapholt, Kolding

OPPOSITE, RIGHT
136. Hans Wegner (1914–2007)
Armchair, 1949
Made by Johannes Hansen (Denmark)
Oak, caning
Museum of Modern Art, New York

BELOW
137. Hans Wegner (1914–2007)
***Halyard* chair**, 1950
Made by Getama (Denmark)
Steel, wood, cord, cloth upholstery
Private collection

RIGHT
138. Finn Juhl (1912–1989)
***Chieftain* chair**, 1949
Made by Niels Vodder (Denmark)
Teak, leather
Musée des Arts Décoratifs, Paris

OPPOSITE
139. Grete Jalk (1920–2006)
Chair, 1963
Made by Poul Jeppensen (Denmark)
Bent teak plywood
Museum of Modern Art, New York

FINN JUHL, THE ARTIST

When Finn Juhl presented the designs he had developed in collaboration with cabinetmaker Niels Vodder at the Cabinetmakers' Guild exhibitions of 1937 and 1940, they were received with a certain incomprehension. His first armchairs and heavily stuffed sofas—sculptural in form—verged on the world of the abstract artists Juhl so admired: Jean Arp and the Dane Erik Thommesen. A critic called them, somewhat nastily, "tired walruses."

Juhl, who originally wanted to be an art historian, was trained as an architect but self-taught as a furniture designer. He did not adhere to Klint's school and approached furniture in a very different manner. Instead of reinterpreting existing models, he sought to create new forms, while still respecting the functionality of the object. In 1941 and 1942, he built his own house in Charlottenlund, near Copenhagen, and developed furniture to meet his personal needs and desires. Passionate about wood, Juhl conceived of chairs with a visible frame from which the seat and back could be detached. Consequently, his massive wooden chairs, as well as his tables, can appear to be very light. During the 1940s, Juhl designed his most famous pieces, the *45* chair (plate 135) and the *Chieftain* chair (plate 138).

Juhl's collaboration with Vodder was very fruitful, allowing him to experiment with new forms and assembly techniques. The complexity of Juhl's designs and the sophistication of his finishes did not seem to be practical for mass production, but in 1950, an American firm, Baker Furniture, began to manufacture his pieces successfully. Shortly after, production started in Denmark as well.

Juhl was a prolific interior designer; a notable example of his work in this field is the Trusteeship Council Chamber at the United Nations in New York. With a career that was both local and international, he taught at the Danish School of Interior Design and was its director from 1945 to 1955. His influence on designers and on Danish decorative art was considerable. During the 1950s, Juhl also designed several exhibitions, including *Good Design* for his friend Edgar Kaufmann Jr. at the Museum of Modern Art in New York and the Danish pavilion at the Triennale di Milano of 1954 and 1957.

Challenged and criticized when he began, Juhl quickly earned recognition from his peers and the public, forming a style and school of his own. Indeed, by the end of the 1950s, his furniture had been so widely copied that the production of his own designs stopped in certain cases. Juhl paved a new pathway in Danish and Scandinavian design, one characterized by sculptural form and the use of teak. An all-around designer, he also worked with glass and porcelain, and made fine wooden bowls for the manufacturer Bojesen.

Today Danish teak furniture from 1940s and '50s is sought out and appreciated for its formal and material qualities. When Juhl began using teak in 1945, it was mostly reserved for exterior use, but he was fond of it because it did not require varnish. It was also dense (allowing it to be precisely shaped), soft to the touch, and relatively light compared to other exotic woods. Circumstances favored its widespread use, as a large stock of teak was to be found in Denmark after the war. The furniture designer Grete Jalk made several pieces from this material, including an innovative set in plywood with a teak veneer (plate 139).

RIGHT
140. Arne Jacobsen (1902–1971)
***Ant* chair**, 1952
Made by Fritz Hansen (Denmark)
Chrome-plated steel, molded laminated teak
Musée des Arts Décoratifs, Paris

THE MASTERS OF STEEL:
ARNE JACOBSEN AND POUL KJÆRHOLM

The architect Arne Jacobsen also approached furniture as sculpture in space. Although his architecture is mostly rectilinear and functionalist, his design is organic and intuitive above all. The *Ant* chair, presented at an exhibition of his work at the Danish Museum of Decorative Art in January 1953, is made from a molded plywood seat placed atop three steel legs (plate 140). The curving forms of the back refer to the animal world, but also to the work of modern artists like Alexander Calder and Henry Moore. Employing the technique of molded plywood developed during the war and adopted by the American designers Charles and Ray Eames as early as 1946, Jacobsen's *Ant* and *Series 7* chairs are simple and very functional (plate 141). They respond to the need for light, stackable seating that is still comfortable and attractive. Manufactured by Fritz Hansen, they were sold by the millions, helping to make Scandinavia a world leader in furniture exports. Founded in 1872, Fritz Hansen has specialized in the production of steam-bent wooden chairs since the 1920s. Its collaborations with modern designers, and with Jacobsen in particular, gave it an international profile and turned it into the Danish leader it still is today (plates 142, 143).

With his *Egg* and *Swan* chairs of 1958, Jacobsen went further toward the sculptural. The prototypes were shaped and reshaped at actual size in plaster until the designer was satisfied. They were then manufactured in a single shell of styropor, a kind of polystyrene invented by the Norwegian designer Henry Klein. Covered in natural fabric or leather and placed on a central steel and aluminum pedestal, these armchairs perfectly contrast the straight lines of Copenhagen's Royal Hotel, of which Jacobsen was the architect (plate 144). He also made steel tableware, lamps, door handles, faucets, clocks, glass, and textiles, all with clean, rigorous forms. A student of Kaare Klint, Jacobsen was the inventor of organic modernism in Denmark, breaking from traditional cabinetmaking.

Poul Kjærholm's 1952 diploma exhibition at the School of Arts and Crafts included a chair in flat steel, with a seat and back in woven cord. This object already revealed the characteristics of the work of this designer, then twenty-three years old: perfect formal mastery, a minimalist simplification of structure, and the use of steel strips combined with natural materials. This chair, which would later take on the name *PK25*, was manufactured by the entrepreneur Kold Christensen, with whom Kjærholm would collaborate on all his major designs. Although he was an accomplished cabinetmaker, Kjærholm came to favor stainless steel after exploring its potential for industrial production. He followed Klint in the sense that he reinterpreted classical or existing models. For example, his *PK41* stool is inspired by a folding stool from antiquity; his *PK22* chair, by the Greek klismos and Mies van der Rohe's *Barcelona*; and his *PK24* chaise longue, by that of Le Corbusier (plate 145). Kjærholm's pieces have a great aesthetic purity and inscribe themselves into space like light, elegant signs.

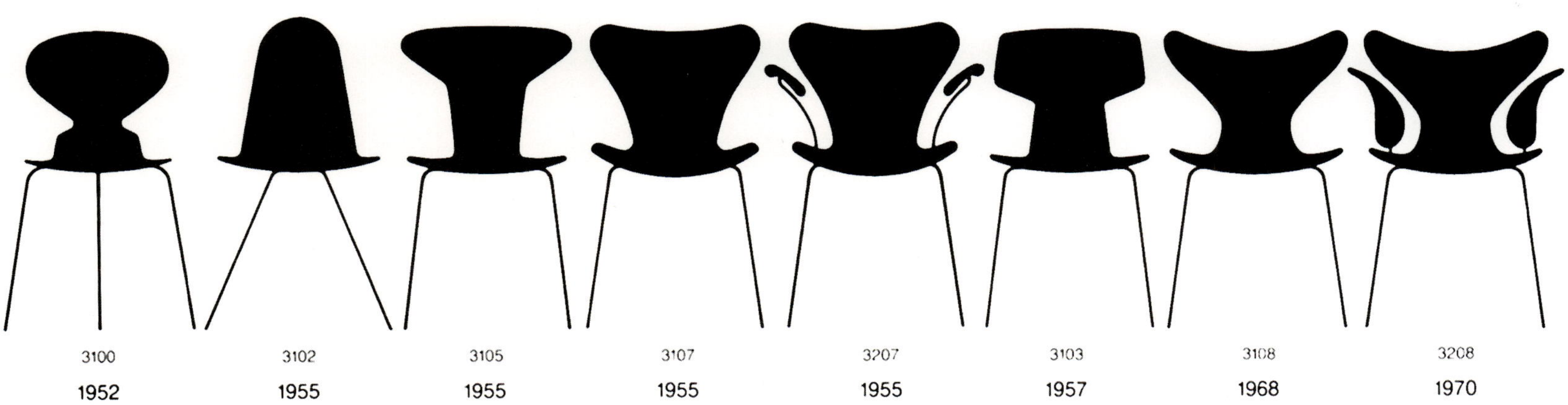

TOP LEFT
141. Arne Jacobsen (1902–1971)
***Series 7* chair**, 1952
Made by Fritz Hansen (Denmark)
Chrome-plated steel, molded plywood
Musée d'Art Moderne, Saint-Étienne Métropole

TOP RIGHT
142. Arne Jacobsen (1902–1971)
***Grand Prix* chair**, 1957
Made by Fritz Hansen (Denmark)
Chrome-plated steel, molded laminated teak
Musée des Arts Décoratifs, Paris

ABOVE
143. Arne Jacobsen (1902–1971)
Silhouettes of chairs designed between 1952 and 1970

OPPOSITE
144. Arne Jacobsen (1902–1971)
***Egg* and *Swan* chairs in a room of the SAS Royal Hotel in Copenhagen**, 1958
Chairs made by Fritz Hansen (Denmark)
Cast aluminum, polyurethane foam, leather

BELOW
145. Poul Kjærholm (1929–1980)
***PK24* chaise longue**, 1965
Made by Kold Christensen (Denmark)
Flat chrome-plated steel, woven rattan, leather
Centre National des Arts Plastiques/Fonds National d'Art Contemporain, Paris, on loan to the Musée des Arts Décoratifs, Paris

BELOW
146. Arne Jacobsen (1902–1971)
***Cylinda Line Classics* tea and coffee service**, 1967
Made by Stelton (Denmark)
Satin-polish stainless steel
Musée des Arts Décoratifs, Paris

OPPOSITE, LEFT
147. Jacob Jensen for Sigvard Bernadotte (1907–2002) and Acton Bjørn (1910–1992)
***Margrethe* bowl**, 1950
Made by Rosti (Denmark)
Melamine
Designmuseum Danmark, Copenhagen

OPPOSITE, RIGHT
148. Erik Magnussen (b. 1940)
Thermos, 1977
Made by Stelton (Denmark)
ABS plastic, glass lining
Musée des Arts Décoratifs, Paris

FROM TABLEWARE TO TOYS

Danish manufacturers draw on a long tradition of fine work in silver and porcelain. The Georg Jensen company, founded in the nineteenth century, produces highly sophisticated modern silver, designed notably by Henning Koppel, Tias Eckhoff, Piet Hein, and Erik Herløw (plate 149). Its boutique on Fifth Avenue in New York is an important showcase of good Danish design overseas. The Royal Copenhagen porcelain manufactory, established under royal patronage in the eighteenth century, continues to produce classic models, including the *Flora Danica* and *Blue Fluted* services, while also collaborating with contemporary designers. In 1965, for example, Grethe Meyer created the *Blåkant* dishware line in white earthenware decorated with a thin blue line.

In 1950, the Swedish designer Sigvard Bernadotte founded an industrial design studio with the Dane Acton Bjørn. Their products notably include the *Margrethe* series of stackable plastic bowls, designed by the Norwegian Jacob Jensen (plate 147). Stelton, established in 1960, began manufacturing modern yet timeless steel objects with Arne Jacobsen's 1967 *Cylinda Line* (plate 146). During the 1970s, Erik Magnussen developed a series of steel and plastic thermoses for Stelton that would become icons of Scandinavian design (plate 148). Magnussen, who was trained as a ceramist, also designed pieces for the porcelain manufacturer Bing & Grøndahl, including original teapots. Wood, and in particular teak, was used in a variety of everyday objects—for example, the salad servers developed in Kay Bojesen's studio in 1949, or Jens Quistgaard's ice bucket (1960) for his company Dansk. Bojesen also designed toy animals in wood, which are very popular.

During the 1950s and '60s, Poul Henningsen continued to develop his modern lights with the manufacturer Louis Poulsen. His *PH Artichoke* lamp from 1957 is a sophisticated reinterpretation of the first *PH* lamps, with the metal leaves now split to add an organic dimension to the functional design (plate 150).

Children play an important role in Scandinavian society. The Danish company Lego, which began manufacturing wooden toys in 1932, diversified into plastic building blocks after the war. The famous Lego blocks have been only slightly modified since, and are still an excellent creative toy appreciated worldwide. Kompan has been manufacturing playground equipment since the 1970s and is now a world leader in this sector. Simple, whimsical, and colorful, its products are rigorously safe in both their materials and their construction.

OPPOSITE
149. Henning Koppel
(1918–1981)
Pitcher *n° 992*, 1952
Made by Georg Jensen
(Denmark)
Silver
Musée des Arts Décoratifs,
Paris

ABOVE
150. Poul Henningsen
(1894–1967)
***PH Artichoke* pendant lamp**, 1957
Made by Louis Poulsen
(Denmark)
Copper, steel
Museum of Modern Art,
New York

RIGHT
151. Verner Panton (1926–1998)
***Cone* chair**, 1958
Designed for the Kom-Igen restaurant in Langesø, Denmark
Made by Fritz Hansen (Denmark)
Chrome-plated steel, latex foam, cloth upholstery
Centre National des Arts Plastiques/Fonds National d'Art Contemporain, Paris, on loan to the Musée des Arts Décoratifs, Paris

OPPOSITE
152. Verner Panton (1926–1998)
***Panton* chair**, 1967
Made by Herman Miller (U.S.)
Polyester reinforced with fiberglass
Victoria and Albert Museum, London

THE COLORISTS:
VERNER PANTON AND NANNA DITZEL

When the *Cone* chair—which looks like an ice-cream cone on a steel pedestal—was exhibited in a New York store window in 1959, it drew such crowds that the police ordered its removal (plate 151). This story attests to the originality and innovative—one might say provocative—spirit of the chair's Danish designer. After studying architecture, Verner Panton collaborated from 1950 to 1952 with Arne Jacobsen, among others, on the development of the *Ant* chair. In 1955, he opened his own design agency and joyfully explored furniture, lighting, textiles, new types of housing, and exhibition displays. The late 1950s marked the beginning of his collaboration with the young Plus-Linje company, with which he developed the *Cone* and *Heart* chairs, and also with Louis Poulsen, for which he designed lights. From 1957 to 1962, he worked on his design for a chair made from a single piece of plastic, inspired by Gerrit Rietveld's *Zig-Zag* chair. After a brief period in Cannes, his research into manufacturing techniques led him to Basel in 1963, where he worked out the details of his plastic chair with Vitra and where he would stay from that point on. The *Panton* chair was finally ready in 1967 and immediately gained wide recognition (plates 152, 154). It set a technical standard that would endure until the appearance of new plastics in the 1990s.

Starting in the 1960s, Panton developed a number of interior designs that were strongly inspired by op art and by research into new ways of inhabiting and moving through space. His *Fantasy Landscape* for Bayer, presented at the *Visiona II* exhibition at the Cologne furniture fair in 1970, was a colorful environment whose sculptural forms, made from foam, blurred the boundaries between floor, wall, ceiling, and furniture in a rather psychedelic fashion (plate 153). With his joyful creative prowess, Panton contributed more than any Nordic designer to the transition from craft traditions to the modern world of synthetic materials.

The milieu of postwar Danish cabinetmakers and furniture designers was essentially male; Nanna Ditzel, along with Grete Jalk, was one of the rare women to break through. It was at the School of Arts and Crafts in Copenhagen, where she studied furniture design, that Nanna Hauberg met her future husband Jørgen Ditzel. They exhibited together at the Cabinetmakers' Guild exhibition as early as 1944, and then collaborated on furniture, textile, and jewelry projects, which earned

BELOW
153. Verner Panton (1926–1998)
***Fantasy Landscape*, in the *Visiona II* exhibition at the Cologne furniture fair**, 1970
Vitra Design Museum, Weil am Rhein, Germany

OPPOSITE
154. **Herman Miller catalog presenting the *Panton* chair**
Vitra Design Museum, Weil am Rhein, Germany

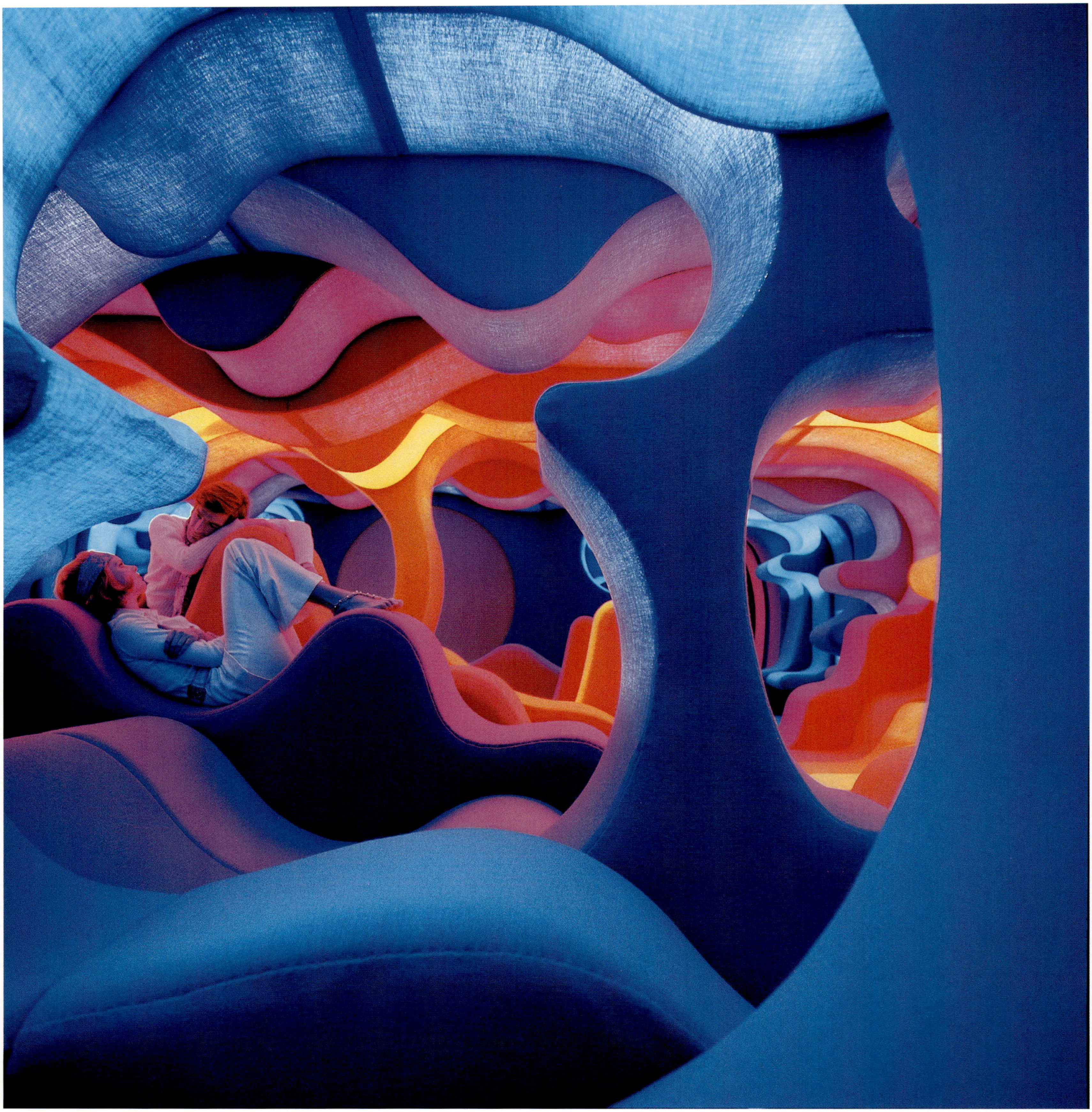

Europe, Africa, Near East
Herman Miller AG
Aeschenvorstadt 4, CH-4010 Basel
(Schweiz)

USA
Herman Miller Inc.
140 McKinley Street, Zeeland
(Michigan)

Deutschland
Fehlbaum GmbH
Kanderstraße 9, D-7858 Weil/Rhein

Schweiz
Fehlbaum AG
Klunenfeldstraße 20,
CH-4127 Birsfelden

France
Interform S.A.
162, Bd Voltaire, F-75 Paris 11e

Italia
ICF de Padova
Via Padana Superiore 280,
Vimodrone-Milano

Great Britain & Scandinavia
Herman Miller United Kingdom
Lower Bristol Road, Bath BA2 3 ER

Herman Miller Information Centre
33 Heddon Street, London W1 R 7 LL

Nederland
A. M. Bakker, »Aiglon«
Teilingen 3, Amsterdam

Belgique
IS. van Brande S.p.r.l.
38, rue Fossé aux Loups
B-1000 Bruxelles

Österreich
Herman Miller AG
Aeschenvorstadt 4, CH-4010 Basel
(Schweiz)

Argentina
Colección S.A.
Buenos Aires

Australia
Herman Miller (Aust.) Pty.
Glen Waverley (Victoria)

Brasil
Moveis Teperman, S.A.
São Paulo

Canada
Herman Miller of Canada, Ltd
Toronto

Colombia
Modulineas Ltda.,
Bogotá

Ecuador
Foresta S.A.
Quito

Japan
Modern Furniture Sales Co., Ltd.
Tokyo

Uruguay
NF Asociados S.R.L.
Montevideo

Venezuela
Decodibo S.A.
Caracas

int. Mod. reg. IR 42288

herman miller international collection **Panton Chair**

OPPOSITE
155. Nanna Ditzel (1923–2005) and Jørgen Ditzel (1921–1961)
Hanging egg-shaped chair, 1957
Made by R. Wengler (Denmark)
Woven rattan

BELOW
156. Nanna Ditzel (1923–2005) and Jørgen Ditzel (1921–1961)
High chairs for children, 1955
Made by Kold Savværk, Kerteminde (Denmark)
Oregon pine

them numerous prizes and awards, notably the Lunning Prize in 1956. The preceding year, they designed the classic children's high chair manufactured by Kold (plate 156) and, in 1957, wicker furniture including a famous hanging seat shaped like an egg (plate 155). The couple's jewelry for Georg Jensen is highly sculptural. After the premature death of her husband in 1961, Nanna continued to operate their design studio, with a particular emphasis on textiles. She was interested in new materials and developed a system of geometric elements in foam that she used to create seats and colorful multilevel interiors. She then moved to London, where she remained for fifteen years, founding the Interspace International Furniture and Design Centre there.

In 1986, at age sixty-three, Ditzel returned to Copenhagen and began a second career. Her furniture made from ultrathin strips of plywood with silkscreened motifs, like the 1989 *Bench for Two* manufactured by Fredericia, once again attracted attention and awards. In her *Trinidad* chair from 1993, openwork rays in the plywood seat and back create a striking visual effect through simple means. Her *Hallingdal* upholstery fabric, distributed by Kvadrat since 1965, today is produced in more than one hundred colors.

BELOW
157. David Lewis (1939–2011)
***BeoSound 8* speakers**, 2010
Made by Bang & Olufsen (Denmark)

OPPOSITE
158. Hans Sandgren Jakobsen (b. 1963)
***Rockable/Unrockable* stool**, 1999
Made by Orebo Traeindustri A/S (Denmark)
Beech

TECHNOLOGICAL EXCELLENCE

The electronics company Bang & Olufsen was founded in 1925 by two young engineers, Peter Bang and Svend Olufsen. Having developed the first radio to work with alternating current, the company diversified into other audio products and then, in the 1950s, into television sets and hi-fi equipment. Since the mid-1950s, the brand has devoted itself to modern design and ease of use as a way of highlighting the advanced technology of its products.

After World War II, Bang & Olufsen worked with designers including Sigvard Bernadotte and Acton Bjørn, but it was Jacob Jensen in particular who would help the firm establish a sophisticated, minimalist aesthetic at the service of exceptional quality. In the face of competition from Asia, the Danish company was able to survive thanks to this high-tech image. Starting in the 1990s, David Lewis developed spectacular and innovative high-end products for Bang & Olufsen, like the *BeoSystem*, *BeoSound*, and *BeoVision* (plate 157). Placed like sculptures in an interior, these devices convey an aura of inventiveness and exclusivity to a demanding international clientele.

THE HEIRS OF DANISH MODERN

It is not easy for younger generations to stand out against such a legacy of furniture design. In 1981, the annual Cabinetmakers' Guild exhibition (Snedkerens Efterårsudstilling, or SE) was relaunched in order to encourage innovation and experimentation within the framework of quality craftsmanship. At the end of the 1990s, the *Walk the Plank* project and competition encouraged collaboration between designers and cabinetmakers. At the beginning of the twenty-first century, manufacturers like Gubi, Hay, and Muuto have been actively promoting contemporary Danish design, while also initiating Nordic and international collaborations. Hans Sandgren Jakobsen operates in the tradition of the designer-cabinetmakers. After studying the Shakers in the U.S. and collaborating with Nanna Ditzel, he began his own design practice in 1997. Well known for his *Gallery* stools of 1998, made from molded plywood, and the sculptural stool *Rockable/Unrockable* for the 1999 *Walk the Plank* (plate 158), he has since developed a more simple and functional style. Jakobsen works on the international level and has received numerous prizes and awards, the most recent being the prestigious Finn Juhl Prize in 2009.

Another recipient of the Finn Juhl Prize is Kasper Salto, the grandson of the Danish ceramist Axel Salto. A cabinetmaker and industrial designer by training, he designed the *Runner* office chair in steel and plywood

LEFT
159. Cecilie Manz (b. 1972)
***Caravaggio* P2 pendant lamp**, 2005
Made by Lightyears (Denmark)
Steel, high-gloss enamel

BELOW
160. Poul Christiansen (b. 1947) and Boris Berlin (b. 1953)
***Gubi* chair**, 2003
Made by Gubi (Denmark)
Plywood

OPPOSITE
161. Cecilie Manz (b. 1972)
***Essay* table (with the *Series* chairs by Arne Jacobsen)**, 2009
Made by Fritz Hansen (Denmark)
Oak, beech, or walnut

for Peter Staerk in 1997. His collaboration with Fritz Hansen has resulted in such pieces as the *Ice* chair of 2002, the *Little Friend* adjustable table (2005), and the nylon *NAP* armchair (2010). *Ice* is made from aluminum and molded synthetic material and suited for indoor and outdoor use. Salto and his partner Thomas Sigsgaard won the contract to provide new furniture for the UN council chamber originally designed in 1952 by Finn Juhl.

Cecilie Manz, who studied in both Denmark and Finland, works in a minimalist vein, creating light, clever solutions. Her designs include furniture for Fritz Hansen (plate 161) and Fredericia; glass for Holmegaard; porcelain for Khler, inspired by Scandinavian designs of the 1950s; textiles for Georg Jensen; and the *Beolit 12* wireless speaker for Band & Olufsen. She has also designed several lamps for the Danish manufacturer Lightyears, including the *Caravaggio*, which has enjoyed great success ever since it was launched in 2005 (plate 159).

The duo from Komplot Design, the Dane Poul Christiansen and the Russian Boris Berlin, presented their innovative *Gubi* chair in 2003 (plate 160). This was the first piece of furniture to be produced by a new technique for molding plywood in three dimensions. Made from a sheet half the thickness of ordinary plywood, this chair, with its soft and "friendly" forms, is found in many restaurants in the Danish capital, making it the heir of Jacobsen's *Ant* and *Series* 7 chairs.

Christian Flindt works in a vein that is even more experimental and futuristic, similar that of international designers like Marc Newson. Two of his chairs may serve to illustrate the variety of his work: On the one hand, the *Rainbow* chair of 2005, made from semitransparent Repsol glass and available in ten colors, employs rectangular forms and can be stacked to form a bench or wall. The *Orchid* armchair of 2008, on the other hand, has an organic form. Made from polished fiberglass, it is a kind of sculpture that is at once sensual and cold, because of its slick and shiny surface.

The Anglo-Danish Louise Campbell also breaks through the divide between design and sculpture. Her creativity and joyful irreverence are striking; witness her chairs in felt (*Bless you*, 1999), openwork wood (for *Walk the Plank*), laser-cut steel (*Veryround*, 2006, and *Spiderwoman*, 2009), and paper cord assembled in biomorphic shapes (*It's complicated*, 2011). Also designing lamps, glass, and porcelain, she has mastered Nordic minimalism as well as more decorative modes, playing with transparency and the integration of space.

BELOW
162. Alvar Aalto (1898–1976)
Door handle, 1950s
Bronze

OPPOSITE
163. Alvar Aalto (1898–1976)
Lamps for the dining room of the Maison Louis Carré, Bazoches-sur-Guyonne, France, 1956–59

FINLAND

The top producer of wood in Europe, Finland has a long tradition of architecture and everyday objects made from natural materials. With its vast forests, 180,000 lakes, and severe winters, the country has always been close to nature, and has developed a particular minimalism drawn from a sense of the essential. Remaining for a long time under foreign domination—Swedish until 1809 and Russian until 1917—the Finnish sought their own cultural identity in their vernacular tradition and their national epic, the *Kalevala*. Art nouveau, imprinted with a romantic nationalism, flourished there, in the work of the architects and designers Eliel Saarinen, Herman Gesellius, and Armas Lindgren as well as the painter Akseli Gallén-Kallela. Weakened at its borders and much impacted by World War II, Finland rebuilt itself not only through architecture but also through the decorative arts and design.

ALVAR AALTO, THE HUMANIST

Alvar Aalto was an active participant in the reconstruction of Finland and endowed his country with many administrative, cultural, educational, and religious facilities. Although his career as an architect—in Finland and internationally—preceded his work as a designer, he completed several pieces of furniture and a number of light fixtures during the 1940s and '50s. His 45 chair, with a laminated birch frame, met the need for light and comfortable seating in his public buildings (plate 165). While simple, this armchair is meticulous in its detail, with armrests wrapped in rattan or leather for comfort. Aalto developed his prototypes with the help of loyal collaborators at Artek, which continues to manufacture his designs. In 1946, he perfected a new furniture leg, the *Y* leg, which he would employ in stools, chairs, and tables. In the mid-1950s followed another, even more sophisticated leg, the *X* or fan-shaped leg, its form practically plantlike. These legs were both based on the *L* leg, patented in 1933, which was made from solid wood reinforced by laminates on the curve of the foot. The *H* leg, or socket leg, from 1956, allowed for the assembly of tables and beds. This furniture was generally framed in birch, with seats or surfaces in wood or other natural materials, like rattan, leather, linen, or glass.

In addition to designing pieces for mass production, Aalto continued to create custom furniture and interiors for his various buildings, both in Finland and abroad. For example, he made successive variations of his bronze door handles and light fixtures, each time inventing new configurations (plates 162, 163). The 1950s were a rich period for his lighting design: he

developed his prewar models, with their indirect illumination, into ceiling fixtures, pendant lamps, and floor lamps. Made in painted metal, they often were lined with brass or adorned with an openwork brass rod in order to diffuse a warmer glow. Lighting was an integral component of his interiors, and his pendant lamps, which hang rather low, helped to articulate the space (plate 164).

Aalto and his second wife, the architect Elissa Aalto, also designed textiles with geometric motifs in primary colors, which are still distributed by Artek. Employing natural materials, Aalto's design work takes the user as its starting point and contributes to a humanist and organic vision of the built environment.

TAPIO WIRKKALA, THE POET OF MATERIALS

When the *Chanterelle* vases were introduced by the Iittala glassworks in 1946, the Finnish discovered the talent of the thirty-one-year-old designer Tapio Wirkkala. After studying sculpture at the Central School of Industrial Arts in Helsinki and working as a commercial artist in an advertising agency, Wirkkala won a competition sponsored by Iittala and began a collaboration that would last his whole life. The *Chanterelles*, with their very thin walls and fine engraved lines, are an abstract and poetic transposition of Finnish wild mushrooms into glass (plate 167). Wirkkala would continue to draw on the inspiration of nature to create modern objects that retain the capacity to transmit emotion and magic. Whether working in glass, wood, ceramic, or metal, Wirkkala greatly respected and truly understood his materials. He knew every technique and developed his designs in collaboration with craftspeople and workers.

In 1951, Wirkkala designed the Finnish pavilion at the Triennale di Milano. It was a great success and marked both the recognition of Finnish design abroad and Wirkkala's own fame. His leaf dish, made of laminated wood and silver, was named "the most beautiful object of the year" by the American magazine *House Beautiful* (plate 166). Danish and Swedish furniture and decorative art were already enjoying international respect, but because of its difficult postwar years, during which means and materials were scarce, Finland was not yet part of the Scandinavian style in the minds of Europeans and Americans. The exhibition of Finnish art and design that traveled the U.S. in 1952 and the Finnish pavilion at the 1954 Triennale di Milano, again designed by Wirkkala, consolidated the country's reputation.

A versatile designer, Wirkkala also created silverware and ceramics for the German firm Rosenthal and glass for the Italian firm Venini (plate 168). In Finland, he collaborated with the manufacturer Asko to make sophisticated pieces in laminated wood and created his own version of the traditional hunting knife, *Puukko*. But it was in glass that he especially excelled and in which he expressed his purist, poetic vision of nature. Through the 1960s and '70s, the objects he designed for Iittala often seemed to evoke the ice of Lapland winters, as does the bottle he created for Finlandia vodka. With his full beard, his pipe, and the hands of a craftsman, Wirkkala himself symbolized the creative genius linked to the spirit of natural materials.

OPPOSITE
164. Alvar Aalto (1898–1976)
Furniture and lamps in the living room of the Maison Louis Carré, Bazoches-sur-Guyonne, France, 1956–59

LEFT
165. Alvar Aalto (1898–1976)
Armchair n° 45/73W, 1946–47
Made by Artek (Finland)
Lacquered birch, rattan, black webbing
Musée d'Art Moderne, Saint-Étienne Métropole

BELOW
166. Tapio Wirkkala (1915–1985)
***Leaf* dish**, 1951
Laminated wood, silver
Museum of Modern Art, New York

BELOW
167. Tapio Wirkkala
(1915–1985)
***Chanterelle* vases**, 1948–54
Made by Karhula-Iittala
glassworks (Finland)
Blown glass with wheel
engraving
Musée des Arts Décoratifs,
Paris

OPPOSITE
168. Tapio Wirkkala
(1915–1985)
***Bolle* bottles**, 1966
Made by Venini (Italy), 1995–96
Blown glass, incalmo technique
Musée des Arts Décoratifs,
Paris

BELOW
169. Kaj Franck (1911–1989)
***Kartio* carafe and glasses**,
1958
Made by Nuutajärvi glassworks (Finland)
Colored mold-blown glass
Museum of Modern Art, New York

OPPOSITE
170. Gunnel Nyman (1909–1948)
Vase, 1947
Made by Nuutajärvi glassworks (Finland)
Glass with bubbles, blown in a spiked mold
Victoria and Albert Museum, London

FUNCTIONAL AND ELEGANT TABLEWARE

There is a long tradition of glassmaking in Finland. In the mid-twentieth century, the country had three large glassworks: Iittala, Nuutajärvi, and Riihimäki. Along with the Arabia ceramics manufactory, these glassworks invited young designers into their modern plants, both to create everyday objects and to experiment with artistic processes.

Like Tapio Wirkkala, Kaj Franck was a multidisciplinary designer. Trained in furniture design, he first worked on interiors and textiles. After the war, it was in ceramics and glass that he made the greatest contributions, leaving a mark on daily life in Finland that still endures. His best-known service for Arabia is the earthenware *Kilta* collection, designed between 1948 and 1952. He conceived of it as a durable, multipurpose dish set, with geometric elements in four colors that could be combined in different ways. (In the early 1980s, he reworked this service, which enjoyed a renewed popularity under the name *Teema*.) At the same time, Franck designed glass objects for Iittala, which he left in 1950 to become the artistic director of Nuutajärvi. There he continued to make simple objects in pressed glass, including the handsome *Kartio* carafe and glasses (1958)—still manufactured today by Iittala (plate 169). Convinced that designers have a social responsibility, Franck adopted functionalist precepts to meet the needs of everyday life in the postwar era. He also made art glass in delicate colors, and at the end of his life, surrendered to the allure of Pop.

The career of the glass, furniture, and lighting designer Gunnel Nyman was short but influential, particularly in the area of Finnish glass. Starting in 1932, she worked for the Riihimäki glass company, where she perfected a technique that releases bubbles to the

surface of the glass (plate 170). Her vases of the 1940s represent the first examples of glass sculpture in Finnish design. Nyman also worked briefly for Iittala and Nuutajärvi, from 1946 to her premature death in 1948, at the age of thirty-nine. Her vases, in particular the *Serpentini* and *Calla*, anticipated the evolution of Finnish glass toward the sculptural and organic forms of the 1950s.

Timo Sarpaneva designed everyday objects, but his work is primarily artistic. A precocious talent, he joined Iittala in 1950, at the age of twenty-four. He invented the steam-blowing technique, which he used to

BELOW
171. Timo Sarpaneva
(1926–2006)
Cooking pot, 1959
Made by W. Rosenlew and Co.
(Finland)
Cast iron, teak
Museum of Modern Art,
New York

OPPOSITE
172. Timo Sarpaneva
(1926–2006)
***Lancet* vase**, 1952
Made by Iittala (Finland)
Transparent and white
opaline glass
Museum of Modern Art,
New York

create his sculptural *Lancet*, *Kayak*, and *Orchid* vases in the vein of Nyman (plate 172). Honored at the Triennale di Milano in 1954, these remarkable pieces propelled him onto the international scene. His delicately tinted *i-line* series of the mid-1950s was situated between housewares and art glass. In 1964, he presented his *Finlandia* vases, blown in wooden molds that were gradually charred by the molten glass. Rawer in facture, each of these objects is slightly different and bears the trace of nature, recalling wood but also ice. A pioneer in this domain, Sarpaneva influenced the "frosted" aesthetic of the 1970s, which others including Wirkkala drew on. The starting point of Sarpaneva's *Festivo* (1966) and *Archipelago* (1981) candlesticks is transparent and expressive glass.

A multifaceted designer, Sarpaneva also designed textiles, porcelain for Rosenthal, and a famous cast-iron pot for W. Rosenlew & Co. (plate 171). Introduced in 1960 and reissued by Iittala in 2003, this piece revisits the traditional cooking pot, giving it a wooden handle that can be detached to allow the removal of the lid. Until his death in 2006, Sarpaneva also created many glass pieces that were either unique or issued in limited editions. The forms of these pieces are pure and geometric, attesting to both his great technical mastery and his artistic talent. In 2011, Design House Stockholm reissued his *Timo* glasses in an edition called *Timo Termo*, made of thermally resistant glass decorated with thin strips of colored silicone, so as to lessen the feeling of heat or cold.

173. Ilmari Tapiovaara
(1914–1999)
***Domus* chairs**, 1946
Made by Keravan Puuteollisuus (Finland)
Solid wood and plywood
Private collection

FURNITURE IN WOOD AND STEEL

The interior designer Ilmari Tapiovaara worked for Finmar, the London-based branch of Artek, in the mid-1930s. In 1946, he designed his famous *Domus* stackable chair, made of solid wood and plywood, for Keravan Puuteollisuus (plate 173). From the start, Tapiovaara was concerned with the functional and pragmatic side of design. He used local materials and sought to rationalize manufacturing techniques for the purpose of large-scale production. His furniture is easily disassembled, stackable, and inexpensive to transport. The 1950s were a fertile period, in which he created a wide range of furniture in wood, metal, and plastic. Tapiovaara also concerned himself with interiors (for the home, airliners, and exhibitions), products, and graphics; he designed lights, silverware, toys, textiles, and wallpaper. Between 1951 and 1964, he won six gold medals at the Triennale di Milano for his work.

Tapiovaara's aesthetic draws on that of Alvar Aalto in its use of bent wood and plywood, as well as in its research into function. But Tapiovaara quickly developed his own style, characterized by the continuous line between the short armrests and the back legs of his chairs, a structural detail that lends them lightness and elegance. Applied for the first time in the *Domus* armchair of 1946, this principle was one he drew on throughout his career. The *Pirkka* chairs (1954) are based on traditional wood furniture, and the stool version, with its thick seat and thin legs, has furnished countless saunas. A great traveler, Tapiovaara was also receptive to influences from outside Finland, and he sometimes recalls the Danish designer-cabinetmakers in his elegant detail and use of exotic woods (see, for example, the *Tale* stool of 1953; plate 174). Through his teaching at the University of Industrial Art in Helsinki as well as through his designs, Tapiovaara played an important role in the history of forms and materials in Finland, bringing to it a vision at once pragmatic and poetic.

Antti Nurmesniemi earned his degree in interior design in 1950. His first piece to attract attention was a sauna stool (1952) for the Palace Hotel in Helsinki. Shaped like a horseshoe and made in birch and teak, this stool reinterprets the traditional Finnish sauna chair (plate 175). Nurmesniemi worked in a modernist vein and was open to foreign influences, especially Italian; he spent six months with a design firm in Milan. In 1957, he designed for Wärtsilä the classic *Fidel* coffeepot in enameled steel, whose bright colors were chosen by his wife, the ceramist and textile designer Vuokko Eskolin-Nurmesniemi (plate 176).

Winner of the Lunning Prize in 1959, Nurmesniemi designed the Finnish pavilions at the Triennale di Milano in 1960 and 1964, which earned him the *gran premio*. For the first one, he designed the *Triennale* chair, an elegant object with a black leather shell atop two T-shaped steel legs. He developed the steel-legged *001* chaise longue of 1968 in collaboration with his wife, who was responsible for the various striped and colored fabrics in which it can be upholstered. Nurmesniemi created a variety of other furniture for Artek, Tecta, and Cassina, as well as numerous interior and exhibition designs. He also worked on kitchen utensils, doorknobs, the carriages of the Helsinki metro (in collaboration with Börje Rajalin), and the *Antti* telephones in colored plastic for the Japanese company Fujitsu. Nurmesniemi was an industrial designer in the modern sense of the word, with a pragmatic and international spirit, who nevertheless remained loyal to the values of Nordic living.

TOP LEFT
174. Ilmari Tapiovaara
(1914–1999)
***Tale* stool**, 1953
Made by Aero (Finland)
Solid wood and plywood
Private collection

TOP RIGHT
175. Antti Nurmesniemi
(1927–2003)
Sauna stool, 1952
Made by G. Soderstrom
(Finland)
Birch, teak
Museum of Modern Art,
New York

ABOVE
176. Antti Nurmesniemi
(1927–2003)
***Fidel* coffeepots**, 1957
Made by Arabia,
Wärtsilä Ab (Finland)
Enameled steel
Museum of Modern Art,
New York

BELOW
177. **Bedroom with furniture by Eero Aarnio**, 1971

OPPOSITE
178. Maija Isola (1927–2001)
***Citrus* fabric**, 1956
Made by Printex Oy (Finland)
Cotton
Victoria and Albert Museum, London

EXPRESSIVE TEXTILES AND POP ENVIRONMENTS

To address the scarcity of fabric and clothing at the end of the war, Viljo and Armi Ratia founded the print fabric company Printex in 1949. Two years later, they also established the retail brand Marimekko with the aim of encouraging the use of Printex textiles in interior design. Armi Ratia wanted different motifs than those available at the time and hired as a designer Maija Isola, a young artist who had studied painting at Helsinki's Central School of Industrial Arts. Isola created abstract, African-inspired motifs in bright colors, which were silkscreened on simple cotton canvas. In 1953, Vuokko Eskolin-Nurmesniemi joined the firm and designed several patterns, including the *Everyboy* for men's shirts, which became a kind of national uniform. Despite a positive reception, Marimekko's beginnings were difficult. However, with the brand's participation in fairs abroad—notably Expo '58 in Brussels—it started to gain international attention. It obtained its first representation in the U.S. a year later, and in 1960 Jacqueline Kennedy advertised the brand by making a purchase—heavily reported—of several Marimekko dresses. That same year, Annika Rimala replaced Eskolin-Nurmesniemi, who left the company to found her own, called Vuokko.

During the 1960s, Isola designed her best-known patterns for Marimekko, including *Unikko* (poppy), *Melon*, and *Kaivo*. These oversize motifs in bursting colors, which emit an energy and joie de vivre approaching the realm of the visual arts, were revolutionary in that epoch of textile design (plate 178). Rimala also created geometric patterns and introduced broad stripes on jersey fabric. The style of Marimekko clothing—simple,

BELOW
179. Eero Aarnio (b. 1932)
***Pastil* chair**, 1967
Made by Asko International, later Adelta International (Germany)
Molded polyester reinforced with fiberglass
Centre National des Arts Plastiques/Fonds National d'Art Contemporain, Paris

OPPOSITE
180. **Young woman in Eero Aarnio's *Balloon* chair**, 1968

full, and unisex—promoted an image of liberation and freedom. (The brand also produces linens and accessories.) After some difficult years in the 1980s, Marimekko has returned to the spotlight with the reissue of iconic patterns from the 1960s and the production of derivative products. One of the latest uses of the *Unikko* flowers is on the outer hull of Finnair planes, as a symbol of the national culture.

Most Finnish furniture designers chose not to adopt synthetic materials, and Eero Aarnio also began by designing wood pieces in the 1950s. However, at the beginning of the 1950s, when he established his own studio, he became interested in fiberglass (plate 177). His first chair in this material was the *Ball* of 1962 (plate 180). An offspring of Jacobsen's *Egg* and the Ditzels' hanging chair, the *Ball* takes the concept further by creating a space within a space. Its spherical shell of white fiberglass has an orange-upholstered interior, and Aarnio even installed a telephone in the first prototype. Asko, the producer of his wood pieces, worked with him to develop this chair and presented it in 1966 at the Cologne furniture fair, where it garnered a great deal of attention. Aarnio also designed a hanging version in transparent acrylic called the *Bubble*, launched in 1968.

At the 1968 Cologne furniture fair, Aarnio showed another design, the *Pastil* rocking chair, made in several colors of fiberglass (plate 179). Slightly smaller than the opening of the *Ball*, it can therefore be shipped inside it. A winner of the American Industrial Award, the *Pastil* can be used outside or in and lends itself to playful activity, like floating on water or sliding on snow. Aarnio's designs belong to pop culture and share the optimistic, playful spirit of the 1960s. In 1973, he designed further synthetic chairs, including the *Tomato* and the *Pony*, shaped like its namesake (plate 181).

These were complemented by other chairs and tables during the 1990s and 2000s. Aarnio's pieces, now manufactured by Adelta, have sparked renewed interest, and a Norwegian telephone company has even placed *Bubble* chairs in its offices to allow its clients to use their cell phones more privately. The prototype of 1962 has thus found a new application a half-century later.

Yrjö Kukkapuro opened his design studio around the same time as Aarnio, at the end of the 1950s. A

BELOW
181. Eero Aarnio (b. 1932)
***Pony* chair**, 1973
Made by Adelta International (Germany), 2001
Steel, molded polyurethane foam, stretch fabric (89% polyester and 11% Lycra)
Centre National des Arts Plastiques/Fonds National d'Art Contemporain, Paris, on loan to the Musée des Arts Décoratifs, Paris

BOTTOM
182. Olof Bäckström (b. 1922)
***O* series scissors**, 1967
Made by Fiskars (Finland)

OPPOSITE
183. Yrjö Kukkapuro (b. 1933)
***Karuselli* (Carousel) armchair**, 1965
Fiberglass, leather, chrome-plated steel
Made by Avarte (Finland)
Private collection

student of Tapiovaara, he remained loyal to functionalist precepts throughout his career. After making furniture in plywood and steel, Kukkapuro drew attention in 1965 for his imposing *Karuselli* (Carousel) armchair (plate 183). Molded on his own body, the fiberglass shell is upholstered in leather and stands on a steel base. The following year, this chair was handsomely enshrined on the cover of *Domus* magazine. Accompanied by a footrest, the *Karuselli*, a Nordic reinterpretation of the Eameses' celebrated 1956 lounge chair, is a coveted classic today. During the 1970s, Kukkapuro took an interest in ergonomics and developed innovative office chairs. A teacher at and director of the Central School of Industrial Arts in Helsinki, he went through a brief postmodern period is the 1980s. First manufactured by Haimi, his pieces are now produced by Avarte.

It was in 1967 that the famous *O* series of scissors was introduced by Fiskars (plate 182). The industrial designer Olof Bäckström drew inspiration from tailor's scissors in developing the series. Perfectly ergonomic, the orange handles in ABS are meant for right-handed users, and the red, for left-handed ones. Fiskars, one of the oldest companies in Finland, was founded in 1649 in the village of the same name, near the country's southern coast. Traditionally a manufacturer of cutlery and kitchenware, Fiskars also manufactures a full range of tools for gardening and forestry and has become a leader in dependability and ease of use.

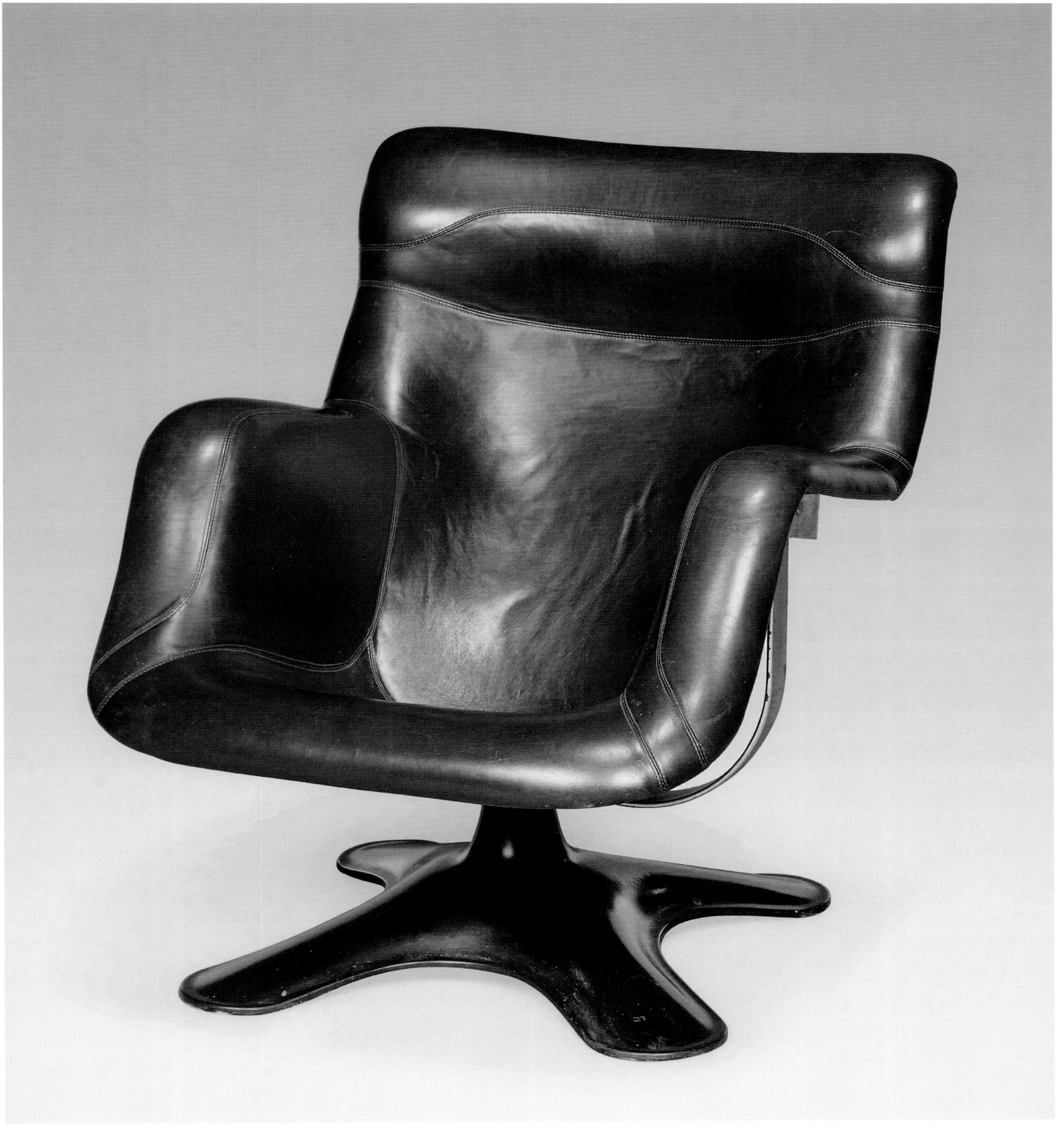

BELOW
184. Frank Nuovo (b. 1961)
***GS 10* mobile phone**, 2002
Made by Nokia (Finland)

OPPOSITE, TOP
185. Ilkka Suppanen (b. 1968)
***Flying Carpet* sofa**, 1997
Made by Cappellini (Italy)
Flexible steel, aluminum finish, felt
Centre National des Arts Plastiques/Fonds National d'Art Contemporain, Paris

OPPOSITE, MIDDLE
186. Harri Koskinen (b. 1970)
***Block* lamp**, 1996
Made by Design House Stockholm (Sweden)
Glass block in two parts
Museum of Modern Art, New York

OPPOSITE, BOTTOM
187. Terhi Tuominen (b. 1979)
***Fold* chairs**, 2008
Made by Dayground (Finland)
Ash

NOKIA PHONES

The Nokia company, originally a paper pulp mill, was founded in 1865. In 1966, it merged with two other companies, one a producer of rubber and the other, of telephone cable. After having launched the first mobile car telephone in the 1980s—weighing more than twenty-two pounds (10 kg)—the group began to focus exclusively on mobile phones in 1992. A pioneer in the sector, Nokia remained the top manufacturer of cell phones worldwide until 2012, the year when competition from Asia took the lead. The American Frank Nuovo, head of Nokia's design team from 1995 to 2006, emphasized simple and solid design, but without strong styling (plate 184). His successor Alastair Curtis attempted to contend with the smartphones being introduced by other brands. In 2011, the Finn Marko Ahtisaari took the helm of the design department and developed models with colorful and sleek cases. A major player in the "Finnish economic miracle" of the 2000s, the Nokia group has wagered on design in its attempt to take back leadership in the cell phone industry.

BETWEEN TRADITION AND GLOBALISM

During the 1980s, Finnish design stood on the sidelines. It did not find a place of its own in postmodernism, and a number of companies focused on reissuing classics. The early 1990s was marked by the economic crisis resulting from the fall of the Soviet Union and the consequent collapse of many Finnish manufacturers' main market. Around 1995, initiatives promoting education and creativity began to bear fruit, while international trends turned again toward simplicity and authenticity. The generation of designers born at the end of the 1960s and the beginning of the 1970s was freed from the weight of past glory; they placed themselves on an international stage and used every technique at their disposal.

Ilkka Suppanen was a founding member of the Snowcrash cooperative in 1997 and later belonged to the Valvomo collective. At the Milan furniture fair, he drew notice for his *Flying Carpet* sofa, which was subsequently distributed by Cappellini (plate 185). The thick felt seat seems to float since it is held up only by thin steel rods. His *Catherine* armchair of 2009 for Vivero adds a free and asymmetrical note to lounge furniture. Suppanen has designed for many companies in Finland and abroad. His work is light and forward-looking, while remaining anchored in the Scandinavian heritage.

In 1996, Harri Koskinen attracted attention with his *Block* lamp for Design House Stockholm (plate 186). The bulb is simply inserted between two blocks of glass, recalling Sarpaneva's or Wirkkala's frosted sculptures. Koskinen has gone on to design glass for Iittala and Venini; speakers for the Finnish company Genelec; watches for Issey Miyake; wood and steel furniture for Artek, Montina, Cassina, and Woodnotes; textiles for

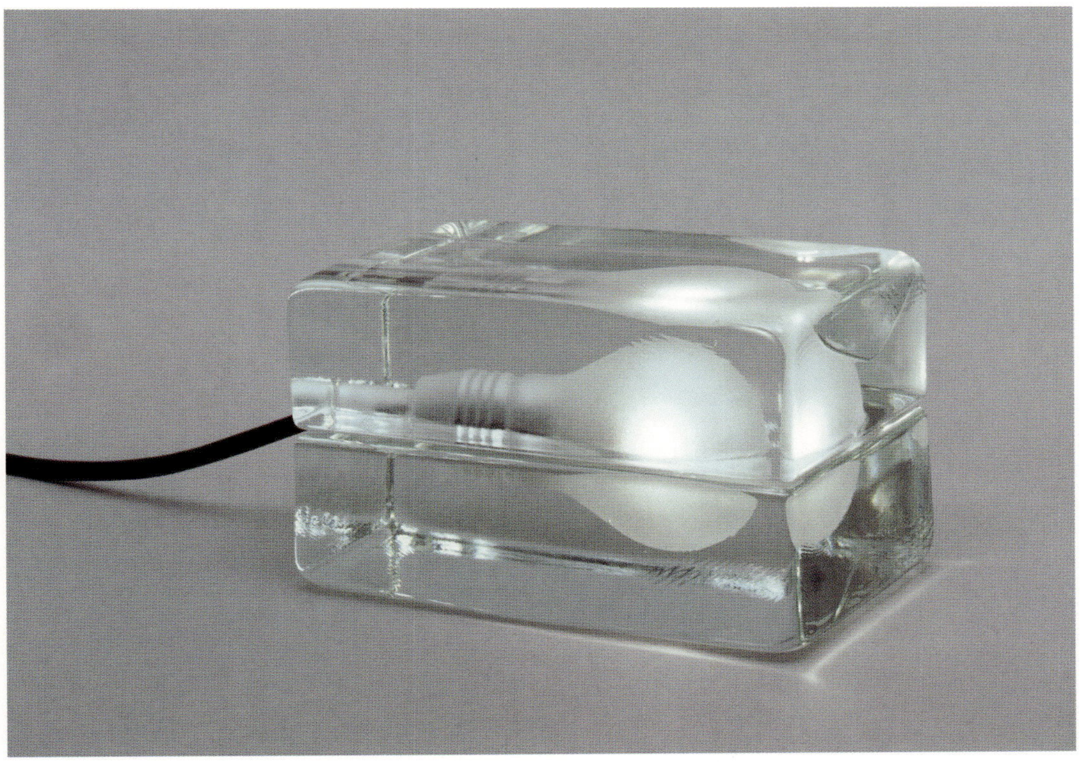

BELOW
188. Timo Salli (b. 1963)
***Helsinki Lighthouse* pendant lamp**, 2006
Made by Saas Instruments (Finland)
Acrylic tubing and fiber optics

OPPOSITE
189. Seppo Koho (b. 1967)
***Secto* pendant lamps**, 1999
Made by Secto Design (Finland)
Laminated birch

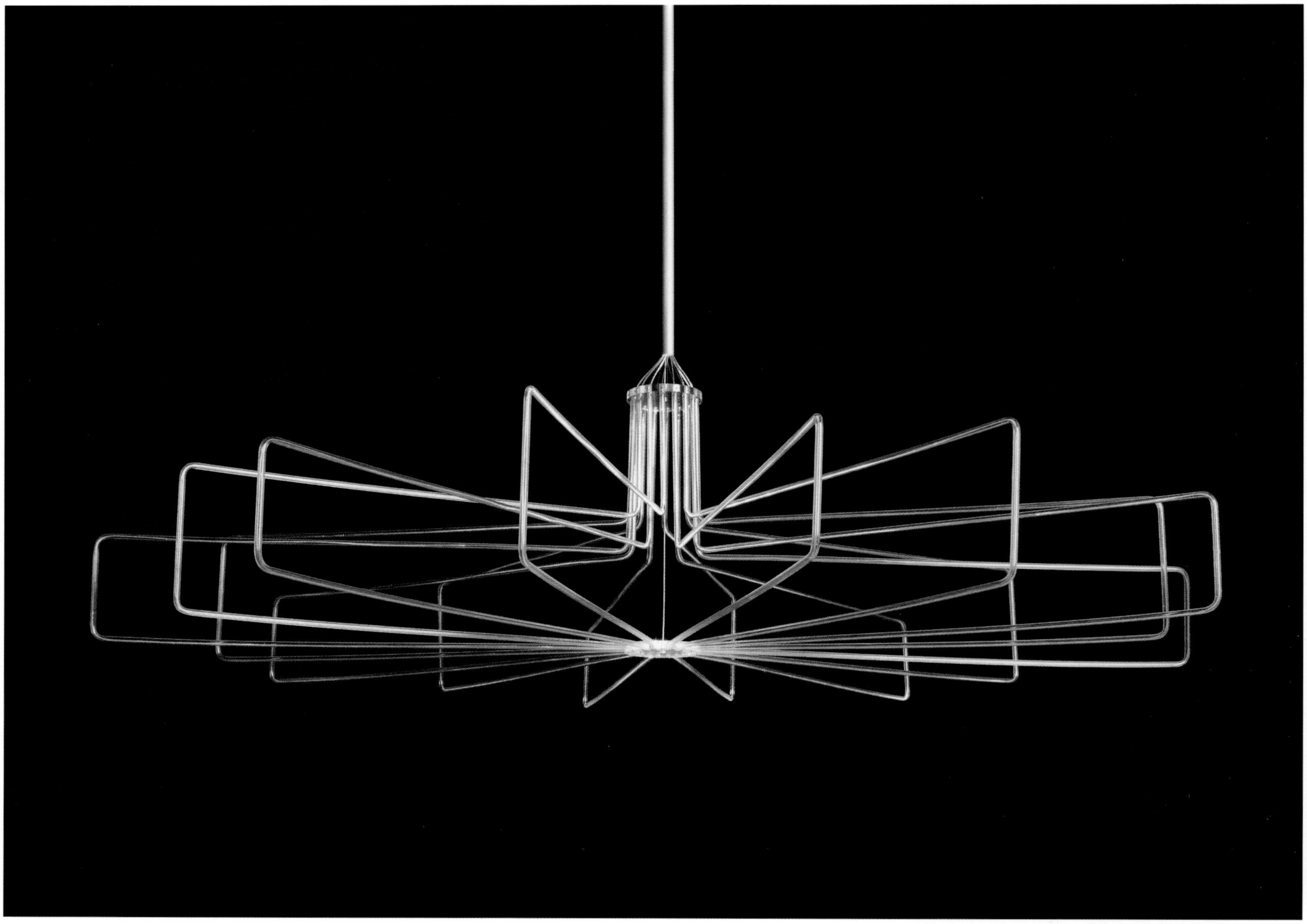

Marimekko; lights for Muuto; and tableware for Hackman, Arabia, and Alessi. He is a direct descendant of Nordic functionalism, with a remarkable sense of proportion and materials.

Named young designer of the year in 2009, Terhi Tuominen has a fresh approach to the objects she makes. In her *Blackbird* chair of 2006, a very thin metal mesh rests on a steel frame painted white or black. Its originality lies in its angular forms, reminiscent of a bird's beak. The stackable *Fold* chair of 2008 displays folds in the blond plywood, creating the effect of a three-dimensional surface (plate 187). Tuominen's structures readily combine an apparent fragility with true compositional strength.

In the technological vein, there is Timo Salli's *Helsinki Lighthouse* lamp (2006; plate 188), sold by Saas Instruments and made of acrylic tubes and fiber optics that generate a soft halo of light. Many designers have attempted to work with LED technology. The *Tunto* lamp by Mikko Kärkkäinen (2009) integrates LEDs into an elegant plywood structure; it can be turned on by a touch. Since 1999, Seppo Koho has designed the fine *Secto* lamps in laminated birch, handmade by Finnish craftspeople and exported throughout the world (plate 189). In 2012, Secto Design presented its *Aspiro* pendant lamp, with a birch spiral illuminated by an LED. These lamps illustrate the trend among young Finnish designers of combining traditional Nordic materials and forms with the newest technologies, or, in other words, of marrying the local and the global.

RIGHT
190. Sveinn Kjarval (1919–1981)
***Horseshoe* chair**, 1960
Made by Nyvirki (Iceland)
Steam-bent and laminated wood, hide

OPPOSITE
191. Halldór Hjálmarsson (1927–2010)
Interior design of Café Mokka in central Reykjavik, 1958
Reykjavik Museum of Photography

ICELAND

A rocky volcanic island, Iceland is poor in natural resources such as wood. For centuries, Icelandic interiors were characterized by the economy of their furnishings, which were generally immobile and made of driftwood or imported wood. Craft traditions focused on wool—abundant and of good quality—as well as silver. Linked to the other Nordic countries through history and trade (and being a Danish colony until 1944), Iceland gradually adopted their lifestyle in the nineteenth and twentieth centuries by importing the necessary materials. In 1930, on the occasion of the Icelandic parliament's millennial celebration, a romantic national awakening was expressed in literature and products inspired by the golden age of the Vikings. At the same time, Scandinavian functionalism became the norm in architecture, accompanied by the production of simple and practical furniture. Occupied by the British and then the Americans during World War II, Iceland emerged from the conflict prosperous and independent. Sheltered by restrictions on furniture imports, the local market saw the emergence of partially automated workshops and some rather fine products in the 1950s and '60s.

DENMARK AS SCHOOL

After World War II, the majority of Icelandic architects and designers continued to study in Copenhagen, as they had before. There they worked under the direction of the leading Danish designers and cabinetmakers and learned their methods. Icelandic design after the war was therefore naturally influenced by the Danish school. Limited means and a small market often ended up stifling the ambition of young designers who returned to Iceland. The first to come home was Sveinn Kjarval, following an apprenticeship with a cabinetmaker and a diploma from the School of Arts and Crafts in Copenhagen. In 1955, he founded the small company and workshop Nývirki, partnering with the carpenter Gísli Ásmundsson. During the 1960s, they produced small batches of several of Kjarval's designs, including the *Horseshoe* chairs (1960; plate 190). Made in laminated, steam-bent oak, it recalls the other chairs with spindle backs, inspired by Windsor chairs, that were popular in postwar Scandinavia for their light weight and low cost. The peculiarity of the *Horseshoe* is that the upper part of the back is steeply angled and widens to form the armrests. The seat is covered in Icelandic wool or hide. Kjarval also designed a lighter version to accompany a dining room table. His rocking armchair of the same period, in beech or oak with a horsehide seat, recalls a design by his teacher Hans Wegner and enjoyed great popularity in Iceland.

Kjarval's peroba chairs were included in the exhibition *Formes Scandinaves* in Paris in 1958. He also designed furniture in steel and plastic for Sindrasmiðjan, starting in 1957. At the beginning of the 1970s, Kjarval once again left Iceland to live in Denmark. Today his furniture is counted among the classics of Icelandic design and greatly sought after.

Halldór Hjálmarsson returned from Copenhagen in 1956 with a degree in furniture design. He developed several chairs in plywood on a steel base and gave them names drawn from Icelandic culture. The *Skata* (Ray) chair, an Icelandic version of Arne Jacobsen's *Ant*, has been reissued by the Museum of Design and Applied Art in Iceland. Like a number of his colleagues, Hjálmarsson also designed interiors. In 1958, he created the interior of the little Mokka café in the center of Reykjavik (plate 191). A beloved destination for artists and intellectuals in the capital ever since it opened, the café has kept its decor of banquettes, fixed tables, and pendant lamps almost fully intact.

Hjalti Geir Kristjánsson did not study in Copenhagen, but rather in Zurich, Stockholm, and New York. When he returned to Iceland in 1951, he worked with his father Kristján Siggeirsson, a carpenter who had founded a furniture factory in 1919. Kristjánsson modernized production and introduced contemporary Scandinavian design. His *ST114* chair of 1958 in teak and leather, designed for mass production, furnished many homes in Iceland. Commissioned by the University of

BELOW
192. Gunnar Magnússon
(b. 1933)
***Milan* armchair**, 1964
Made by Skeifan (Iceland)
Teak, wool upholstery

Iceland to make a chair for its ceremonial hall in 1965, he designed a simple seat, upholstered in light-colored leather. When Iceland joined the European Free Trade Association (EFTA) in the early 1970s and taxes on imports gradually disappeared, competition with imported furniture became quite a bit harsher, and the company began to focus on office furniture.

GUNNAR MAGNÚSSON, THE GEOMETER

The son of a shipowner from northern Iceland, Gunnar Magnússon learned cabinetmaking under Guðmundur Guðmundsson, known as the "Blind Man," in his Víðir furniture factory in Reykjavik. From 1959 to 1963, he studied furniture design at the School of Arts and Crafts in Copenhagen. During those years, he was selected to participate in the Cabinetmakers' Guild exhibition on several occasions, and his pieces were manufactured by the cabinetmakers Søren Horn, Christensen & Larsen, and Anders Svendsen, and by the Skalma company. Magnússon quickly distinguished himself with the clean lines of his geometric designs. At the time, Finn Juhl and Hans Wegner's organic forms dominated the Danish scene, but Magnússon tended increasingly toward the style of Børge Mogensen.

His first pieces included an innovative set of nesting tables, allowing for the assembly of three tables of the same height. He also designed a teak bed called the *Rockefeller,* since Nelson Rockefeller bought one at the famous Den Permanente boutique in Copenhagen. The legs and armrests of his oak *Inka* armchair are formed by two rectangles, joined by an outer frame that supports the back. A sofa and coffee table complete the set. Magnússon's designs were very well received by the public and critics alike. In 1963, he was one of the winners of an important international competition sponsored by the London *Daily Mirror*, and his bedroom furniture was produced by the Archie Shine company.

Despite this success, Magnússon decided to return to Iceland upon finishing his studies. Beginning in 1963, he adapted some of his Danish pieces for the Icelandic company Skeifan and the Meiður carpentry workshop. The *Milan* sofas and armchairs, with their clean lines and upholstery in bright blue Icelandic wool, are a variation of his *Blanca* series (plate 192). In 1968, he created the *Apollo* series in Oregon-pine plywood, inspired by the circular forms of a traditional butter churn (plate 194). Magnússon designed numerous interiors, notably those of the Holt Hotel in Reykjavik between 1963 and 1992, parts of which can still be seen. In 1972, he designed the impressive table in mahogany, leather, and marble that was used for the famous chess match between Boris Spassky and Bobby Fischer in Reykjavik. Starting in the early 1980s, he focused on office furniture for Kristján Siggeirsson. Today, some of Magnússon's models are being reissued, and he is recognized as the most original Icelandic designer of the postwar period, bringing together vernacular inspiration and Scandinavian refinement (plate 193).

RIGHT
193. Gunnar Magnússon
(b. 1933)
Pendant lamp, 1975
Made by Málmsteypa Ámunda
(Iceland)
Painted metal

BOTTOM
194. Gunnar Magnússon
(b. 1933)
***Apollo* sitting room**, 1968
Made by Kristján Siggeirsson
(Iceland)
Oregon pine, wool upholstery

RIGHT
195. Einar þorsteinn Ásgeirsson (b. 1942)
***Bucky* stool,** 1980
Made by the designer
Plywood with metal fasteners

BOTTOM
196. Dieter Roth (1930–1998) and Manfreð Vilhjálmsson (b. 1928)
Kúlan (The Ball) boutique in Reykjavik, 1962
Reykjavik Museum of Photography

OPPOSITE
197. Högna Sigurðardóttir-Anspach (b. 1929)
Interior in Gardabær, Iceland, 1965

BRUTALIST AND EXPERIMENTAL FURNITURE

An interesting interlude took place in Reykjavik when Swiss artist Dieter Roth moved there with his Icelandic wife between 1957 and 1964. He became friends with the architect Manfreð Vilhjálmsson, and showed him his way of making furniture from salvaged materials. In 1962, they and some fellow artists founded a boutique called Kúlan (The Ball), where Roth presented furniture in raw wood and metal, jewelry, books, and toys. Vilhjálmsson contributed chairs, stools, and tables that he made from painted metal tubes and sailcloth. The experiment, which lasted only three months, in some ways anticipated today's inexpensive furniture that is sold disassembled (plate 196). Vilhjálmsson, who was one of the leading Icelandic architects of the twentieth century, continued to use repurposed materials—notably rake handles for shelves and racks in his own house. In 1969, for the interior of a nightclub, he designed some thirty stools made from red buoys encased in painted metal frames.

Another architect, Högna Sigurðardóttir-Anspach, designed a few houses in Iceland in the 1960s before moving to France (plate 197). All the furniture—including the beds, shelving, and tables—was conceived in raw, unfinished concrete, and the light fixtures were built into the walls. This approach reflected a new conception of the home in a brutalist aesthetic, as a sculpture to live in.

The *Bucky* stool, designed by the architect and mathematician Einar þorsteinn Ásgeirsson in the early 1980s, is an investigation into geometric forms (plate 195). In an homage to Buckminster Fuller, it is made from four hexagonal sheets of plywood, assembled with metal fasteners. Ásgeirsson later moved to Berlin, where he continued to study and construct geometric forms, often on behalf of the Icelandic-Danish artist Ólafur Elíasson.

BELOW
198–99. Pétur B. Lúthersson (b. 1936)
***Stacco* stackable chairs**, 1980
Made by Labofa (Denmark)
Steel, upholstered seat

OPPOSITE
200. Valdimar Harðarson (b. 1951)
***Sóley* folding chairs**, 1982
Made by Kusch & Co (Germany)
Steel, lacquered wood

FURNITURE FOR EXPORT

After Iceland joined the EFTA, there was an attempt to encourage the export of Icelandic design. The models presented at international fairs often piqued interest but generally were not backed by production capacity or marketing. In a market as small as Iceland—where the population was not even three hundred thousand at the time—designers nevertheless had to look abroad to make a living for their work. According to Pétur B. Lúthersson, if the Icelandic people had invested the same energy in selling furniture as they did in selling fish, their furniture industry would be in much better shape today. A furniture designer since 1964, Lúthersson established his own relationships with foreign distributers, and his office furniture is produced in Denmark, Germany, Italy, Great Britain, the Netherlands, and the U.S. The Danish distributor Labofa sold several hundred thousand units of his stackable *Stacco* chair of 1980 (plates 198, 199).

The architect Valdimar Harðarson offered his folding *Sóley* armchair (named after an Icelandic flower) to the German firm Kusch & Co. in 1982 (plate 200). Unique in that it could be folded in a single motion and hung on the wall, *Sóley* was named "furniture of the year" in 1984 and also produced in Japan until 2003. In 2012, it was reissued in a wide variety of colors and finishes.

Erla Sólveig Óskarsdóttir studied in Denmark and collaborated with several distributers there and abroad. Onecollection, the company that reissued Finn Juhl's furniture, has distributed her chairs since 1997. The folding *Ames Einn* chair from 2001 is a good example of her style, at once rigorous and fluid. Óskarsdóttir is today one of Iceland's most prolific furniture designers.

BELOW
201. Dögg Guðmundsdóttir (b. 1970) and Rikke Rützou Arnved (b. 1973)
***Fifty* chair**, 2012
Made by Ligne Roset (France)
Steel, cord

BOTTOM
202. Dögg Guðmundsdóttir (b. 1970)
***Stjaka* candelabra**, 2011
Made by Epal (Iceland)
Aluminum

OPPOSITE
203. Katrín Ólína (b. 1967) and Michael Young (b. 1966)
***Tree* coatrack**, 2001
Made by Swedese (Sweden)
Lacquered MDF

INNOVATIVE TECHNOLOGY

At the 2012 Olympic Games in London, the disabled South African sprinter Oscar Pistorius competed against able-bodied athletes for the first time in an official competition. His prosthetic legs were developed by the Icelandic company Össur, founded in 1971 and today a global leader in orthopedics. Design contributes not only to the performance but also to the high-tech, even stylish image of these prostheses that some must wear on a daily basis.

Perhaps more predictably, Icelanders are at the technological forefront of the fishing industry. Since its founding in 1983, Marel, originally a small manufacturer of fish scales, has grown to become the top supplier of high-tech equipment and solutions for the food processing industry. Vaki, founded in 1986, is also a leader in the development, manufacturing, and marketing of high-tech equipment for fish farming and environmental monitoring. While design is not a top concern for these companies, their products have an entirely Nordic simplicity and efficiency.

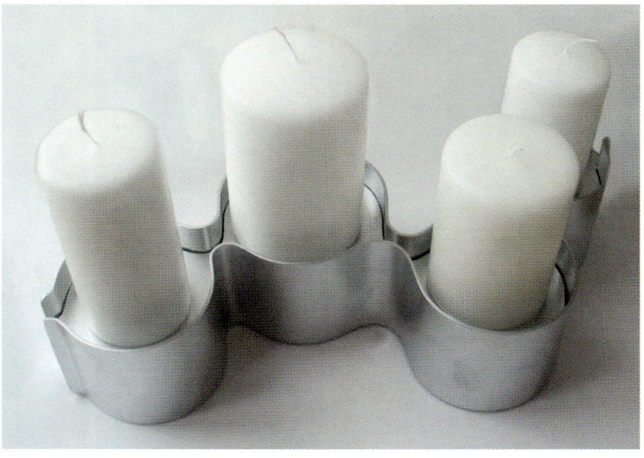

FEMININE ENERGY

While the design profession in Iceland has remained essentially male for decades, the generation born around 1970 includes several women who are distinguishing themselves through their work. Dögg Guðmundsdóttir studied in Milan and Copenhagen before moving to the latter city, where she opened Dögg Design. Guðmundsdóttir's work stems from Scandinavian functionalism but introduces a dose of humor and playfulness. Her *Kite* lamp (1998), designed with Carlo Volf and produced by Bsweden, is made from a sheet of polycarbonate attached to a metal plate with magnets, allowing for various configurations. The *Rocky Tre* rocking chair in laminated wood is inspired by the work of Alvar Aalto and has been manufactured by the Icelandic company Sólóhúsgögn since 2009. Her *Stjaka* candelabras in aluminum for Epal (2011) also have a fine organic quality in the Nordic tradition (plate 202). Guðmundsdóttir works with international distributers and has designed a number of objects for Ligne Roset, Cinna, Christofle, Finn Frogne, and Elimenti. Her most recent piece for Ligne Roset is the *Fifty* armchair, designed in collaboration with Rikke Rützou Arnved (plate 201). Inspired by a metal and rope chair by Hans Wegner from the 1950s, *Fifty* is a contemporary object with a sculptural appearance, both light and imposing.

After studying in Milan and England, Tinna Gunnarsdóttir opened her design studio in Reykjavik in 1993. The daughter of designer Gunnar Magnússon, she has a poetic approach to product design. Her work is characterized by a minimalism that draws from the cultural heritage of her country. Her garden gnome of 2002 is a simple ball of Icelandic lava, evoking fairies and nature myths. She also designed a system of lava tiles in which can be inserted garden tables or cubical stools whose chrome finish reflects the surrounding environment. Gunnarsdóttir often incorporates ornamental motifs into her geometric structures, for example in the openwork of her rubber place mats and doormats.

The *Tree* coatrack by Katrín Ólína and the Briton Michael Young has inspired several designers since its introduction by Swedese in 2001 (plate 203). Ólína—a graduate of the ESDI in Paris—has gone on to develop a very personal universe, populated with characters dear to her imagination, in a nature magical and sublimated. Her unique graphic designs are incorporated in her interiors—for example, in her *Cristal Bar* in Hong Kong (2008)—and have also decorated products such as snowboards for Elan-Colorsurf System, figurines and dishware for Rosenthal, and area rugs for Ege. In 2012, she designed *Friends of Steel*, a line of furniture and housewares made in painted steel.

Since the economic crisis struck Iceland in 2008, design has undergone a resurgence in the country. Tradition and the vernacular often serve as a source of inspiration, combined with a healthy dose of humor and self-mockery. Contemporary Icelandic design is both sophisticated, drawing on the latest technology, and anchored in the material and tactile, with a lively creativity.

BELOW
204. Gerhard Berg (b. 1927)
***Berg* lamp**, 1967
Made by Northern Lighting (Norway), reissued 2006
Acrylic, aluminum

OPPOSITE, LEFT
205. Hans Brattrud (b. 1933)
***Scandia* stackable chair**, 1957
Sold by Norge (Stordal, Norway)
Made by Hove Mobler
Bent laminated wood, steel
Nasjonalmuseet for Kunst, Arkitektur og Design, Oslo

OPPOSITE, RIGHT
206. Gerhard Berg (b. 1927)
***Nor* chair**, 1955
Made by Tynes Møbelfabbrikk (Norway)
Mahogany, pine, brass, wool upholstery
Nasjonalmuseet for Kunst, Arkitektur og Design, Oslo

NORWAY

With its deep fjords and villages nestled in breathtaking nature, Norway has a rich tradition of artistic craftsmanship that varies from one region to the next. After a period of romantic nationalism characterized by the "dragon style"—Viking in inspiration—the Norwegians adopted international art nouveau starting in 1905, the year the country celebrated its independence. During the 1920s and '30s, functionalism gained ground, in tandem with a gradual industrialization. In 1937, the engineer Jacob Jacobsen adapted the design of an English office lamp, the *Anglepoise* by George Carwardine, and manufactured it through his own company, Luxo. More than twenty-five million of these articulated office lamps have been sold worldwide under the name *Luxo L-1*, and they have notably served as a model for Pixar Animation Studios' mascot, a little hopping desk lamp. After World War II, Norway benefited from the "Scandinavian design" label, distinguishing itself above all in tableware.

ORGANIC FURNITURE

During the 1950s and '60s, Norwegian furniture designers adopted the soft forms and exotic woods of the Danish cabinetmakers. Hans Brattrud's *Scandia* stackable chair (1957), in bent laminated wood and steel, is among the classics of Norwegian design (plate 205). But the country's greatest contribution to organic Scandinavian design came from the interior designer Henry Klein. After studying in Denmark under Finn Juhl, Klein returned to Norway in 1953 to experiment with polystyrene chair shells. He produced his first chair using this technique in 1955, and the following year he sold licenses and several designs to Nordic manufacturers including the NK stores in Sweden, Asko in Finland, and Fritz Hansen in Denmark. His *1007* armchair probably served as the model for Arne Jacobsen's *Egg*, distributed by Fritz Hansen in 1958. In 1960, Klein began a fruitful collaboration with the Danish company Bramin.

Gerhard Berg studied at the Oslo National Academy of the Arts with the architect and modernist designer Arne Korsmo. As early as 1955, he was designing teak and metal furniture for Langlos, Stokke, and Vatne Lenestolfabrikk. He also explored fiberglass shells, for example in his *Giska* armchair, which was distributed by Hareid Bruk starting in 1958. Instead of pairing a synthetic seat with a metal base, Berg often used an exotic wood base, thereby connecting traditional cabinetmaking practices with new materials (plate 206). In 1967, he designed the sculptural *Berg* lamp, made from rectangular acrylic elements assembled in an irregular, organic fashion (plate 204).

BELOW
207. Grete Prytz (1917–2010)
Casserole dish, 1962
Made by Cathrineholm (Norway)
Enameled metal
Norsk Folkemuseum, Oslo

OPPOSITE, TOP LEFT
208. Tias Eckhoff (b. 1926)
***Maya* cutlery**, 1962
Made by Norsk Stålpress (Norway)
Stainless steel

OPPOSITE, RIGHT
209. Arne Jon Jutrem (1929–2005)
***Arne Jon* glass**, 1954
Made by Hadeland glassworks (Norway)
Glass
Nasjonalmuseet for Kunst, Arkitektur og Design, Oslo

OPPOSITE, BOTTOM LEFT
210. Willy Johansson (1921–1993)
***Multe* salad bowl**, 1965
Made by Hadeland glassworks (Norway)
Glass
Nasjonalmuseet for Kunst, Arkitektur og Design, Oslo

TABLEWARE

The winner of the second Lunning Prize, in 1952, was Grete Prytz. Having studied silversmithing, she worked in the family business, J. Tostrup, where she perfected the technique of enameling, a Norwegian specialty since the beginning of the twentieth century, both for jewelry and for large silver dishes. She also worked with her husband, the architect and modernist designer Arne Korsmo, and with him formed the iconic design duo of postwar Norway. At the 1954 Triennale di Milano, both won awards, she for her enamel work, he for his silver cutlery. Prytz also designed textiles, ceramics, and glass for Hegeland. After Korsmo passed away, she took the name of her second husband, Kittelsen. In the 1960s, she made a set of enameled metal kitchenware that was distributed by Cathrineholm. Decorated with a lotus motif designed by Arne Clausen, these inexpensive objects, simple in shape and cheerful in color, were hugely popular in Norway (plate 207). In 2008, the National Museum of Art, Architecture and Design mounted a retrospective of the work of the woman who is often referred to as the queen of Scandinavian design.

Tias Eckhoff was one of the most influential figures of postwar Norwegian design. A ceramist by training, he began working with the Porsgrund porcelain manufactory in 1949, while still a student at the Oslo National Academy of the Arts. From 1953 to 1960, he served as the artistic director of this porcelain works, the country's principal one, which had begun making modern products in the 1920s, under the aegis of his predecessor Nora Gulbrandsen. Among Eckhoff's best-known designs are the white tea and coffee set *Det riflete* (The Fluted), designed in 1949 and introduced in 1952, and the *Regent* service of 1962, which bears no decoration whatsoever. Eckhoff's work, which possesses a refined elegance expressed in soft, delicate lines, won awards at the Triennale di Milano in 1954, 1957, and 1960.

Eckhoff also designed the *Cypress* silver cutlery set for Georg Jensen, which won the Lunning Prize in 1953. His best-known set of flatware is *Maya*, designed in 1962 for Norsk Stålpress (later Norstaal, then acquired by Stelton). Composed of geometric forms in stainless steel, it was developed for mass production and enjoyed considerable success (plate 208). A true industrial designer, Eckhoff also made glassware, doorknobs, and furniture. His stackable *Tomi* chairs of the early 1980s consist of a plastic shell atop a steel base; the last version to date, the *Bella* of 1995, has a seat in bent laminated wood.

Hadeland, founded in 1762, is the largest glassworks in Norway. It produces both utilitarian glassware and a limited number of artistic designs. In 1936, the young Willy Johansson joined the company as an apprentice. The son and grandson of glassblowers, he quickly learned the trade and complemented his training with academic studies in the 1940s. In 1947, he became the director of the Hadeland design team and, along with Herman Bongard and Arne Jon Jutrem, built a solid reputation for Norwegian glass, which was honored at the Triennale di Milano and awarded the Lunning Prize. During the 1950s and '60s, they made everyday pieces in thin glass with simple shapes (plates 209, 210), as well as art glass in sophisticated colors and patterns.

OPPOSITE, TOP LEFT
211. Peter Opsvik (b. 1939)
***Tripp Trapp* chair**, 1972
Made by Stokke (now Variér) (Norway)
Ash

OPPOSITE, TOP RIGHT
212. Terje Ekstrøm (b. 1944)
***Ekstrem* chair**, 1973
Made by Stokke (now Variér) (Norway)
Metal, cloth upholstery
Stiftinga Sunnmøre Museum, Ålesund

OPPOSITE, MIDDLE LEFT
213. Peter Opsvik (b. 1939) and Hans Christian Mengshoel (b. 1946)
***Variable Balans* chair**, 1979
Made by Stokke (now Variér) (Norway)
Ash, cloth or leather upholstery
Stiftinga Sunnmøre Museum, Ålesund

OPPOSITE, MIDDLE RIGHT
214. Bjorn Refsum, Hilde Angelfoss, and Bard Eker
***Xplory* stroller**, 2005
Made by Stokke (now Variér) (Norway)
Aluminum and polyester

OPPOSITE, BOTTOM LEFT
215. Andreas Engesvik (b. 1970)
***Break* sofa**, 2004
Made by Norway Says (Norway)
Steel, cloth upholstery

OPPOSITE, BOTTOM RIGHT
216. Andreas Engesvik (b. 1970)
***I'm Boo* carafe,** 2006
Made by Norway Says (Norway)
Glass

ERGONOMIC AND SOCIALLY CONSCIOUS DESIGN

Although its design generally followed the pattern of the other Scandinavian countries in the postwar era, in the 1970s Norway stepped into the international spotlight with original ergonomic solutions to the problems of modern society. In 1972, the furniture manufacturer Stokke introduced the *Tripp Trapp* high chair, designed by Peter Opsvik (plate 211). Stable and adjustable, the *Tripp Trapp* follows the growth of the child from six months to adolescence, and even into adulthood. First made in polished birch, this symbol of a humanistic and pragmatic approach to childhood is now offered in various woods and colors. More than three million have been sold, and the design has been widely copied. At the Copenhagen furniture fair of 1979 there appeared another innovative design by Opsvik, the *Variable Balans* chair (plate 213). Developed in collaboration with Hans Christian Mengshoel, it is the fruit of in-depth research into seating ergonomics. Meant to relieve the back by transferring part of the body's weight onto the knees, *Variable Balans* revolutionized the way we sit. Produced by Stokke, it underwent many variations and has also been much copied.

Stokke, which has always been willing to explore types of seating, launched the *Ekstrem* chair by Terje Ekstrøm in 1973 (plate 212). With a metal frame covered in foam and fabric, it is a kind of sculpture that accommodates a variety of postures. Recalling the work of Verner Panton or Olivier Mourgue, the *Ekstrem* is a Norwegian nod to the 1970s and postmodernism.

A focus on childhood, the disabled, safety, and social responsibility is apparent in Norwegian industrial design, notably in the work of the K8 studio. Established in 1998, K8 drew attention with its *Xplory* stroller, winner of an international competition on the theme of city-friendly baby transportation, sponsored by Stokke (plate 214). A bit like the *Tripp Trapp* chair, the *Xplory* grows with the child and allows for multiple configurations. K8 members Marius Andresen, Rolf Blomvågnes, Olivier Butstraen, and Erik Lanuza have also designed solar chargers, an innovative stretcher, a wheelchair, and water bottles. The serious aspects of their work do not bar a playful and ecologically conscious spirit.

NEW PATHWAYS

In the early 2000s, Norwegian architecture attracted international attention with the design of the Bibliotheca Alexandrina by the young firm Snøhetta. At the same time, design seemed to revive in the wake of the success of the Norway Says collective at the 2000 Milan furniture fair. Since oil drilling began in the 1970s, Norwegian authorities had somewhat neglected design as a source of revenue and exports. Aware that they were lagging behind neighboring countries, they began encouraging and supporting the creative initiatives and marketing efforts of designers and studios. Norway Says developed furniture, light fixtures, and tableware in a neo-modernist minimal style with playful and sensuous touches (plate 216). With its asymmetry and cutout seat, the *Break* sofa from 2004 creates a level of intimacy in public places (plate 215). When Norway Says disbanded in 2009, two of its members, Torbjørn Anderssen and Espen Voll, founded the Anderssen & Voll studio. The third, Andreas Engesvik, pursued his own career in collaboration with distributors in Norway and abroad.

Johan Verde's design possesses a sculptural quality that distinguishes him from his neo-modernist colleagues. His 2002 *Loop* and *Peel* chairs, designed with

BELOW
217. Johan Verde (b. 1964), Olav Eldøy (b. 1948), and Ole Petter Wullum (b. 1965)
***Loop* chair**, 2002
Made by Fora Form (Norway)
Chrome-plated steel or polished aluminum, elastic fabric

OPPOSITE
218. Johan Verde (b. 1964)
Gravy boat from the *Opera Verde* service, 2008
Made by Porsgrund Porselen (Norway)
Porcelain
Nasjonalmuseet for Kunst, Arkitektur og Design, Oslo

BELOW
219. Cathrine Kullberg (b. 1971)
***Norwegian Forest* pendant lamp**, 2007
Made by Northern Lighting (Norway)
Birch

OPPOSITE, TOP
220. Daniel Rybakken (b. 1984)
***Daylight Entrance* installation**, 2010
Commissioned by Vasakronan AB, Stockholm
Corian, LED

OPPOSITE, BOTTOM
221. Daniel Rybakken (b. 1984)
***Counterbalance* lamp**, 2012
Made by Luceplan (Italy)
Steel, aluminum, LED

Olav Eldøy and Ole Petter Wullum, present themselves as objects in space, inviting the human body to place itself differently (plate 217). The series of dishes and objects he designed for Oslo's new opera house revives Scandinavian tradition with new materials and with more expressive forms (plate 218).

Tradition reasserts itself in Cathrine Kullberg's *Norwegian Forest* lamps (plate 219). Inspired by Scandinavian lampshades in openwork birch from the 1960s, these pieces are decorated with laser-cut forest and animal motifs. The birch shade is assembled by hand and sewn to the lamp using leather strips. In this way Kullberg brings together vernacular materials, craft processes, and images of Nordic nature with high technology and very contemporary design.

Among young Norwegian designers there is a trend toward poetry and immateriality. In 2011, the two founders of SHE Design, Silje Søfting and Eva Marit Tøftum, launched the *Myk* clock, a circle of stretched white canvas that presents the time in a peaceful, minimalist fashion. Daniel Rybakken, one of the most promising Norwegian designers, works mostly with light. In 2007, at the age of twenty-three, he drew attention with his *Daylight Comes Sideways* lamp, which creates the illusion of natural light with an array of LEDs behind a semitransparent screen. In 2010, he designed an installation for an office building in Stockholm that simulates sunlight in a windowless stairwell (plate 220). Rybakken also works with mirror effects for Ligne Roset and Galerie Kreo, among others. His *Counterbalance* lamp, presented in Milan in 2011 and distributed by Luceplan in 2012, is a clever rendition of Jean Prouvé's sconce (plate 221). A gear and magnet mechanism allows for the position of its lightweight LED to be precisely controlled. Rybakken's work proves that design, at its highest level, has the ability to introduce poetry and joy into daily life.

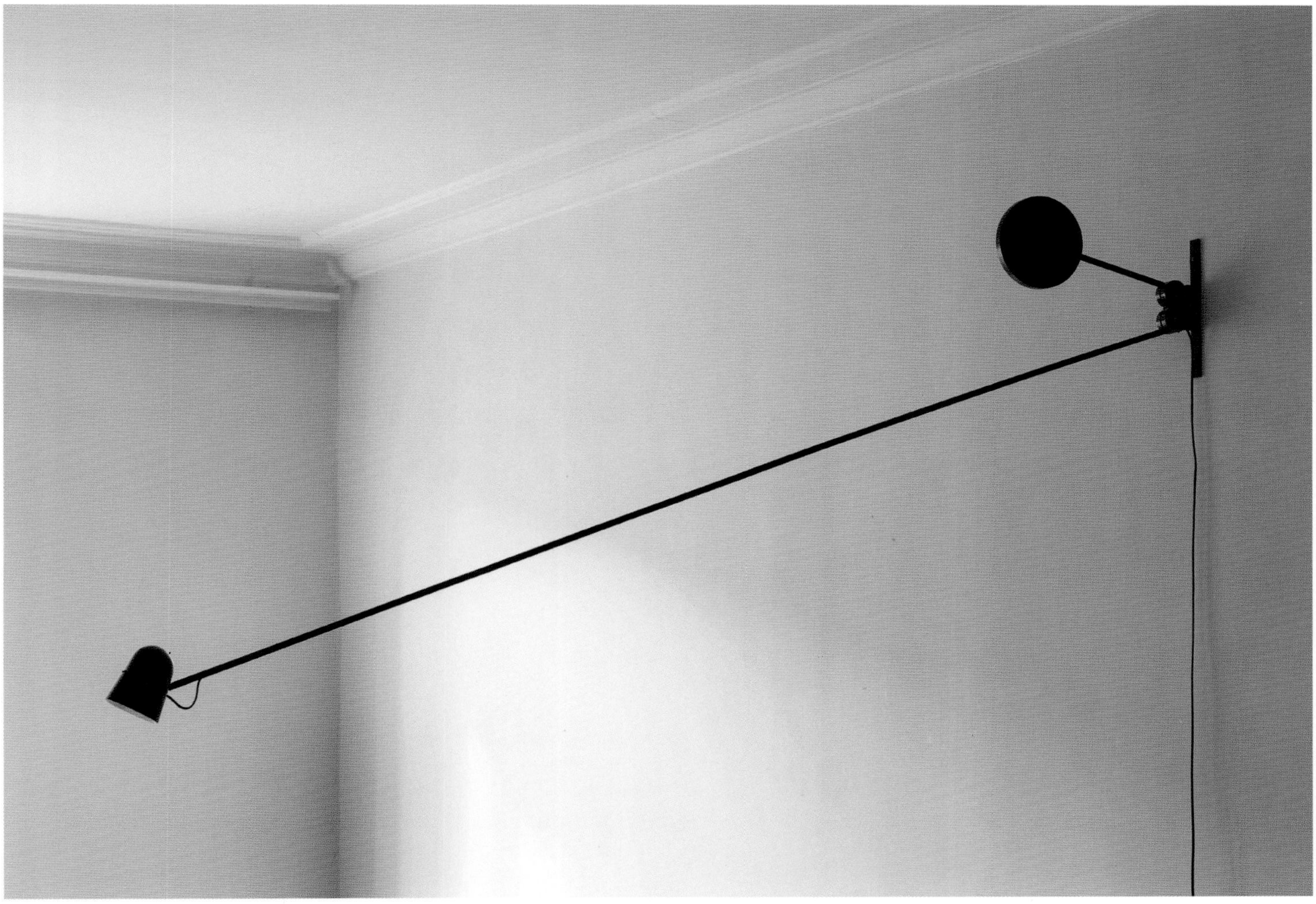

BELOW
222. Josef Frank (1885–1967)
***Bow* textile**, c. 1960
Made by Svenskt Tenn (Sweden)
Cotton, linen
Cooper-Hewitt, National Design Museum, New York

OPPOSITE
223. Bruno Mathsson (1907–1988)
Chaise longue and footrest, 1944
Made by Karl Mathsson (Sweden)
Bent birch, jute
Saint Louis Art Museum

SWEDEN

Situated in the heart of Scandinavia and having close political contact with the European continent, Sweden has always absorbed and assimilated international influences. The Gustavian style of the eighteenth century was in fact a lighter, "Nordic" reinterpretation of the Louis XV style. After the adaptation of the English Arts and Crafts style by Carl and Karin Larsson and a dalliance with the organic art nouveau of the continent, the Swedish decorative arts—glassmaking in particular—were distinguished by an elegant classicism. At the 1925 Paris International Exposition, the English critic P. Morton Shand coined the phrase "Swedish grace," widely adopted thereafter. The most influential furniture designer and teacher in this vein was Carl Malmsten, who drew his inspiration from the vernacular and the Gustavian style. He opposed the international functionalism of the 1920s and '30s, which was represented by Gunnar Asplund and Sven Markelius in particular. Nevertheless, they each in their own way put into practice the concept of "beauty for all" promoted by Ellen Key and Gregor Paulsson, and sought to create attractive, functional objects that were accessible to the greatest number of people.

As the only neutral and unoccupied country in Scandinavia during World War II, Sweden was much better off than its neighbors at the outcome of the conflict. The welfare state launched a massive housing construction program, requiring all kinds of fixtures and furnishings. The *H55* exhibition in Helsingborg in 1955 presented the best of Scandinavian design at the time and offered an attractive showcase for Sweden's own productions.

A SOFTER FUNCTIONALISM

Bruno Mathsson is one of the best representatives of Swedish organic modernism. His "lounge," "easy," and "work" chairs (the last known as the *Eva*) of 1934–35 are made from laminated wood and webbing. The production techniques and soft curves recall the work of Mathsson's colleague Alvar Aalto. During the 1940s, Mathsson developed variations of these chairs, including some with armrests and a chaise longue called *Pernilla*. The son of a cabinetmaker, he worked in, and produced his first designs for, the family business, Karl Mathsson, which later became Mathsson International (plate 223). He studied in the U.S. for six months in 1948 and on his return set out to build glass homes that were innovative in terms of energy efficiency. His growing interest in metal in the 1960s is exemplified by his *Jetson* armchair. A collaboration with the Dane Piet Hein yielded the *Superellipse* tables, with novel metal legs, distributed by Fritz Hansen.

Voted the best Swedish furniture design of the twentieth century, the *Lamino* armchair, designed by Yngve Ekström in 1956, also showcased organic lines. Its ergonomic back conforms to the curve of the spine and draws an elegant line in space. Originally from Hagafors, in the south of Sweden, Ekström began working in the country's oldest furniture factory at age thirteen. In 1945, he founded his own company with his brothers, which was initially called ESE, and later Swedese. Today one of the main distributors of Swedish furniture, Swedese put *Lamino* back in production in 2003. As opposed to their Danish colleagues, the Swedish designers of the 1950s and '60s generally preferred blond wood to exotic wood and simple forms to the technical feats of the cabinetmakers.

An atypical figure who nevertheless had a certain influence on Scandinavian design was the Austrian Josef Frank. Fleeing Nazism, he settled in Sweden in 1934 and worked closely with the interior design company Svenskt Tenn, founded by Estrid Ericson ten years earlier. Frank had been involved with the Wiener Werkstätte and participated in the exhibition of modern architecture at Stuttgart in 1927. He brought a softness and refinement to Swedish functionalism, often going against prevailing trends. His textiles in particular, with their exotic colors and motifs, would have a lasting influence on the Scandinavian aesthetic (plate 222). In the end, his work accorded well with the natural and somewhat romantic spirit of the country that took him in.

LEFT
224. Stig Lindberg (1916–1982)
Pitcher from the *Berså* service, 1960
Made by Gustavsberg (Sweden)
Porcelain
Victoria and Albert Museum, London

BELOW
225. Ingeborg Lundin (1921–1992)
***Äpple* vase**, 1957
Made by Orrefors (Sweden)
Glass
Private collection

OPPOSITE
226. Nils Landberg (1907–1991)
***Tulip* glasses**, 1956
Made by Orrefors (Sweden)
Glass
Private collection

GLASS AND CERAMICS

However, postwar Swedish design distinguished itself less in the area of furniture than in that of tableware, particularly glass. Since the 1910s, Simon Gate and Edward Hald had been developing sophisticated techniques at the Orrefors glassworks. In 1954, Sven Palmqvist presented his series of *Fuga* bowls, created by an innovative centrifuging method. Ingeborg Lundin was the first woman to join Orrefors as a glass designer, in 1947. Winner of the Lunning Prize in 1954, she exhibited her *Äpple* vases at *H55* (plate 225). These very fine glass objects, simple in form and delicate in color, have become classics of Nordic design. Lundin also created thicker pieces with colorful and artistic patterns, drawing on the *Ariel* and *Graal* techniques perfected by her predecessors. Her colleague Nils Landberg was known above all for his *Tulip* glasses, which were blown in a single piece—proof of his technical virtuosity (plate 226). Symbolizing "Swedish grace," they received the gold medal at the Triennale di Milano in 1957 and remained in production at Orrefors until 1981.

The Kosta glassworks, founded in 1742, concentrated on utilitarian glassware before recruiting designer Vicke Lindstrand in 1950. Lindstrand, who had worked at Orrefors, introduced art-glass techniques at Kosta and developed sculptural pieces there in the 1950s and '60s. A former illustrator, he readily employed line as a decorative motif. In the 1970s, he designed monumental glass sculptures, including the thirty foot (9 m) tall *Green Fire* for the city of Umeå. Following a merger, Kosta took on the name Kosta Boda in 1976, and then was bought by Orrefors in 1989.

The best-known ceramics manufactory in Sweden is Gustavsberg, founded in 1825. Wilhelm Kåge, its artistic director from 1917 to 1949, set the company on a modern path, particularly with the *Praktika* and *Grey Stripes* services. During the 1950s, he made stoneware with geometric motifs and art brut accents. His successor, Stig Lindberg, designed both organically shaped pieces, like the *Pungo* vase of 1953, and everyday services. His *Terma* service in black stoneware was long a staple in Swedish kitchens, and his *Berså* service in white porcelain decorated with a leaf pattern continues to be very popular (plate 224). Also a textile designer and an illustrator, Lindberg left a mark on his times with his joyful and colorful visual universe. Gustavsberg manufactured plastic objects as well, designed by Lindberg, Carl-Arne Breger, and Sven-Eric Juhlin, the founder of Ergonomidesign, among others. One of the primary Scandinavian producers of sanitary equipment since the 1950s, the company today belongs to the Villeroy & Boch group and focuses on that sector.

BELOW
227. Hugo Blomberg (1897–1994), Ralph Lysell (1907–1987), and Gösta Thames (1916–2006)
Ericophone, 1954
Made by Ericsson (Sweden)
Museum of Modern Art, New York

OPPOSITE, TOP
228. Sixten Sason (1912–1967)
Saab *92*, 1947
Made by Saab (Sweden)

OPPOSITE, BOTTOM
229. Nils Bohlin (1920–2002)
Three-point seat belt, 1959
Made by Volvo (Sweden)

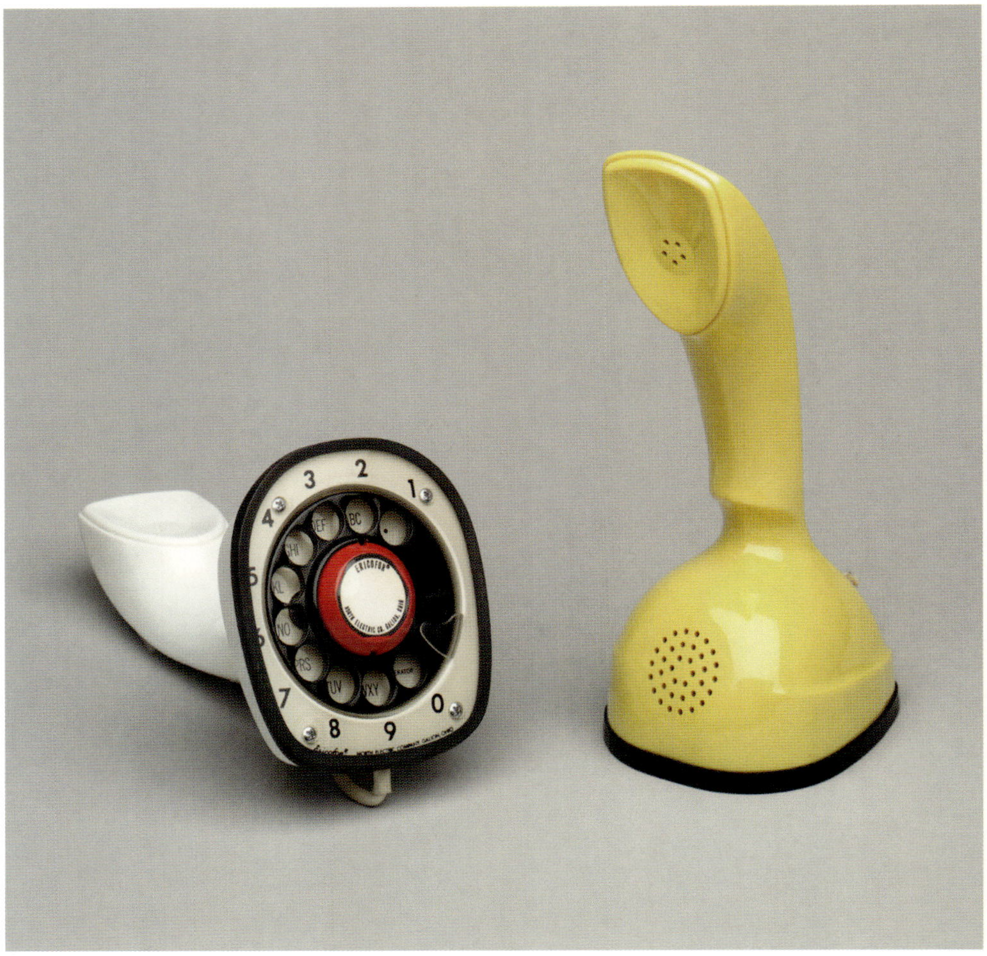

DESIGN FOR GROWING INDUSTRIES

Although it industrialized relatively late, in the twentieth century Sweden led the other Scandinavian countries in manufacturing. As early as the 1920s, the industrial sector contributed to Sweden's economic expansion and rising standard of living. The electrical appliance manufacturer Electrolux has been at the forefront of innovation since it was founded in 1919. Electrolux was the first company to offer households a lightweight, portable vacuum cleaner (the *Model V*, on metal runners, in 1921) and a refrigerator (1925), and in the 1930s it employed the pioneering American designer Raymond Loewy. World War II marked the beginning of a collaboration with the Swedish designer Sixten Sason, who created many aerodynamically shaped devices for the firm. Today the owner of brands like AEG, Zanussi, Arthur Martin, Husqvarna, Faure, and Frigidaire, the Electrolux group is the global leader in electrical appliances, with a strong interest in ecological materials and energy efficiency.

Sason was one of the first Scandinavian industrial consultants. A silversmith and pilot by training, he collaborated with numerous Swedish companies, notably Saab, for which he was artistic director from 1945 to his death in 1967. A manufacturer of military airplanes when it was founded in 1937, Saab turned to automobiles after World War II. The aerodynamic Saab 92, introduced in 1947, presented numerous innovations, including an aerodynamic monocoque body (plate 228). It was Sweden's first small car. Sason also designed the world's first single-lens reflex camera with interchangeable parts, the *1600F*, for Hasselblad. During the 1950s, he created several products for Husqvarna, including a drill and a sewing machine. Husqvarna is also known for its bicycles and motorcycles, and its chain saws, which are much appreciated by professional foresters.

Volvo put its first series of cars on the market in 1927. Making safety a top priority, the brand introduced the three-point seat belt, designed by Nils Bohlin, in 1959 (plate 229); it was the first carmaker in the world to make this belt a standard feature. In the decades that followed, Volvo's innovative safety measures and solid bodywork entailed robust, square designs. In 2004, the *YCC* concept car, designed by a team of eight women, reimagined the automobile from a female perspective. The Volvo Car Group was part of the Ford Motor Company from 1999 to 2008, when it was sold to the Chinese automaker Geely.

The telecommunications company Ericsson was founded in 1876 and was at the cutting edge of the European telephone industry by the early twentieth century. In 1954, Hugo Blomberg, Ralph Lysell, and Gösta Thames designed the original *Ericophone*, taking advantage of new materials that had been used in the war, like plastic and nylon (plate 227). Originally intended for bedridden patients, it was the first telephone to combine the dial and the handset in a single unit. In the 1980s, Ericsson entered the mobile phone market, operating a joint venture with Sony in this sector from 2001 to 2012.

BELOW
230. Ergonomidesign (now Veryday)
***Speedglas* welding helmet**, 1996
Made by 3M (U.S.), reissued 2005

OPPOSITE, LEFT
231. Gillis Lundgren (b. 1929)
***Billy* bookcase**, 1979
Made by Ikea (Sweden)

OPPOSITE, TOP RIGHT
232. Nicholai Wiig Hansen
Storage unit from the *PS* collection, 1999
Made by Ikea (Sweden)

OPPOSITE, BOTTOM RIGHT
233. **Bowls, cups, and utensils for children**
Made by Ikea (Sweden)
Plastic

ERGONOMICS AND SAFETY

During the 1970s and '80s, ergonomics and the needs of social or medical minorities were among the primary concerns of Swedish design. In a country that prioritizes safety and equality, public authorities and private companies have worked together to develop policies and products that accommodate the disabled. In fact, the familiar blue and white wheelchair icon, the International Symbol of Access, was designed during a conference of design students in Stockholm in 1968.

The design studio Ergonomidesign, which was founded in 1971 and merged with Designgruppen in 1979, worked with the Swedish Handicap Institute and the health care industry to develop products adapted to their specific needs. Over time, the studio has expanded its practice to other aspects of life, including childhood, work, sports, sustainable development, and biomimetics. In 1984, two of its members, Maria Benktzon and Sven-Eric Juhlin, designed a new coffeepot for the airline SAS. To replace the old steel pots, which had caused muscle strain among flight crews, they created a light and ergonomic object with a perfectly calculated center of gravity. Today this coffeepot is used by more than thirty airlines worldwide. The *Speedglas* welding helmet, introduced in 1996, offers unprecedented comfort for professional welders with its light weight, its auto-darkening lens available in several thicknesses, and its safety features (plate 230). Ergonomidesign also takes into account less sophisticated needs and in 2010 designed the *EzyStove* cookstove for developing countries. It draws the maximum heat from firewood, limits carbon dioxide emissions, and can be manufactured on-site and transported flat. In 2013, the company changed its name to Veryday.

IKEA'S GLOBAL REACH

The Ikea company was founded in 1943 by seventeen-year-old entrepreneur Ingvar Kamprad. Its name is composed of Kamprad's initials and those of his family farm (Elmtaryd) and native village (Agunnaryd), and is therefore solidly anchored in his personal history. After having sold various products through catalogs, Kamprad began offering furniture in 1953, and introduced original designs two years later. His first designers were Gillis Lundgren and Bengt Ruda for furniture, and Viola Gråsten and Göta Trägårdh for textiles. Seeking to produce modern and attractive objects at a low price, Ikea adopted flat packaging and home assembly in the late 1950s. In the 1960s and '70s, the Ikea style adapted to changes in society and lifestyles, and made greater use of plastic and chipboard panels (plate 231). (In terms of the brand's relation to Scandinavian design, the line between inspiration and copying has sometimes seemed hazy.) Success came quickly: stores opened in Scandinavia in the 1950s and '60s, in Switzerland in 1973, and then across the world; the first U.S. Ikea opened in Philadelphia in 1985.

Aside from lamps designed by the Finn Tapio Wirkkala in 1963 and the Italian Vico Magistretti in 1970s, Ikea's designers long remained in the shadows. However, the company's *PS* (Post-Scriptum) project, conceived by the entrepreneur Stefan Ytterborn and the designers Thomas Eriksson and Thomas Sandell, was intended to promote the up-and-coming generation of Scandinavian designers (plate 232). The first *PS* collection, launched at the 1995 Milan Furniture Fair, took as its title an Ikea slogan, "democratic design," which reflects the company's mission of making design affordable to the masses.

Today Ikea is the top furniture manufacturer in the world, and every year its stores are visited by hundreds of millions of people. Its approach, relying on the mass production of relatively ephemeral products, contributes to the globalization of design and to the export of "Scandinavian good taste" (plate 233).

"All you
need
love"
IKEA

BELOW
234. Mårten Claesson (b. 1970), Eero Koivisto (b. 1958), and Ola Rune (b. 1963)
***Isola* chair,** 2012
Made by Tacchini (Italy)
Steel, marble, cloth upholstery

OPPOSITE, TOP
235. Thomas Eriksson (b. 1959)
***Red Cross* medicine cabinet**, 1992
Made by Cappellini (Italy)
Metal

OPPOSITE, BOTTOM
236. Thomas Sandell (b. 1959)
***TS* chair**, designed for the Moderna Museet, Stockholm, 1990
Made by Asplund (Sweden)
Birch

MINIMALISM AND THE OFFBEAT

In response to the morose ambience that followed the extravagance of the 1980s, Stefan Ytterborn presented the work of twelve young Swedes at the 1992 Milan Furniture Fair. This immediately drew attention to contemporary Swedish design, whose light and simple objects in blond wood broke with Latin postmodernism. Two years later, Ytterborn founded the company Swecode, which collaborated with young distributors like Asplund and David Design to bring Swedish design to an international stage. Other producers, including Lammhults, Källemo, Swedese, Offecct, and Design House Stockholm, followed suit, creating a favorable climate for designers and studios.

Thomas Sandell is emblematic of this generation. His *TS* chair, distributed by Asplund for the Moderna Museet in Stockholm in 1990, is simple and stackable, and seems to float atop its natural wood feet (plate 236). Sandell founded the Sandellsandberg company in 1995 and focused increasingly on architectural projects. Thomas Eriksson, who like Sandell worked for Ikea and other Swedish and international retailers, designed the *Red Cross* medicine cabinet for Cappellini in 1992 (plate 235). Conceived as a sign, this small cabinet literally embodies the first-responder symbol. Pia Wallén also uses the cross form in her textiles and jewelry in a very contemporary manner. The architecture and design studio opened by Mårten Claesson, Eero Koivisto, and Ola Rune in 1995 brought a refined neo-minimalism to Scandinavia. Their recent *Isola* armchair, soft and organic in form, incorporates a small table that can hold work, a glass, or a book (plate 234).

The milieu of Swedish design today is as much female as it is male. Gunilla Allard came to notice in 1994 with her *Cinema* armchair, developed with Lammhults. *Cinema Sport*, introduced in 1999, is a lighter version, inspired by 1950s sports cars. Carina Seth-Andersson uses very minimalist forms for her glass and tableware (plate 238). The cylindrical bowls she designed for the Finnish retailer Hackman in 1998 have a double wall of stainless steel to maintain the temperature of the

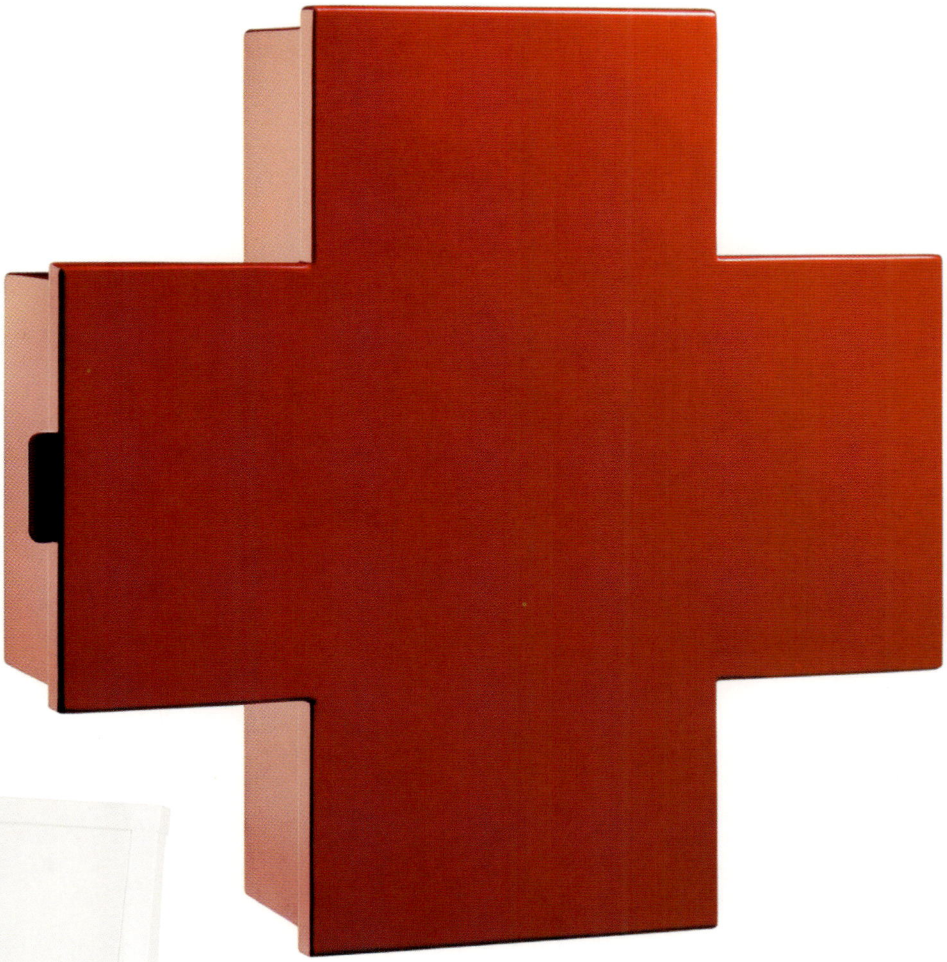

237. Jonas Bohlin (b. 1953)
***Concrete (Béton)* chair,** 1981
Made by Källemo (Sweden)
Concrete, metal

contents, whether hot or cold. Seth-Andersson may be compared to the grande dame of Swedish ceramics and glass, Ingegerd Råman. The latter, who has been creating timeless forms with simple decoration since the 1960s, launched a beautiful collection of glass for Orrefors in 2000.

However, contemporary Swedish design goes beyond minimalist objects. A playful and unique touch has been introduced, blurring the line between design and art. Jonas Bohlin turned heads when he exhibited his *Concrete* armchair, made of concrete and metal, in 1981 (plate 237). He first produced it himself before Källemo began distributing it as a limited edition. A provocative postmodernist, Bohlin has also designed porcelain for Rörstrand and several restaurant interiors, and has taught at Beckmans Academy. His sculptural and emotional approach has added a certain levity to Swedish design.

The Front collective, founded by Sofia Lagerkvist, Charlotte von der Lancken, Anna Lindgren, and Katja Sävström in 2003, caused a sensation with its animal-made collection, which included wallpaper gnawed by rats and a table decorated with the flight paths of insects. Sävström has left the group, but Front continues its playful and offbeat work with, for example, the *Surface Tension* lamp, presented by the Dutch company Booo at the 2012 Milan Furniture Fair (plate 239). Fueled by an LED light source, the lamp makes soap bubbles that grow and pop, before starting again. This poetic and living light contrasts the long duration of the LED with the ephemeral and fragile quality of the bubbles.

The Scandinavian countries share a Protestant, democratic, humanist, and egalitarian culture. The climate favors the development of domestic interiors and inspires the search for what is essential. A proximity to nature encourages a knowledge of, and respect for, materials, while a late industrialization enabled the preservation of craft skills. All these commonalities justify the discussion of Scandinavian design as a single entity. However, as the preceding pages indicate, each of the five countries has its own specific qualities, tied to its geography and history.

By the end of the twentieth century, borders were no longer what they had been, particularly when it came to business. In the face of globalization, multinational corporations have been formed, and the flagship companies of one nation or another have passed under the control of groups headquartered in a neighboring country. For example, Artek, founded by Alvar Aalto in Finland, now belongs to the Swiss furniture maker Vitra; the Swedish glassworks Orrefors and Kosta Boda were part of the Danish group Royal Scandinavia before being resold to the New Wave Group based in Stockholm; Volvo's car division was bought successively by Ford and Geely; and Ericsson partnered with Sony. Even Ikea is controlled by a Dutch foundation. Young Nordic designers are also very mobile, often working with international retailers. In this context, can we still speak of Scandinavian design? I believe so. Designers today draw on the heritage of Scandinavian modernism in creating their own contemporary forms, which stem also from globalization and ecological awareness, and are supplemented by a dose of humor and poetry that is decidedly "Nordic."

RIGHT
238. Carina Seth-Andersson (b. 1965)
Bowls, 1998
Made by Iittala (Finland)
Glass, stainless steel

BELOW
239. Front collective
***Surface Tension* lamp**, 2012
Made by Booo (Netherlands)
LED, soap bubbles

GERMANY AND SWITZERLAND

JEREMY AYNSLEY

PRECEDING PAGE
240. Konstantin Grcic (b. 1965)
***Chair_ONE*, 2003**
Made by Magis (Italy)
See plate 289.

BELOW
241. Wilhelm Kienzle (1886–1958)
Watering can, 1935
Reissued by Mewa Blattmann (Switzerland), 1991
Museum für Gestaltung, Zurich

OPPOSITE
242. Wilhelm Wagenfeld (1900–1990)
***Kubus* stackable glass containers**, 1938
Made by Vereinigte Lausitzer Glaswerke (Germany)
Molded glass
Cooper-Hewitt, National Design Museum, New York

From a contemporary perspective, it is clear that postwar design in Germany and Switzerland fulfilled the promise of early twentieth-century modernism while also questioning some of its basic principles and offering new ways of thinking. Both these countries had established reputations for design that put them in a good position to revive traditions after the trauma of World War II. The situation varied considerably between the two, however. In Germany, modern designers negotiated the political disruption, censorship, and fatal consequences of the Third Reich and ensuing war only to find themselves in a divided country in 1949, with the founding of the Federal Republic of Germany (FRG) allied to Western forces and the Soviet-oriented German Democratic Republic (GDR) in the east. By contrast, Switzerland, as a neutral country during the war, offered greater possibilities of cultural continuity, becoming a vital center for design during the troubled early postwar years. Poised between the design capitals of Paris and Milan, Switzerland attracted leading figures who taught in its art and design schools while working for national and international companies. At the beginning of the twenty-first century, in the context of a completely changed political map of Europe, a growing digital economy, and globalized markets, design in Germany and Switzerland is still commanding significant critical attention and acclaim.

It can be notoriously difficult and, at times, even dangerous to make generalizations about national styles or tendencies in design; unhelpful stereotypes may thus be created or reinforced. National character in design is often a construct developed through rhetoric, discourse, promotion, and advocacy as much as reality. Writing in 2007, the Swiss curator Claude Lichtenstein was well aware of this problem, and resisted too singular or simplified a characterization of his country's design, instead embracing the terms "paradox" and "contrast." He observed that Swiss design is rarely marked by "effusive indulgence" and tends to employ an "intelligent economy of means," not "expensive solutions." Swiss designers, he contended, "avoid theatricality" while still taking creative risks, which he saw as symptomatic of nonconformism.[1]

One design that fits such a definition is by Wilhelm Kienzle, a furniture designer who studied at the Schule für Gestaltung Basel from 1901 to 1905 and served as director of the interior design courses at Zurich's Kunstgewerbeschule until 1951. His *Zierpflanzen* (*Cactus*) watering can, originally designed in 1935, was reissued in 1991 (plate 241). Kienzle took into account the limited space of the narrow windowsill that such an apparently mundane object usually occupies, stressing verticality over horizontality. He enjoyed the idea that the watering can stood out "like a bird sitting in the bushes," made all the more striking through a vibrant use of color. With its extended neck, the *Cactus* revealed an early awareness of the ergonomics of pouring, yet its anthropomorphic wit and sense of self-deprecation also predicted postmodern design strategies, which no doubt added to its renewed appeal in the 1990s.[2]

BELOW
243. ***Die gute Form*, Schweizerische Werkbund (SWB) exhibition held in Zurich during the Zurich Festival**, June 1950
Museum für Gestaltung, Zurich

OPPOSITE
244. **Poster for another exhibition called *Die gute Form*, at the Gewerbemuseum Winterthur am Kirchplatz**, 1958
Museum für Gestaltung, Zurich

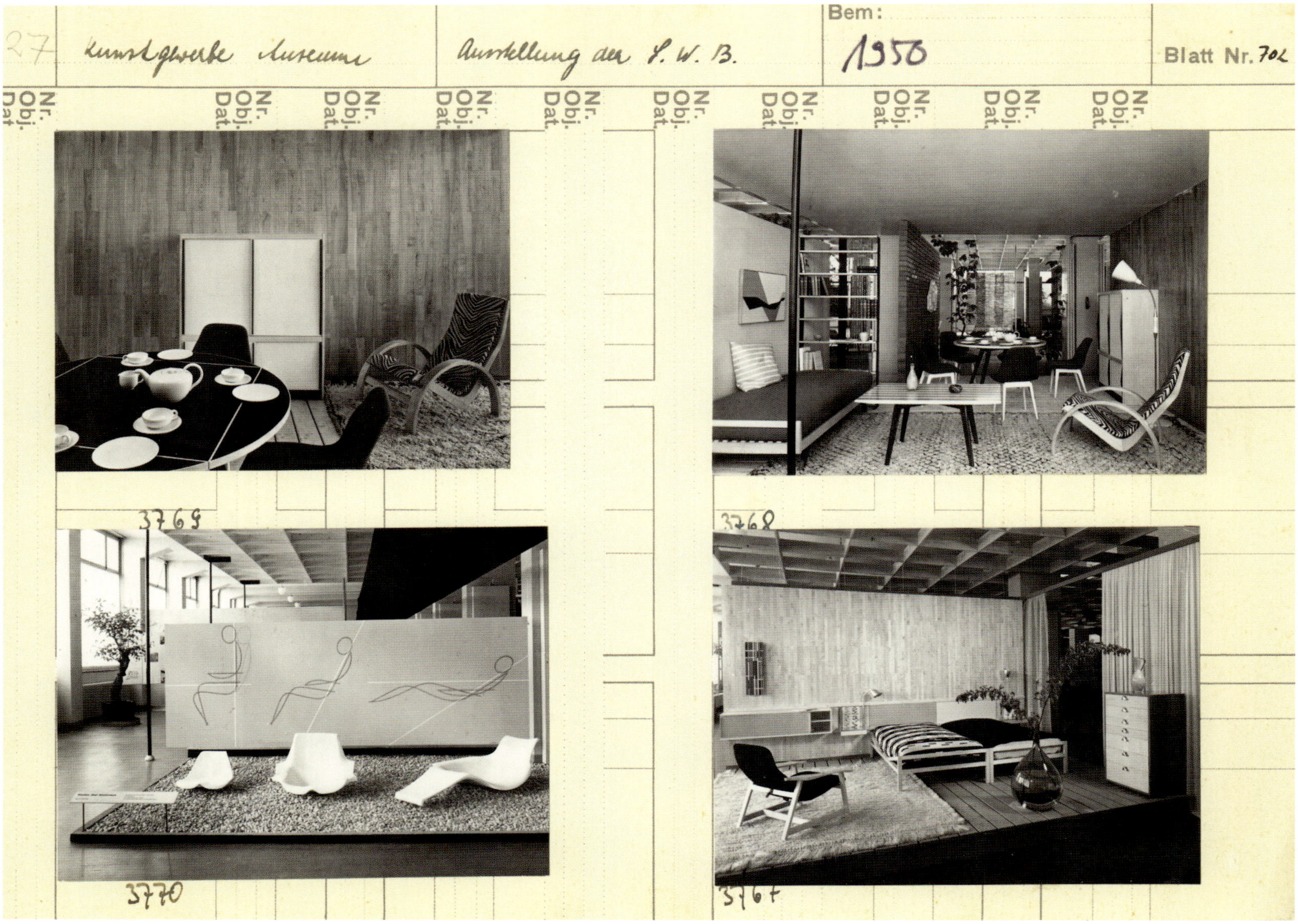

In the case of German design, numerous commentators have stressed how the legacy of the Bauhaus and Weimar functionalist traditions was realized, revived, or extended after the war, particularly in West Germany (plate 242). To take one instance, design writer Gwendolyn Ristant has commented that, at this time, officially sanctioned design in the Federal Republic generally followed from a belief in objectivity, or *Sachlichkeit*—frequently used as a word of approval in design writing. German style, she suggested, was "modest in appearance, functional in use, matter-of-fact, with neat right angles, in white, gray or black, without ornamentation, with no more than precise, technically necessary details."[3] The best-known exponent of such an approach is Dieter Rams, whose work for the Braun Electric Company from the mid-1950s onward epitomizes Ristant's description. Through Rams and others, "Design in Germany" became associated with well-engineered products of high-quality manufacture—marked by values such as durability and, later, sustainability—that earned their place in the canon of history of design, as well as museum collections worldwide.

A shared feature of Germany and Switzerland is that neither country has a single major center for design. Instead, both countries' political organization in the second half of the twentieth century favored federal and regional initiatives over national ones. In the case of Germany, this federal structure encourages regional capitals to function as cultural centers and draws designers to them. It is impossible to locate design solely in Berlin; Munich, Stuttgart, Hamburg, and Leipzig, to mention just a few, also function as important hubs. In Switzerland, a small country composed of twenty-six cantons, the major cities of Basel, Lausanne, and Zurich each have important design cultures, as do many smaller cities. Swiss identity is further complicated by the coexistence of four official languages, including Romansh. Inevitably, the French-, German-, and Italian-speaking regions have much in common with the country of their language as well as with each other.

die gute Form
Gewerbemuseum Winterthur am Kirchplatz
17. Mai bis 29. Juni 1958
Werktags 14-18 Uhr, Mittwoch u. Freitag auch 19-21 Uhr
Sonntag 10-12 und 14-17 Uhr
Entwurf Ruder
Druck Gewerbeschule Basel

BELOW
245. Max Bill (1908–1994)
Kitchen timer, 1956–57
Made by Junghans AG (Germany)
Musée des Arts Décoratifs, Paris

OPPOSITE
246. Gerd Alfred Müller (1932–1991)
Braun *KM3* food mixer with attachments, 1957
Made by Braun AG (Germany)
Museum of Modern Art, New York

PROMOTING "GOOD DESIGN"

For much of the twentieth century, one of the strongest preoccupations in official circles was to select examples of what was considered "good design" (*gute Form*). This term was used to distinguish design that followed the officially sanctioned principles of functionalism and modernism. Many educational and professional bodies engaged in the debate about "good design." The longest-standing design organizations, the German and Swiss Werkbund, were established respectively in 1907 and 1913 to nurture and promote good relations between designers, manufacturers, retailers, and consumers. Traveling exhibitions put out the message that design would contribute to future economic prosperity, citizen well-being, and cultural identity (plates 243–44). For instance, the exhibition *Wie Wohnen?* (How to Live?), shown in Stuttgart and Karlsruhe, was drawn from a competition organized by the Ministry of Finance in Baden-Württemberg. The first postwar design exhibition in Germany, it opened in December 1949. At a time of reconstruction, the catalog claimed, "the task set . . . was to design functional, durable, tasteful, and above all cost-effective furniture making the best possible use of the materials."[4]

Similar traveling exhibitions were organized by Max Bill, an important figure who long acted as a conduit for avant-garde design ideas between his native Switzerland and Germany. In the 1920s, Bill had studied at the Zurich Kunstgewerbeschule before enrolling at the Dessau Bauhaus. He spent the 1930s and '40s in Zurich, but was called back to Germany in 1953 to become the first rector of the Ulm Hochschule für Gestaltung (HfG). As a leading advocate of avant-garde and constructivist art, Bill believed design could improve the quality of life. His reduction of form to pure geometry and surface simplicity, grounded in European abstraction, was intended to shun associative values or accusations of "style" (plate 245).[5]

Dieter Rams, one of Germany's most prominent designers, also had close associations with the Ulm HfG, realizing its major principles in a long career as a product designer and important spokesperson for design. Rams changed the international face and reputation of German industrial design more than any other designer during the second half of the twentieth century, and his achievements have been recognized with countless awards and exhibitions. Joining Braun as an interior designer and architect in 1955, Rams became head of its design department by 1961 (plates 246–50). Rams has written about his design philosophy throughout his career, notably in a ten-point "manifesto": good design is innovative; it makes a product useful, aesthetically pleasing, and understandable; it is unobtrusive, honest, long-lasting, thorough down to the last detail, environmentally friendly, and as little designed as possible.

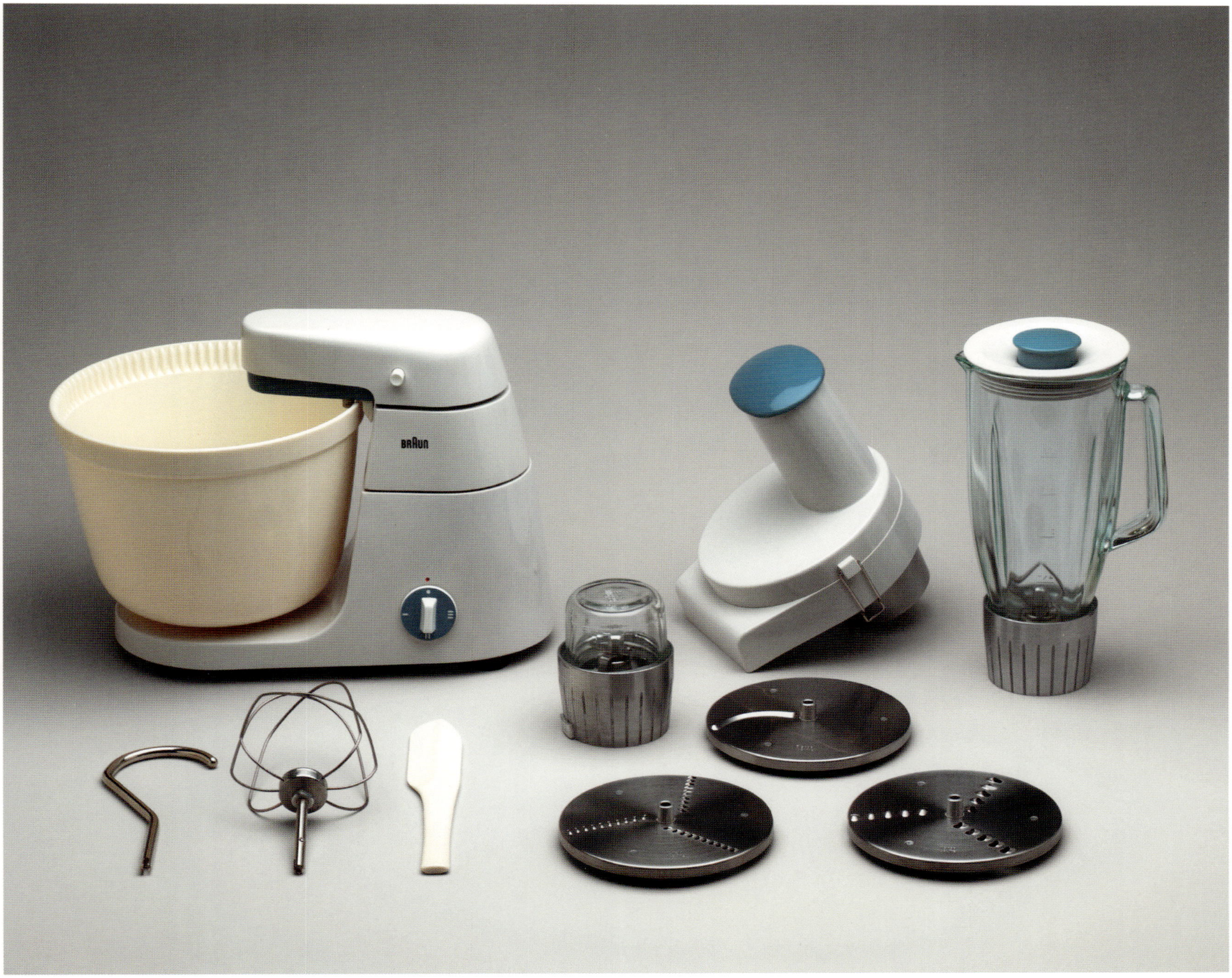
BRAUN

BELOW
247. Richard Fischer (b. 1935) and Dieter Rams (b. 1932)
***Sixtant 6007* razor**, 1973
Made by Braun AG (Germany)
Musée National d'Art Moderne, Centre Georges Pompidou, Paris

BOTTOM
248. Reinhold Weiss (b. 1934)
***KSM 1/11* coffee grinders**, 1967
Made by Braun AG (Germany)
ABS plastic, polystyrene
Musée d'Art Moderne, Saint-Étienne Métropole

Adapting Mies van der Rohe's famous earlier dictum of "less is more," Rams concluded that "less is better."[6]

Through the 1950s, '60s, and '70s, Werkbund initiatives included exhibitions strongly didactic in orientation and favoring modern functionalism. Some consisted of cases of exemplary objects that traveled to colleges and schools to teach students—the country's future citizens—to recognize, understand, and appreciate good design. Inevitably, the enthusiastic zeal of these organs of approval came to be challenged by the late 1960s counterculture, the alternative political movements and ecological critique of the '70s, and the myriad consumer lifestyle choices of the '80s.[7] As public interest in design broadened, official taste came to be considered overly reductive or prescriptive. Design selection lost its appeal as it gave way to a more inclusive approach that interpreted design culturally, as a matter of personal preference, social engagement, or consumer enjoyment, rather than merely aesthetic judgment. A case in point is the designer Volker Albus, an important member of the new German design movement of the early 1980s, whose multimedia installations explored color, narrative, ornament, and glamour—all properties ostensibly denied by modernism.[8]

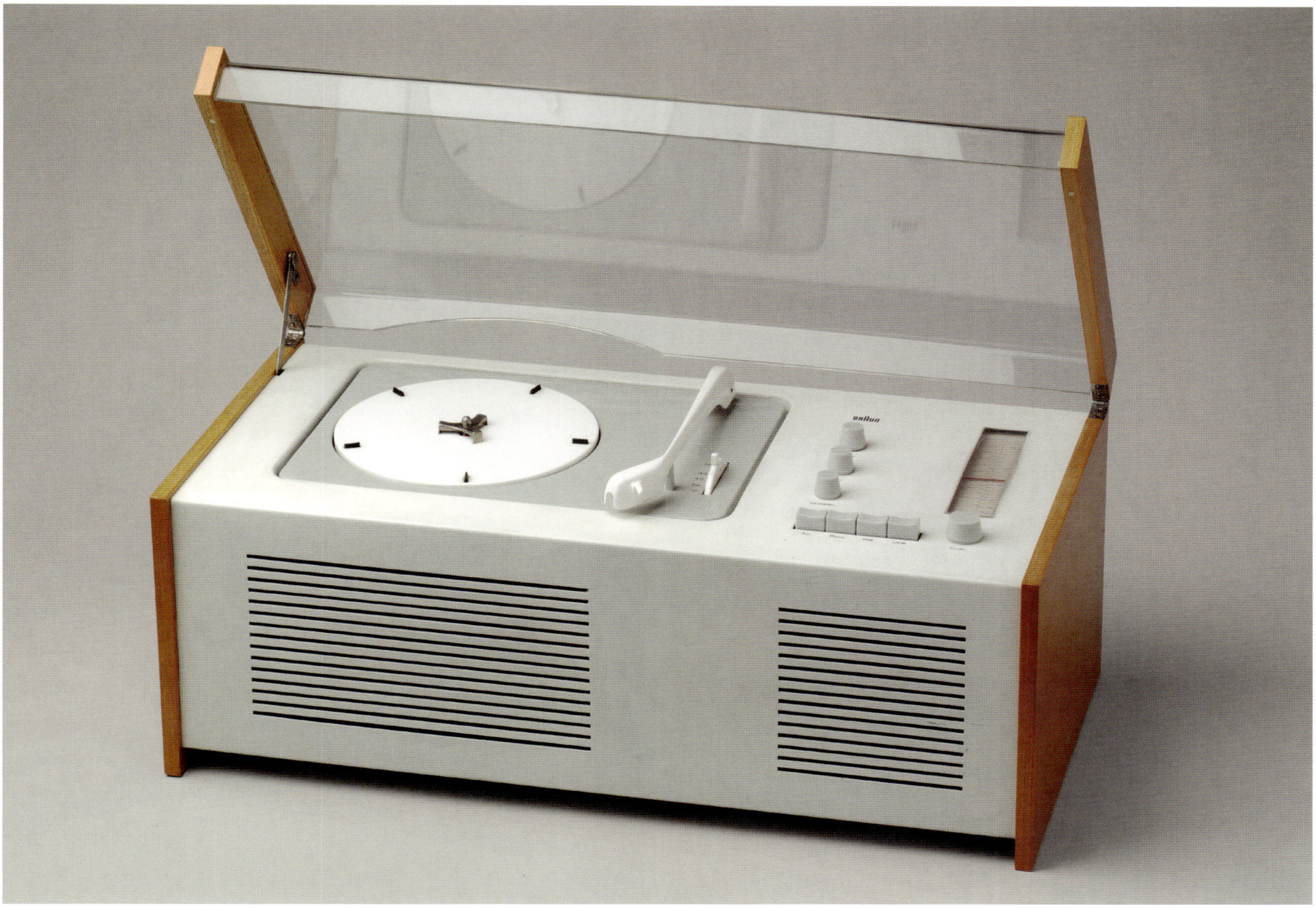

ABOVE
249. Hans Gugelot (1920–1965) and Dieter Rams (b. 1932)
***SK 4* turntable with radio**, 1956
Made by Braun AG (Germany)
Lacquered metal, plastic, wood
Museum of Modern Art, New York

LEFT
250. Dieter Rams (b. 1932)
***TFG 2* table lighter**, 1968
Made by Braun AG (Germany)
Plastic, painted and chrome-plated metal
Museum of Modern Art, New York

BELOW
251. **Cover of the first issue of *Form*, with a drawing by Le Corbusier and a photograph of a chair by Gerrit Rietveld**, 1957

OPPOSITE
252. **View of the spring trade fair in Leipzig**, March 1967

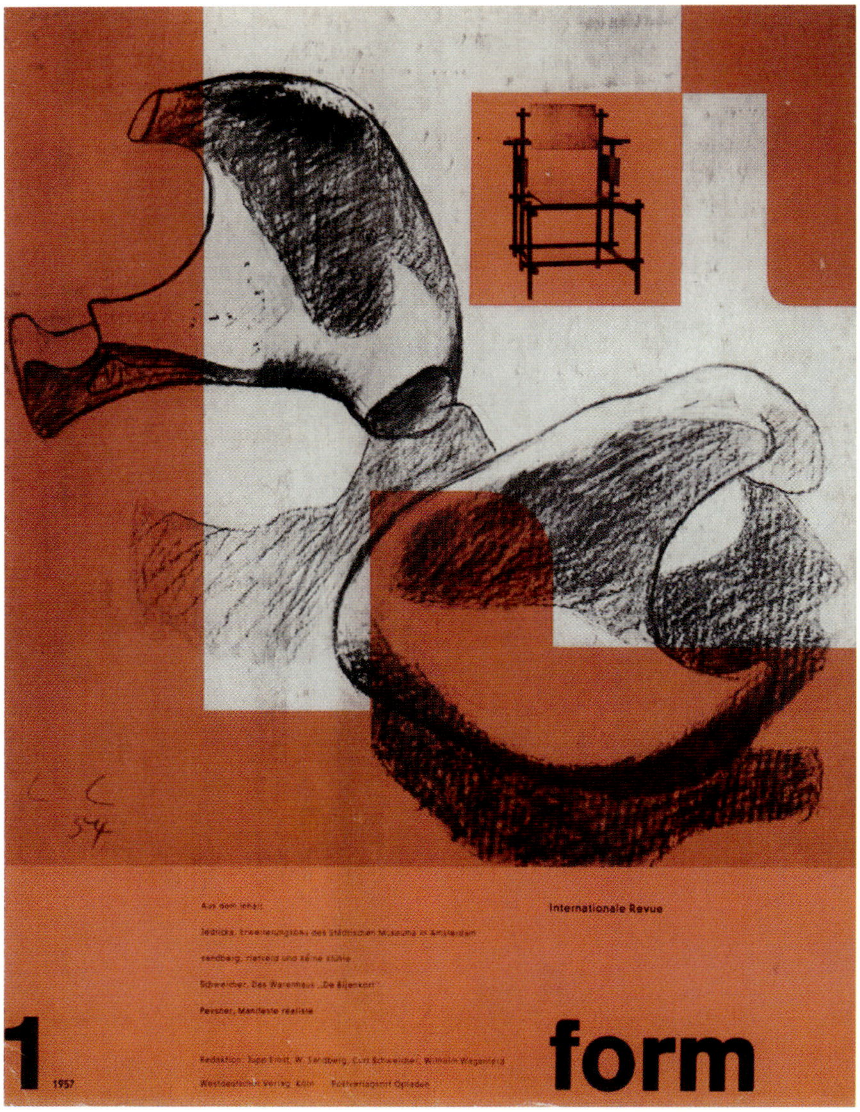

DISSEMINATING DESIGN

The second half of the twentieth century witnessed a significant growth in commentary on design in the form of magazine journalism. Among the titles that established Swiss and German design on the international stage were *Form* (1957– ; plate 251); *Werk und Zeit* (1952–2007), an organ of the German Werkbund; the official industrial design magazine of East Germany, *Form und Zweck* (later *form+zweck*; 1956–); and *Design Report* (1987–), published by the German Design Council (Rat für Formgebung).

In tune with changing attitudes and a rising popular interest in design, the late twentieth century also saw the emergence of the design museum as an institution distinct from the traditional museum of arts and crafts, or of decorative arts.

Two examples offer different perspectives on the question of curating modern design. The first, the Bauhaus Archiv, cares for the legacy of this most famous art and design school. The archive's research and exhibition program illuminates the teaching and design activities of the staff and students of the Bauhaus, and its continuing importance to contemporary design. The first Bauhaus Archiv was founded by Hans Maria Wingler in the Ernst-Ludwig-Haus on Mathildenhöhe in Darmstadt in 1960, based on bequests of former staff. Following its transfer to Berlin in 1971, the archive moved into a purpose-built gallery of Walter Gropius's own design in 1979. After the German reunification in 1990, a parallel collection-based exhibition and publication program developed at Dessau, and the school buildings and masters' houses there were restored in connection with their classification as a World Heritage Site in 1996. Meanwhile, the Bauhaus Museum Weimar has run a series of exhibitions since 1995, while building its permanent collection as part of the Klassik Stiftung Weimar.[9]

A contrasting curatorial project, located on the Swiss-German border, is the Vitra Museum. The Vitra furniture company was originally established by Willi Fehlbaum in Basel in 1934 and later moved to Weil am Rhein, just across the southwest border with Germany. In 1957, the firm obtained from Herman Miller the license to distribute the furniture of Charles and Ray Eames and George Nelson in Europe. Vitra's own collaboration with contemporary designers began with Verner Panton and his *Panton* chair of 1967. Willi Fehlbaum's son Rolf took the company in several

RFT
VEB CARL ZEISS J
BAUELEMENTE
DDR NACHRICHTEN
mit RFT
weltweitsenden
präzis vermitteln

BELOW
253. Frank O. Gehry (b. 1929)
Vitra Design Museum, Weil am Rhein, Germany, 1989

OPPOSITE
254. **A display at the Vitra Design Museum, Weil am Rhein, Germany**

original directions. In Weil am Rhein, two new factories designed by British architect Nicholas Grimshaw were built in 1981 and 1986. In 1993, a fire station by Zaha Hadid and a conference pavilion by Tadao Ando were completed. Vitra became an international center, running workshops on themes in architecture and design and, most importantly, opening the Vitra Design Museum in a Frank O. Gehry building in 1989, under the direction of Alexander von Wegesack (plate 253). The museum houses part of the Fehlbaums' extensive collection of furniture (plate 254) and has formed the base for original curatorial projects and publications. Alongside this cultural curation, since 1987 Vitra Editions has allowed internationally acclaimed designers to develop projects in laboratory conditions, outside the normal restrictions of the marketplace.[10]

Other types of design museum include the Neue Sammlung, which opened in 2002 in the Pinakothek der Moderne in Munich as a permanent home for a collection already begun in the 1920s, and the Museum für Gestaltung, Zurich, which has adopted the format of the Kunsthalle, drawing on the historical and contemporary collections of the city's Arts and Crafts Museum for a program of innovative thematic exhibitions covering all genres of design.[11]

PRODUCTIVE NATIONS AND INDUSTRIAL DESIGN

In the reconfiguration of Europe under the European Recovery Program of 1947, more commonly known as the Marshall Plan, West Germany was recognized as crucial to the wider economic and political stabilization of the continent and benefited from substantial American financial investment, leading to its economic miracle, or *Wirtschaftswunder*, of the 1950s. Abundant reserves of coal and iron provided for the country's heavy industries,[12] while the federal government invested in engineering and technical education and set great store in design's benefits to industrial growth and the national economy. Consequently, West Germany became a leading producer of capital and consumer goods, with many thriving international companies, a position it maintained after political reunification in 1990. It is impossible to list all the companies that were recognized for their effective design policies, but one litmus test will provide some insight. In 1996, the Bundespreis Gute Form, an annual award for high-quality technical products whose "design combines functionality and aesthetics," honored cars from Audi and Porsche, Lamy pens, the Noblex *135 U Panorama* camera, Schindler elevators, Blanco Med medical equipment, and Rodenstock eyeglass frames.[13]

BELOW
255–56. Walter Zapp (1905–2003)
Minox *VEF Riga* miniature camera (shown open and closed), c. 1936
Made in Latvia 1937–43 and in Germany from 1948
Museum of Modern Art, New York

OPPOSITE
257. Ernst Leitz (1871–1956)
***IIIc* telemetric camera**, 1940–51
Made by Leica (Germany)

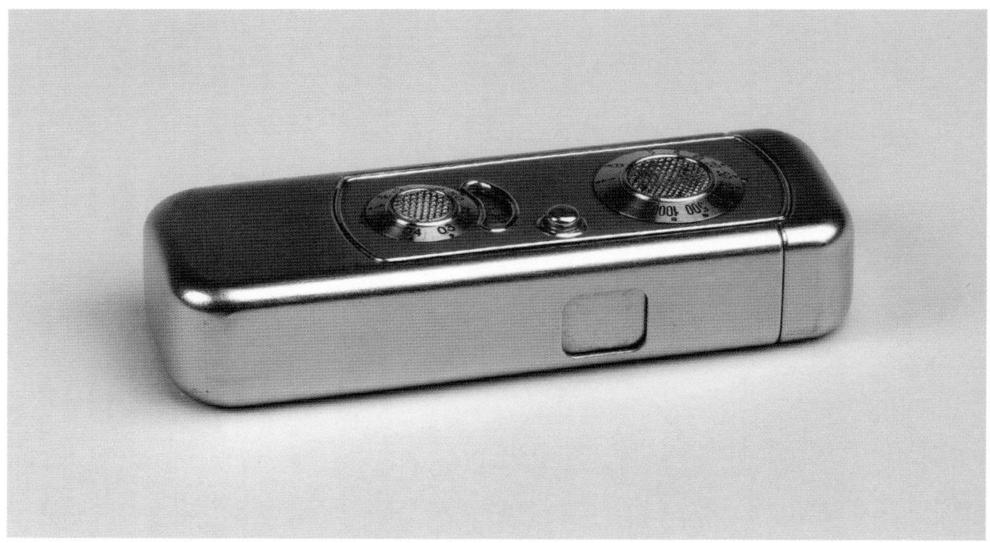

In the political context of the Cold War, the division of Germany meant that two design traditions evolved in ways that revealed both parallels and differences. Design in the Federal Republic of Germany, governed by consensus politics and a welfare-state democracy, was seen as an important means for the country to develop a consumer society. By contrast, under the command economy of the German Democratic Republic, the mission was to establish "real existing socialism," and design was marshaled to the cause of defining a new way of life ideologically and materially distinct from capitalism. East German designers faced decisions about how to combine the legacy of Weimar modernism, to which many were immediately drawn, with contemporary political imperatives. Alongside other Eastern European countries, the GDR established an identity as a productive nation and by the early 1960s was ninth in the league of the world's manufacturing nations.[14]

Many structures for design in the GDR mirrored those of the West. These included a central institute for the promotion of industrial design, founded in 1950; the dedicated design magazine *Form und Zweck*, which began publication in 1956; and annual design awards and exhibitions. Many important design institutes founded in the early twentieth century were also located in the Eastern bloc. The Leipziger Messe, a biannual trade fair, was an important venue to display consumer and capital goods for the export market (plate 252). Products of particular quality included Zeiss cameras and optical equipment, Jena glass for industrial and domestic use, and specialist industrial machinery.[15]

Much debate concentrated on how goods could both look and be "socialist" beyond banal decoration or the immediate iconography and symbolism of politics. Many modern designers risked being charged with "formalism" if their design solutions appeared too similar to those used in the West. Unlike West Germany, the East lacked key mineral resources and, out of necessity, instituted the Chemieprogramm in 1959 to develop its chemical industries, in part to make products for the home. One important debate was how to serve the most needs with the fewest resources. Accordingly, standardized systems of products found favor, such as Franz Ehrlich's 1957 collection of fitted furniture, the *Typensatz 602*, which was manufactured by the VEB Deutsche Werkstätten in Hellerau near Dresden, which had its origins in the distinguished Jugendstil workshop movement.[16]

Another priority in the GDR was to design products for use in the public as well as the private sphere. A notable example is *Rationell*, a stacking coffee service by

№394953
Leica
D.R.P.
Ernst Leitz
Wetzlar

TOP
258. **Assembly line at the Volkswagen plant in Wolfsburg, Germany**, 1952

ABOVE
259. **Volkswagen *Export Type 1* (*"Beetle"*)**, 1953
National Motor Museum, Beaulieu, U.K.

OPPOSITE
260. **Advertisement for the Porsche *356***, 1955
National Motor Museum, Beaulieu, U.K.

The world's most talked-about car

Margarete Jahny, which was produced in one version with a logo decoration for Mitropa, the catering division of the GDR state railway, and in another with a floral decoration for home use.

One particular field of West German prowess was automotive design, which took off when car ownership changed from being a luxury to a reasonable expectation for many middle-class families. Not surprisingly, the 1950s was a boom decade for the West German car, with the number of new owners per annum growing from 216,107 in 1950 to 1,503,000 in 1959.[17] The Volkswagen (literally "the people's car") was the first important small car in Germany, designed to appeal to first-time buyers. It was originally developed by Ferdinand Porsche under National Socialism, as part of the Kraft durch Freude (Strength through Joy) movement. Although initially intended as a response to Henry Ford's *Model T* in the U.S., before 1939 the Volkswagen was used only as a propaganda tool, and its production was diverted to military use. Not until after the war did it become a true consumer model, manufactured in Wolfsburg according to assembly-line principles (plate 258). With its rear engine and endearingly distinctive shape, the car became an international cult object by the 1960s (plate 259). Dubbed the "Beetle" (Käfer), it subsequently starred in Hollywood films. However, for all its popularity, the car's small backseat and rear engine were considered by some to be a disadvantage.[18]

The Volkwagen stood out all the more as most German car manufacturers gradually adopted body styles based on American patterns. The first Opel *Kadett* was manufactured between 1937 and 1940, but it was the second version, the *Kadett A*, introduced in 1961 at the directive of Opel's parent company General Motors, that presented a full-on challenge to the Volkswagen, as manifested in its design (plate 262). Under a development team led by Opel product director Karl Stief, the *Kadett* was styled according to Detroit principles, with a large horizontal windshield, chrome trim, and a streamlined form. An elegant coupé version appeared in 1965. Even if it was smaller than most equivalent American models, the *Kadett*, available in bright colors, became popular with German motorists.[19] The

OPPOSITE, TOP
261. **Advertisement for the Trabant *601 Limousine Standard* and *Universal Standard*,** 1983

OPPOSITE, BOTTOM
262. **Opel *Kadett A* assembly line in Bochum, Germany,** 1962

ABOVE
263. **Parking lot at Marx-Engels-Platz in East Berlin with Trabant and Wartburg cars,** c. 1970

BELOW
264. ***R1200C* motorcycle**, 1997
Made by BMW (Germany)

OPPOSITE
265. ***300SL* "gullwing" coupé**, 1957
Made by Mercedes-Benz (Germany)
National Motor Museum, Beaulieu, U.K.

GDR introduced its own small car in 1958: the Trabant. The *Model 601*, produced between 1964 and 1990, had a body made from Duroplast, a plastic material reinforced by fibers (plates 261, 263).

Germany's reputation as a major automobile producer continues today with Audi, BMW (plate 264), and Mercedes-Benz (plate 265), all known for their carefully engineered high-performance vehicles. At the highest level of styling and performance, the successive generations of the Porsche family continue to lead the field for Germany, with a series of models that have eroded the distinction between the race car and the sports car (plate 260).

Volkswagen achieved another breakthrough when it commissioned Giorgetto Giugiaro, an Italian design star already known for his work with Maserati and Alfa Romeo, to create the *Golf*, which was introduced in 1974 (plate 266). Coinciding with the oil crisis and ensuing concerns about environmental sustainability, the *Golf* represented a further stage in the Europeanization of small cars for the city. The *Golf*'s straight, angular lines distinguished it from its predecessor, the *Beetle*. This first hatchback introduced a body style that would have a profound impact on car design around the world.[20]

While Switzerland may not manufacture automobiles, the country has made important contributions to the expanded field of transport design for road, rail, and flight, involving multidisciplinary teams embracing graphic, architectural, and product design. Ahead of its time when introduced in 1953, the distinctive winged-arrow logo designed by Rudolf Bircher of Zurich and Karl Gerstner of Basel for Swissair asserted a strong visual presence alongside other nations' fleets (plate 268). The design was used not only on the aircraft

SYN 126F

themselves but also on flight-crew uniforms, printed material, advertisements, and accessories, including popular flight bags. In 1978, Gerstner was entrusted with the renewal of the entire corporate identity.[21]

Deutsche Lufthansa, the West German state airline, also distinguished itself through design. In 1955, for instance, it commissioned an elegant in-flight tray service from Wilhelm Wagenfeld, the leading glass designer of the period. The company also employed Otl Aicher and his Gruppe E5 (consisting of students from the Ulm HfG), to develop its first corporate-identity manual in 1962.[22]

A few years earlier, the Swiss Federal Railways (Schweizerische Bundesbahnen, or SBB) had also become known for the high quality of its corporate identity and for its level of design awareness in the field of structural engineering. This involved the building of remarkable bridges, tunnels, and rail networks, providing a vital infrastructure for the country's extreme geography. The station platforms were overseen by Hans Hilfiker, a designer with an engineering background, who sought to provide passengers with the best possible information and amenities. In particular, he introduced a new system of concrete tubes to support light and elegant canopies that had a sculptural appearance and facilitated visibility by avoiding unnecessary columns. Hilfiker's station design incorporated his Swiss railway clock of 1944, to which he now added the famous red second hand. Moreover, all the station clocks were synchronized by a central system, which has come to be seen as a metaphor for the precision and efficiency of Switzerland (plate 267).[23]

A final example of a corporate identity for a complex national network with enormous reputational value is that commissioned by Deutsche Bahn, which was charged with standardizing and modernizing the German rail system after reunification. The company turned to Erik Spiekermann, Germany's foremost designer of typefaces for branding. He worked in collaboration with Christian Schwartz, and their design was broadly acclaimed, winning the Designpreis Bundesrepublik Deutschland in 2007.[24]

OPPOSITE
266. **Advertisement for the Volkswagen *Golf***, 1983
Private collection

TOP
267. Hans Hilfiker (1901–1993)
Swiss Federal Railways station clock, 1944 (redesigned 1952)
Museum für Gestaltung, Zurich

ABOVE
268. Rudolf Bircher (1911–2009) and Karl Gerstner (b. 1930)
Swissair logo, c. 1960

BELOW
269. **Advertising art for Herbert Hirche office furniture**, 1960s
Museum für Gestaltung, Zurich

OPPOSITE, TOP
270. Ludwig Walser (b. 1936)
Garden seat, 1959
Made by Eternit (Switzerland)
Fiber cement
Musée National d'Art Moderne, Centre Georges Pompidou, Paris

OPPOSITE, BOTTOM
271. Willy Guhl (1915–2004)
Garden rocking chair, 1954
Made by Eternit (Switzerland)
Fiber cement
Musée National d'Art Moderne, Centre Georges Pompidou, Paris

IDEAL AND REAL HOMES

In the interwar years, modern designers had frequently turned their attention to mass-housing schemes for the working classes, with furnishings to match, using such catchwords as "functionalism" (*Funktionalismus*), "necessity" (*Bedarf*), or "minimum existence" (*minimum Existenz*). While some designers remained committed to social causes and resisted the pull of the market, the rising standard of living meant that, by the 1960s, design was more often integrated into the modern lifestyle through consumer choice than social programs. For the domestic interior, "modern" could mean furniture manufactured in Scandinavian-inspired woods or chromed steel, with clean lines, fashionably organic shapes, and abstract patterns borrowed from modern art. Design advice manuals often advocated furniture made up of standard modular elements for easy storage. Textiles, ceramics, glassware, and metalware successively referenced concrete art, abstract expressionism, and Pop art. Such tendencies were prominent until alternative lifestyles offered a critique of the pursuit of the "ideal" home. The radical design movement of the late 1960s introduced informal seating and experimental new materials as a response to a generational change in social attitudes. As one indication of this larger shift, Susi and Ueli Baer—of the Swiss ad-hoc design collective Terreno, founded in 1971—created flexible environments for alternative lifestyles, with "inflated, expanded, folded, combined, stacked and knocked down furniture."[25]

Innovative furniture designs are regularly presented at the major trade fairs, such as the storied Triennale di Milano (plate 272) or the more recent Design Miami, as well as many other exhibitions in less expected venues. For example, at the G59 garden show in Zurich, what has been called the "lowest seat in Europe" was exhibited. The chair, designed by Willy Guhl

BELOW
272. ***Ideal Home* exhibition at the Triennale di Milano, with furniture designed by Egon Eiermann**, 1954

OPPOSITE, TOP
273. Egon Eiermann (1904–1970)
***SE 42* chairs**, 1949
Made by Wilde + Spieth (Germany)
Laminated wood
Private collection

OPPOSITE, BOTTOM LEFT
274. Dorothee Maurer-Becker (b. 1938)
***Uten.Silo* organizer**, 1969
Made by Ingo Maurer GmbH (Germany), then by Vitra (Switzerland)
Plastic

OPPOSITE, BOTTOM RIGHT
275. Egon Eiermann (1904–1970)
***E10* chair**, 1949
Made by Heinrich Murmann (Germany)
Rattan
Private collection

in 1954, combined a novel material (fiber-reinforced concrete) with an abstract sculptural form to produce a low rocking chair for the terrace or garden (plate 271).[26]

The World's Fairs also continued to be an important venue for judging the state of a nation's design. For West Germany, the 1958 Brussels World's Fair was crucial, as it marked the country's return as an exhibitor after a break of more than twenty years following the 1937 Paris International Exposition. The choice of Egon Eiermann to design the pavilion represented the continuity of modernism and won international approval. One of Germany's most prominent architects and designers, Eiermann had studied under the first-generation expressionist and modernist Hans Poelzig in Berlin in the 1920s, and was a founding member of the German Design Council in 1951. He was sometimes referred to as the "German Eames" for his experiments in bent-plywood chair construction for the Esslingen company Wilde + Spieth, which ran in parallel with the more celebrated designs from California (plate 273). Eiermann's Brussels pavilion, a structure of black steel and glass, had a simplicity, transparency, and openness that symbolized West Germany's desire to be seen as a "good citizen of Europe." When illuminated at night, it became a showcase for the sculptural forms of a variant of Eiermann's *E10* wickerwork chair (plate 275).[27]

Other landmarks in exhibition design include the West German pavilion at Expo '67 in Montreal, designed by Frei Otto. Along with his contemporary Buckminster Fuller, Otto pioneered tensile and membrane techniques to produce lightweight structures in cable net. He went on to design the much acclaimed roof of the stadium for the 1972 Olympics in Munich. Both structures signaled that German design could embrace the high-tech, the organic, and the futuristic, and break away from the more familiar rectilinear modernist conceptions for which the country had such a deserved reputation.[28]

THE MILLION DOLLAR HOTEL

BELOW
276. Luigi Colani (b. 1928)
***Poly-COR* chair**, 1968
Made by COR Sitzkomfort, Rheda-Wiedenbrück (Germany)
Polyester resin reinforced with fiberglass
Alexander von Vegesack collection

OPPOSITE
277. Helmut Bätzner (1928–2010)
***BA 1171* stackable chair**, 1964–65
Made by Bofinger (Germany)
Polyester resin reinforced with fiberglass
Musée des Arts Décoratifs, Paris

BELOW
278. Walter Gropius (1883–1969)
***Tac 1* teapot**, 1969
Made by Rosenthal Studio Line (Germany)
Black porcelain
Museum für Gestaltung, Zurich

OPPOSITE, TOP
279. Luigi Colani (b. 1928)
Teapot and milk jug from the *Drop* tea service, 1971
Made by Rosenthal Studio Line (Germany)
Porcelain
Musée National d'Art Moderne, Centre Georges Pompidou, Paris

OPPOSITE, BOTTOM
280. Timo Sarpaneva (1926–2006)
***Suomi* service**, 1976
Made by Rosenthal Studio Line (Germany)
Porcelain, stainless steel
Private collection

ART AND DESIGN COLLABORATIONS

The decorative arts have a long and distinguished history in Germany, going back to court manufactories that produced elegant luxury objects with great artistry and skill. Although this emphasis on craftsmanship may seem to sit awkwardly alongside the principles of modern Swiss and German design, with its emphasis on the democratic, the affordable, and the mass-produced, it would be a mistake to regard the two as entirely contradictory.

Among initiatives to combine artistic excellence with modern creativity in the traditional decorative arts, the Rosenthal Studio Line stands out (plate 278).[29] Echoing the Scandinavian tradition of commissioning high-profile designers to develop special product lines, Philip Rosenthal introduced the Studio Line in 1961 to complement his company's reputation for excellence in modern and traditional ceramic services. He went on to work with more than one hundred artists and designers, particularly to encourage new ideas for working in porcelain. To take a case in point, the versatile Swiss-born German designer Luigi Colani was invited to design a tea service for the Rosenthal series in 1973. His *Drop* teapot minimized the profile of the lid and incorporated the handle within the body of the vessel (plate 279). This design allows the pot to be held close to its center of gravity for ease of pouring, and creates a smooth teardrop profile that shares many characteristics with Colani's other product designs, from cars to cameras (plate 276). One of the most acclaimed products of the Studio Line series was the *Suomi* service (1976) by the Finnish designer Timo Sarpaneva, best known for his achievements in glass. For Rosenthal, Sarpaneva developed a teapot, coffeepot, cup, saucer, bowls, and plates in simple, elegant forms (plate 280). In its undecorated version, the *Suomi* service became an acknowledged classic. Part of its elegance came from the treatment of the lids as integral to the oval composition, and of the chrome steel handle as an extension of the silhouette. The service won the gold medal in the International Ceramic Competition in Faenza in 1976. An eclectic group of artists—including Salvador Dalí, Eduardo Paolozzi, and Victor Vasarely—were also invited to use its open and lidded bowls, teapot, and coffeepot as a blank canvas.[30]

A more recent collaboration between artists, designers, and manufacturer came from the less expected quarter of Swiss watch design. Conventionally associated with centuries of precision engineering, fine metals, and decorative excellence, the Swiss watch industry had, by the early 1980s, been fundamentally challenged by the introduction of inexpensive quartz movements produced outside Europe, and Swiss export figures were plummeting. The founder of Swatch,

BOTTOM
281. Helen von Boch (b. 1939)
***Avant-garde* stackable service**, 1971
Made by Villeroy & Boch (Germany)
Vitroceramic
Musée des Arts Décoratifs, Paris

BELLOW RIGHT
282. ***Falling Star* watches**, 2000
Made by Swatch (Switzerland)

OPPOSITE
283. ***Smart* car**, 1998
Made by Daimler AG
Museum of Modern Art, New York

Nicolas G. Hayek, reconceptualized the Swiss watch and was largely responsible for the revival of this threatened industry after 1983.[31] His introduction of the plastic Swatch watch turned tradition on its head. The product's name cleverly combined the idea of a "Second *watch*" with the national prestige of the "*Swiss*," but Hayek insisted its innovation went beyond a marketing ploy. The Swatch's affordable price, achieved by reducing the number of parts, allowed it to be treated as a changeable fashion statement, something that was particularly attractive to younger people. The close association with Swiss identity was reinforced by the incorporation of the national flag in the branding, realized in a crisp graphic style (plate 282).

The Swatch was available in immediately identifiable, highly designed outlets, including modular kiosks at railway stations and airports, mall shops, and flagship stores in world cities. The appeal of the Swatch was enhanced by associations with modern art. The first artist commissioned to design a Swatch watch was Kiki Picasso, in 1984. He was followed by the hip New York artist Keith Haring, whose four Swatch designs were introduced in the U.S., among them the *Milles Pattes* (Centipede) in 1986. Swatch later extended its purview, commissioning watches from fashion designers, filmmakers, and architects, each chosen for their youth appeal and street cred. The company's associations with fine art continued with its sponsorship of the 2011 and 2013 Venice Biennale. At the same time, Swatch aimed to attract a different type of customer by sponsoring the winter sports of freestyle snowboarding and skiing, as well as the summer sports of basketball and volleyball.

By the early 1990s, Hayek had turned his attention to the automotive industry and again challenged many assumptions about product design and consumer behavior, in this case by reconceptualizing the city car. He established an alliance with Mercedes Daimler-Benz AG in 1994, and the resulting *Smart* car made its debut at the 1998 Paris auto show (plate 283). Following a similar pattern to "Swatch," its name combined the first letters of "Swatch" and "Mercedes" with the word "art."[32] The *Smart* was developed as a rear-engine two-seater in various models, under the direction of designer Johann Tomforde. Taking environmental issues into account, it was intended for ease of city driving and parking. In its mode of manufacture and appearance, the *Smart* had as much in common with other types of industrial products as with conventional automobiles.

BELOW
284. Richard Sapper (b. 1932)
***Tizio* desk lamp**, 1972
Made by Artemide (Italy)
ABS plastic, aluminum
Victoria and Albert Museum, London

OPPOSITE
285. Richard Sapper (b. 1932)
Whistling kettle, 1982
Made by Alessi (Italy)
Stainless steel, polyamide, brass
Centre National des Arts Plastiques/Fonds National d'Art Contemporain, Paris

DESIGN STARS AND DESIGN CULTURE

Many German and Swiss designers have looked south to Milan, sometimes forming partnerships with Italian design groups and manufacturers that have resulted in products with considerable flair and imagination. After studying in Munich, the German industrial designer Richard Sapper joined Gio Ponti's studio in Milan in the late 1950s, and ended up staying in Italy permanently. There he created the iconic *Tizio* lamp (1972), kitchenware for Alessi, and other products that placed him in the front rank of postmodern designers (plates 284, 285).[33]

Another leading personality of German design is Ingo Maurer of Munich. Maurer's fascination with the magical and transformative properties of materials—from humble paper to suggestive gold leaf—leads him to create new illuminated environments. In one of his first designs, the *Bulb* series of 1966 (plate 286), Maurer borrowed the form of the incandescent bulb itself to play on scale and cultural significance. In a manner that echoed contemporary Pop sculptors such as Claes Oldenburg, he used the blow-up to create lighting as sculpture in an expanded field.[34]

A new range of possibilities came with developments in electronics. The LED facilitated Maurer's deconstructive approach, through which he questioned conventional distinctions between cable, container, and light source (plate 288). While constantly developing innovative lighting for the domestic environment, Maurer also works on a theatrical scale, as in his lighting schemes for the Munich subway and his conceptual museum installations.

Konstantin Grcic is one of the most successful German industrial designers of his generation, adapting to the changing requirements of his profession in an increasingly globalized context, with fierce competition from Asian manufacturers in particular. Grcic trained in furniture design at Parnham College in Dorset and the Royal College of Art in London before founding Konstantin Grcic Industrial Design (KGID) in his home city of Munich in 1991. This international design group specializes in lighting, furniture, and product design, working with leading European companies (plate 289).[35]

Grcic's design for the *MYTO* chair, which won international acclaim, represents the key elements of his approach (plate 290). Like many of his designs, it is made from a single material. In this case, Grcic was commissioned by the German chemical company BASF to explore the potential of Ultradur High Speed, a resilient plastic usually used in the automotive industry. The strength of the material led to Grcic's design, a modern

LEFT
286. Ingo Maurer (b. 1932)
***Bulb* lamp**, 1966
Made by Ingo Maurer GmbH (Germany)
Chrome-plated metal, glass
Museum of Modern Art, New York

BELOW
287. Stiletto (aka Frank Schreiner, b. 1959)
***Consumer's Rest* armchair**, 1983
Made by Stiletto Studios (Germany)
Varnished metal, plastic
Vitra Design Museum, Weil am Rhein, Germany

OPPOSITE
288. Ingo Maurer (b. 1932)
***Lucellino* wall lamp**, 1992
Made by Ingo Maurer GmbH (Germany)
Glass, brass, plastic, goose feathers
Museum of Modern Art, New York

variant of the cantilevered chair, which he associated particularly with German modernism of Marcel Breuer and Mies van der Rohe. (Grcic has written that he seeks to address "function in human terms.") No mere technical exercise, the *MYTO* chair bears natural as well as cultural associations: Grcic has commented that "the perforations look quite like an animal skin, and the chair has a reptilian quality with a precise outline and roundish, but tense surfaces. Sometimes animals assume that position when they are ready to pounce."[36]

Another way star designers reach a wider public is through design editing. The German brand Authentics, which belongs to the Flötotto furniture company, publishes an online catalog of affordable high-quality design from its headquarters in Gütersloh. Authentics works with a stable of young, mainly European designers, including Grcic, Sebastian Bergne, Edward Barber and Jay Osgerby, Christina Schäfer, Constantin and Laurene Boym, Shin Azumi, Tord Boontje, and Marti Guixé.[37] In tune with contemporary urban lifestyles, the selection concentrates on apparently quotidian items that are both useful and aesthetically pleasing. Authentics calls these objects, which are usually simple and industrially produced, "elementary things." They often transcend conventional boundaries, such as those between living and work space, or inside and outside. With environmental awareness now taken for granted, the Authentics Web site includes a lexicon that describes the properties of the natural and synthetic materials used in its products. One aim is to alter

OPPOSITE
289. Konstantin Grcic (b. 1965)
Chair_ONE, 2003
Made by Magis (Italy)
Painted aluminum, varnished concrete
Musée des Arts Décoratifs, Paris

BELOW
290. Konstantin Grcic (b. 1965)
***MYTO* chairs**, 2008
Distributed by Plank (Italy)
Made by BASF
Plastic

attitudes—for instance, by promoting colorful and durable plastics as a material of choice for items such as wastebaskets and kitchen and bathroom accessories. The belief is that designers can impact everyday life in deceptively unassuming ways while meeting consumers' expectations.

With a similar ethos, the brothers Daniel and Markus Freitag developed a product that has become part of the iconography of design-conscious cities where physical and creative mobility is key.[38] Living on an arterial road on the outskirts of Zurich, they grew up seeing tractor-trailers on a daily basis. Inspired by this experience, they developed their first shoulder bags in 1993, made from recycled heavy-duty truck tarps and modeled on the American bike courier bag. The tarp is cut by hand, washed in rainwater, and photographed. Only then are design decisions made. The Freitag brothers' concept was enormously appealing to young urban dwellers from Tokyo to New York, and has remained commercially successful, despite being quickly emulated. Their company, Freitag, creates a contemporary street-style mix, with product names referring to the club scene and American popular culture. The bags are seen as environmentally friendly but also perform as strong style statements. A small creative firm, Freitag has retained full control of the design, manufacturing, and marketing of its goods. Its products are available only through the company's own outlets or selected retailers who share a similar outlook. Freitag stores are individually designed according to their location. The shop in Zurich, for example, designed by Annette Spillmann and Harald Echsle in 2006, reflects the brand's recycled, industrial aesthetic, as it is made from seventeen stacked freight containers—ready-mades—held in place by tension members (plate 293). The display system by Colin Schälli uses stackable boxes held together by bungee cords. The shop is located on the edge of the city, away from other designer outlets. The Freitag concept is as thoroughly thought out as any of the earlier corporate design schemes for which German and Swiss companies were so renowned. Only the ingredients, the look, and the ethos differ, equipped as they are for the twenty-first century.

OPPOSITE, TOP
291. Axel Kufus (b. 1958)
***Egal* bookshelves**, 2001
Made by Nils Holger Moormann (Germany)
Birch plywood

OPPOSITE, BOTTOM
292. Big-Game (est. 2004)
***Bold* chair**, 2009
Made by Moustache (France)
Steel, polyurethane foam, detachable fabric cover

BELOW
293. Anette Spillmann (b. 1969) and Harald Echsle (b. 1968)
Freitag's flagship store in Zurich, 2006

ITALY

ANTY PANSERA

PRECEDING PAGE
294. Achille Castiglioni (1918–2002) and Pier Giacomo Castiglioni (1913–1968)
***Arco* lamp**, 1962
Made by Flos (Italy)
See plate 343.

OPPOSITE
295. Giacomo Balla (1871–1958)
Coffee set, c. 1929
Made by Atelier Gatti (Faenza, Italy)
Tempera on ceramic

According to one definition, offered in the 1950s by Gillo Dorfles, the preeminent Italian critic and historian of design, the term "industrial design" refers only to cases in which the conception of the object and its production are linked to *aesthetic* values. Such a love for the beautiful seems to be an integral part of the Italian DNA. But the Italians have another innate quality that should not be forgotten: a resourcefulness that allowed them to make something positive from the burdensome restrictions imposed by the peace treaty after World War II and to undertake a program of reconstruction characterized in part by the development and distribution of products made in Italy.

Italian designers have a unique ability to synthesize their traditional artistic culture, the applied arts, and technology. That provides a first clue as to why Italian industrial design has been and continues to be so successful worldwide. Italian designers have always been able to add technical and aesthetic value to a product, and their particular gift for marrying memory and innovation has helped them find ways to breathe new energy into the domestic environment.

THE ROOTS OF ITALIAN DESIGN

It is impossible to speak of industrial design in a country that is not yet industrialized. This process did not take place in Italy until after World War II, although a series of Italian products manufactured by semi-mechanized means had already been displayed in the International Exhibition of Mass Production organized by Giuseppe Pagano at the seventh Triennale di Milano in 1940.

The first artist-designers to focus on utilitarian objects displayed a keen interest in machinery and in nascent industry. These were the Futurists. On March 11, 1915, Giacomo Balla and Fortunato Depero published a first reflection on the relationship between art, craftsmanship, and industry in a manifesto entitled *The Futurist Reconstruction of the Universe*, the official beginning of a long and inventive history. In the first *Futurist Manifesto* (1909), Filippo Tommaso Marinetti had already envisioned the possibility of producing objects in a different way: "It is the solidity of a steel plate in itself that interests us, that is, the incomprehensible and inhuman alliance of its molecules and electrons."

A few years later, Nuove Tendenze (New Trends)—a movement active in Milan from August 1913 to June 1914—brought together painters, artisans, and architects determined to combine pictorial experiments and product design with artisanal practices. One of them was Marcello Nizzoli, who will be discussed later in this chapter, in the context of Olivetti. In designing the shawls, pillows, and handbags produced by his sisters, Nizzoli anticipated the first of the Italian *case d'arte*, the "art houses" that the Italian Futurists opened during the interwar period, from Trentino to Sicily, from Milan (where Cesare Andreoni and Fedele Azari were involved) to Palermo (where Pippo Rizzo was), and also, of course, in Rovereto and Rome. In 1909, Balla had painted a new utilitarian object, the arc lamp, choosing it less for its form than its intrinsic technical value.

In the 1909 *Manifesto*, it is easy to discern (behind the presentiment that human life might someday be governed by the life of the factory) the environment in which Italian industrial design would set down roots. For example, an evident taste for fine craftsmanship marks all the comments about furniture. But the Futurists never sought to develop a mode of production and never designed for industry. Rather, they grasped in aesthetic terms the eminently technological nature

BELOW
296. Fortunato Depero
(1892–1960)
***Ballerina* exhibited at the first Monza exposition**, 1923
Painted wood

OPPOSITE
297. Fortunato Depero
(1892–1960)
Depero Futurista, 1927
Published by Dinamo Azari
Book bound with steel bolts
Private collection

of their age and exalted its speed, its dynamism, its energy. They designed furniture and accessories, but conferred on them the status of art, which makes them not commodities but symbols, feelings (plate 295).

The forms of these objects are deliberately provocative (angular chests of drawers, uncomfortable seats intended for "superfast" use), and the greatest attention is paid to formal and structural solutions. The decorative elements always arise from the very form of the object, which tends to narrate the precise function it is called on to fulfill. The furniture, in bold colors, is generally made of painted wood, completely at odds with bourgeois comfort. The Futurists avoided desks and secretaries, symbols of the academy and of contemplation—and even beds, the burial plots of love. Conversely, they anticipated the use of the sofa bed, the lowboy, the night table, and convertible furniture, multiplying the functions of each element. With the exception of their ceramics, these pieces were one of a kind or produced in limited editions.

Beyond the performances of Marinetti, the Futurists put their stamp on Italian history, especially in the area of furniture. They marked out several essential directions for the future of Italian design, including rationalism and, later on, postmodernism.

THE BIRTH OF AN ITALIAN IDIOM

The predecessor of the Triennale di Milano was the International Exposition of the Decorative Arts,[1] held between 1923 and 1930 at the Villa Reale in Monza. These events, at first taking place every two years, were the birthplace of a specifically Italian idiom, and also offered a unique opportunity to bring Italian creations into dialogue with those of other countries.

The different stylistic grammars of the many regions that constitute the Italian Peninsula—regions attentive to the ideas being developed by their respective European models (London for Sicily, Paris for Piedmont, Vienna for Lombardy and Venetia)—found at the Villa Reale a forum for introducing themselves to the public and, especially, rich soil in which to grow. The historian and theorist Guido Marangoni,[2] who wanted, perhaps, to play the same role in Italy that Hermann Muthesius did in Germany, was the founder and, until 1930, the coordinator of these exhibitions. Even before World War I, he had planned to create an international exposition combined with a school to train designers. That school, the Istituto Superiore Industrie Artistiche (ISIA; Higher Institute for Artistic Industries), opened its doors in 1922 and sought to impose itself as the "Italian Bauhaus." It offered an extremely stimulating work space, thanks to its teaching methods and the renown of its faculty: masters of the decorative arts from Lombardy, such as Alessandro Mazzucotelli, a specialist in Liberty-style wrought iron, and the goldsmith Alfredo Ravasco; artists such as Arturo Martini and Marino Marini; and major figures in the modern movement, such as Giuseppe Pagano, Edoardo Persico, and Marcello Nizzoli. As Gio Ponti remarked, it was a school "where the student and the teacher make and invent

DINAMO-AZARI
DEPERO
FVTVRISTA
EDIZIONE ITALIANA
DINAMO
AZARI MILANO
Via S. Orsola, 6 - Telefono 82520
NEW-YORK-PARIS
BERLIN

BELOW
298. Francesco Menzio (1899–1979)
Confectioner's shop at the third Monza exposition, 1927

OPPOSITE
299. Luigi Figini (1903–1984), Guido Frette (1901–1984), Adalberto Libera (1903–1963), and Gino Pollini (1903–1991) of Gruppo 7, with Piero Bottoni (1903–1973)
Entrance to the Electric House at the fourth Monza exposition, 1930

together." A sort of university campus avant la lettre, it attracted young people from all over Italy, along with a large number of foreign students. Of these first students, we may mention the celebrated "three Sardinians": the graphic designer Giovanni Pintori and the sculptor Costantino Nivola (both of whom eventually work for Olivetti),[3] and the ceramist and sculptor Salvatore Fancello. The ISIA would close its doors for political reasons in 1943, but branches would reopen in the 1970s and '80s, in Venice (briefly), Urbino, Rome, Florence, and Faenza.[4]

The First International Exposition of the Decorative Arts, which opened in May 1923, sought to present the different regional cultures and to begin the search for a specifically Italian style. Its aim was to promote a modernity in tune with the changing spirit of the times and also with regional production capabilities. The local cultures, moreover, were brought face-to-face with fourteen foreign exhibitors, from France, Austria, Hungary, Sweden, and other countries. The exposition addressed the question of how to give shape to everyday life within the domestic environment, by means of the objects that inhabit it.

To be sure, Trentino Hall, designed by Depero and overflowing with artistic objects, elicited the disapproval of critics. They accused Depero of being "only a brass band musician from a country fair." The display by Richard Ginori, the venerable ceramics manufactory that had just been overhauled by Ponti, was a greater success.

Two years later, in 1925, the Monza biennial faced stiff competition from the Exposition Internationale des Arts Décoratifs et Industriels Modernes in Paris. Many exhibitors defected, the Futurists in particular: Balla and Depero opted for Paris. However, Germany chose to exhibit the most significant products of the Deutscher Werkbund at Monza. On the whole, the program for this biennial differed little from that of 1923, although there were a few encouraging signs of vitality: the display, on distinctive mannequins, of silk shawls designed by Nizzoli and embroidered by Piatti of Como; new works by Guido Andloviz, produced in Laveno by the Società Ceramica Italiana (SCI); and the renewed success of the Richard Ginori manufactory.

In 1927, when Ponti became a member of the biennial's arts council, the Novecento movement (as the "return to order" was called in Italy) finally earned its rightful place, being represented by Gio Ponti's and Emilio Lancia's neoclassical furniture. Several artists created stage sets for the biennial, such as a "commercial street" that offered a "very modern slice of city life." Its stylish establishments—a butcher's shop, a bar, a telephone switchboard, a pharmacy, a confectioner's shop (plate 298)—were designed by, among others, the Turinese Gruppo dei Sei (Group of Six), who were gradually approaching the main European current. In the garden of the Villa Reale, Depero created a standalone pavilion for the Bestetti-Tumminelli publishing house, in which the architecture, furniture, and decorative elements were made up of giant three-dimensional letters.

In 1930, the event was renamed the International Triennial Exposition of the Modern Decorative and Industrial Arts, and experimental architecture was again on display in the garden of the Villa Reale. This fourth exhibition, which sought to provide functional and formal solutions to the housing shortage, included three full-size model homes: the *casa del dopolavorista* (workingman's cottage) by Luisa Lovarini—an architect lost to history—as well as Ponti and Lancia's Vacation House and the Electric House by Gruppo 7 (plate 299).

The Vacation House was conceived as a showcase for the Domus Nova line of furniture, which Ponti and

CASA ELETTRICA
SOCIETA EDISON
B.466

OPPOSITE
300. **Kitchen of the Electric House by Gruppo 7 at the fourth Monza exposition**, 1930

LEFT
301. Franco Albini (1905–1977)
Radio, 1938
Receiver and speaker mounted between plates of Securit glass
Single prototype exhibited in the Radio Salon at the seventh Triennale di Milano, 1940

Lancia had designed for the Milanese department store La Rinascente. Intended for the middle classes and offered at very reasonable prices, it could also be purchased by mail: a beautiful catalog featured it under the rubric "modern furnishings." The Edison General Electric Company sponsored the Electric House, designed by Luigi Figini and Gino Pollini to display all the household electrical appliances existing at the time, whether made in Italy or elsewhere. It was filled with functional furniture on the leading edge of domestic comfort, designed by Frette and Libera. Piero Bottoni was responsible for all the kitchen equipment and bathroom fixtures (plate 300). The Electric House served as the manifesto of the rationalists, whose aesthetic would to a great extent characterize the 1930s. It attested to their interest in new materials, particularly metal, and in a closer relationship with industrial production.

At the Triennale of 1930, the Italian productions displayed a certain stylistic unity, set to compete with the transalpine world. On that occasion, Monza also welcomed experiments from foreign schools: the Deutscher Werkbund of Berlin, cofounded by Bruno Paul, and the Viennese branch of the Werkbund, run by Josef Hoffmann. Examples of rigorous functionalism were represented as well: the lamps designed by the Bauhaus of Dessau, and Mies van der Rohe's chair in chrome-plated tubular steel. France sent extraordinary pieces in the 1925 style, as well as the beautiful travel accessories of Louis Vuitton.

At the time, given the success of the event and the difficulties of drawing in people from the Milanese hinterland, there was talk of transferring the Triennale to Milan. The idea even arose (already!) of constructing a light-rail system. Three years later, a building designed by Giovanni Muzio to house the Triennale was unveiled in Parco Sempione, in the very heart of Milan: the Palazzo dell'Arte. This versatile building, illuminated by natural light, already housed various services: a post office, a fire station, a library, a theater, a bar, and a restaurant attached to the garden terrace on the roof. Next to the Palazzo dell'Arte, and in sharp contrast to its Novecento style, Gio Ponti and Cesare Chiudi erected their Torre Littoria, or "Fascist Tower," made entirely of Dalmine steel pipes.

At that Fifth International Exposition of the Modern Decorative and Industrial Arts and of Modern Architecture (architecture now being an integral part of the theme), the concept of industrialization was already integrated into the furniture designs on display (plate 302). These all paid great attention to new materials, such as linoleum and its derivatives, boxwood, Anticorodal (a 98-percent aluminum alloy), and steel—to which an entire pavilion, the Steel-frame House in Parco Sempione, was devoted—as well as metal tubing, which the Columbus company bent into chairs, tables, and desks.[5] These prototypes were analyzed by critics from the specialist magazines but would ultimately have only a modest commercial success, being used primarily in public and collective spaces and in the Italian overseas colonies.

302. Franco Albini (1905–1977), Renato Camus (1891–1971), Paolo Masera, and Giancarlo Palanti (1905–1977)
Sitting room and office on two floors, at the fifth Triennale di Milano, 1933

In accordance with Ponti's principle of the "unity of the arts," painting was also represented, in the form of monumental murals by Mario Sironi, Massimo Campigli, Achille Funi, Giorgio de Chirico, and Gino Severini. Among the many structures erected in the park was a Futurist terminal for a civilian airport, designed by Enrico Prampolini with the collaboration of several "aeropainters" (Andreoni, Depero, Munari).

The sixth Triennale, whose tone of moral responsibility can be attributed to Pagano and Persico, took place in 1936. It sought to promote "an intimate and profound embrace of life by the arts," as Pagano wrote. He was eager to make the Triennale a hub of activities, constructive debate, and permanent research, and for that reason sought to establish a study center there.

Despite the general climate of autarky reigning in Fascist Italy in the years prior to World War II, the Palazzo dell'Arte was open to experiments from outside the country, which would have a significant impact on the culture of Italian design. Max Bill, for example, designed the Swiss pavilion, and Alvar Aalto's work sparked the interest of many designers. But the essence of the sixth Triennale lay in the Salone della Vittoria (Victory Salon), a collaborative effort by Persico, Palanti, Nizzoli, and Lucio Fontana, who created a successful synthesis of architecture and the visual arts, marked by a creativity linked directly to the rationalist experiment. This monumental project was envisioned as an "act of faith in the greatness of pacified Europe," but it was partly mutilated by state censorship, just as the Triennale was deprived of the presence of Gropius and other German architects.

The seventh Triennale di Milano, in April 1940, included the International Salon of Mass Production, organized by Pagano with the aim of "helping to think form as it is thought by the collectivity." In many respects, this salon marked the transition from mechanized craftsmanship, characteristic of "protodesign," to industrial production. On display were the very first Italian products that openly took the industrial process into account as a methodological model for design; these items were manufactured by some of the largest Italian companies (Fiat, Olivetti, Pirelli). The theme of the salon was "standardization"—a subject of debate in Europe since the beginning of the century but still unknown in Italy. For the rationalists, standardization gave concrete form to the principles of quality and social utility, a choice more ethical and pedagogical than productive or economic.

At the same Triennale, Livio and Pier Giacomo Castiglioni made their debuts (as did Luigi Caccia Dominioni) as commissioners of the Radio Salon. About twenty receivers were collected there, each displaying an alternative solution to the traditional radio set concealed in a chest or cabinet. (In 1938, Franco Albini had already exposed the radio receiver and speaker between two plates of Securit glass; plate 301). Office furniture also played a prominent role. The trend was toward efficiency: everything had to be within reach—tools as well as work surfaces—and the materials were to be metallic, lightweight, and strong. Several models of telephone were exhibited, as well as many typewriters that Nizzoli had designed for Olivetti.

The year 1940 was indeed an emblematic one for Italian design. However, as we will see, the birth of a real notion of industrial design[6] and of the figure of the designer—someone who creates prototypes that industry produces on a large scale and by mechanical means—dates to the beginning of the 1950s.

1408.

RIGHT
303. **Cover of the review *Domus*, founded by Gio Ponti,** No. 303, February 1955

OPPOSITE, TOP
304. Gio Ponti (1891–1979) ***Superleggera* chair**, 1957 Made by Cassina (Italy) Lacquered ashwood, cane seat Musée des Arts Décoratifs, Paris

OPPOSITE, BOTTOM
305. **Demonstration of the lightness of Gio Ponti's *Superleggera* chair**

THE AGE OF PIONEERS

The Futurists certainly played a role in the very early days of design, but it was primarily the rationalists who set the stage in the 1930s. The pronouncements of Gruppo 7 (formed in 1926 by the architects S. Larco, G. Frette, C. E. Rava, L. Figini, G. Pollini, G. Terragni, and A. Libra) on the need for a "strict respect for logic and rationality" originally referred only to architecture, but they were quickly applied to furniture and other objects as well. In the words of the critic Edoardo Persico, design was a means for moving "beyond architecture."

It was primarily architects and engineers, in fact, who were advancing toward an integral conception of the home environment. Along with a handful of artists and craftsmen, also attentive to technological developments, they were the pioneers of "protodesign," a neologism that is still useful for defining the relationship that was gradually established between art and industry, design and product. They were able to provide new answers in the area of production but also a poetics: in other words, a new language.

Foremost among them was the Milanese Gio Ponti (1891–1979),[7] a multitalented architect and eclectic creator of forms, ranging from the small curio to the urban plan, who set to work in the early 1920s. It should not be forgotten that he also played an important role in kindling industry's passion for art, by means of the now-legendary review *Domus,* founded in 1928 (plate 303), and a lesser-known magazine, *Stile* (1941–47), intended to assemble everything related to "expression, ornament, or instrument in our lives and beautiful homes." *Stile* also published "free projects," such as Carlo Mollino's furniture patterns. "Industry," said Ponti, "is the style of the twentieth century, its way of creating a relationship between art and industry: art is the species; industry, the condition."

The Novecento movement and its aesthetic, then rationalism, shaped Ponti's idiom. He could be called eclectic, although the synthesis of the different arts was always the guiding thread of his poetics, and his work was always directed toward "the original purity of the relationship between form and function." His first experiment consisted of modernizing—with a touch of classicism and of solemnity mixed with irony—the ceramics of the eighteenth-century Richard Ginori manufactory (plate 306). This was followed, also in the 1920s, by the creation of the Domus Nova furniture collection, with its very clean lines, for the department store La Rinascente (1928). The previous year, he had founded Il Labirinto, which offered objects and furniture to a more refined, bourgeois clientele, and in the early 1930s he established Fontana Arte[8] to produce lamps and glass furniture.

Ponti also made a fundamental contribution to furniture manufacturing after World War II. For Cassina,

RIGHT
306. Gio Ponti (1891–1979)
***La Passeggiata Archeologica* urn**, 1925
Made by Richard Ginori (Italy)
Enameled porcelain
Metropolitan Museum of Art, New York

BELOW
307. **Interior of the Parco dei Principi Hotel in Sorrento, designed by Gio Ponti, with furniture by the Norwegian firm Arnestad Bruk (produced by Cassina)**, 1962

OPPOSITE
308. Gio Ponti (1891–1979)
***Morandiana* bottles**, c. 1950
Made by Venini (Italy)
Hand-blown glass, glass rods

BELOW
309. Bruno Munari (1907–1998) and Enzo Mari (b. 1932)
***Diamante* espresso machine**, 1956
Made by La Pavoni (Italy)

OPPOSITE, TOP
310. Marcello Nizzoli (1887–1969)
***Mirella* sewing machine**, 1957
Made by Necchi (Italy)
Museum of Modern Art, New York

OPPOSITE, BOTTOM
311. Marco Zanuso (1916–2001)
***1100/2* sewing machine**, 1956
Made by Veglia Borletti (Italy)

he designed the famous *Superleggera* chair (1957), still produced today, which revisited the Chiavarina (or Campanina) chair, a pure product of artisanal cabinetmaking (plates 304, 305), developed in Chiavari (Liguria) in the early nineteenth century. The *Superleggera* was characterized by the lightness of its ashwood frame (3.75 pounds, or 1.7 kg) and the translucency of its seat in woven India cane. Chairs, a veritable playground for designers but also a showcase for furniture makers, tell their own story of Italian design. The infinite variety of the solutions imagined is a constant source of surprise.

Ponti also designed for other manufacturers, who sought him out precisely because he was an ardent defender of "industry, the style of the twentieth century" and the "hub of our social organization." He designed sewing machines, for example, such as the *Visetta* of 1948. In the following decade, a number of other designers would reinterpret this essential household appliance, and two would win the Compasso d'Oro for their efforts: Marco Zanuso for the *1100/2*, made by Borletti in 1956 (plate 311); and Nizzoli for the *Mirella*, produced by Necchi in 1957 (plate 310). Ponti also created coffee machines for bars: in 1949, he designed a horizontal model for La Pavoni, with each functional component clearly delineated; and in 1956, through *Domus*, he sponsored a competition to design such a machine, which was won by Bruno Munari and Enzo Mari's *Diamante* model (plate 309). He produced lights and bathroom fixtures that favored physiological rather than geometric forms (especially the range of products distributed by Ideal Standard).

Typically "Pontian" designs include the "wall organizer" (1949), the "headboard-dashboard" (1953), the "furnished window" (1954), and the "domestic cell," a kind of flexible housing called *alloggio uniambientale* in Italian. In the 1950s, when Italy converted the vessels requisitioned for the war back to civilian use, Ponti designed a large number of amenities for cruise ships. He was also responsible for the famous Pirelli Tower (1956), a skyscraper designed in collaboration with Antonio Fornaroli and Alberto Rosselli, which for years was a symbol of modern Milan.

Through *Domus*, Ponti contributed significantly to the renewal of Italian production. For several decades, the review, which initially targeted a broad audience of enlightened amateurs, promoted a new domestic environment. Ponti played an equally decisive role in establishing the Compasso d'Oro prize and the Associazione per il Disegno Industriale (ADI).

In 1928, another magazine made its debut: *La Casa Bella*, founded by Guido Marangoni. Later named *Casabella*, it would undergo profound changes over the years. From 1933 on, it was edited by Pagano,[9] an architect, designer, and theorist active in the field of industrial design, who worked with Ponti on railroad car compartments for the Breda company.

In fact, the first companies in Italy to commission "modern" products from artists and architects included not only ceramics and glass factories, but also aircraft and automobile manufacturers: they wanted "designer-engineers." As Giovanni K. Koenig has shown, the Italians were "widely recognized as the best in the world" at aeronautical design.[10] (It must be said that this new industry was supported by Fascism.) One important aircraft manufacturer was Caproni, founded by the engineer Giovanni Caproni in 1910, the year after the first major Italian aeronautical exhibition in Milan. All Caproni planes had a distinctive, functional, and very elegant architecture, and the company enlisted the design services of Cesare Pallavicino[11] (beginning in 1933)

NECCHI
mirella

BORLETTI

BELOW
312. Marcello Dudovich (1878–1962)
Poster for the Fiat *508 Balilla*, 1934

OPPOSITE, TOP
313. **Advertisement for the launch of the new Fiat *500*,** 1957

OPPOSITE, BOTTOM
314–15. Pinin Farina (1893–1966), with Piero Dusio (1899–1975), Giovanni Savonuzzi (1911–1987), and Dante Giacosa (1905–1996)
***Cisitalia 202 GT*,** 1946–48
Made by Pininfarina (Italy)
Museum of Modern Art, New York

and Secondo Campini, who would design the first Italian jet in 1940. By 1917, the Breda company had also constructed large hangars outside Milan, at Sesto San Giovanni and Bresso, where it produced civilian and military aircraft. Filippo Zappata, having returned to Italy after a productive experience at Blériot Aéronautique, designed the four-engined Breda *BZ308*, which was completed in 1948. In addition to being on the cutting edge of technology, the *BZ308* was characterized by its extraordinary aesthetic quality, the result of a harmonious synthesis between design, construction technique, and aerodynamic research.

Since its inception at the beginning of the twentieth century, the Italian auto industry has succeeded in combining artisanal traditions, entrepreneurial know-how, and creativity to close the technological gap that once separated Italy from more industrialized nations.[12] The city of Turin in particular has become synonymous with automotive design. It was there, in 1899, that Giovanni Agnelli founded Fiat (Fabbrica Italiana Automobili Torino). The mass motorization of Italy began with the Fiat *508 Balilla*, exhibited at the Milan Trade Fair in 1932 (plate 312). Among its designers was the engineer Dante Giacosa,[13] who, from 1929, dedicated himself to Fiat's small cars: the *Topolino* in 1936, followed, after World War II, by the *600* (1955–56) and the *Nuovo 500* (1957), "the Italian's first car," destined to replace the bicycle and the motor scooter (plate 313).

Also exercising his talents in the Piedmont capital was Giovanni Battista Farina, nicknamed "Pinin Farina." After on-the-job training, a visit to Detroit in 1920 (where he was welcomed by Henry Ford), and various successes as a racer, he opened the Carrozzeria Pinin Farina in 1930. Even before 1939, he had begun to develop the revolutionary forms that would culminate, after the war, in the Cisitalia *202* (plates 314, 315). With this little coupé, he established a new style in which the volumetric shape of the car body was no longer merely the result of assembling two-dimensional panels. As we will see below, the Pininfarina company went on to create many other influential designs for Italian and foreign carmakers.

Lombardy would also make its contribution to the history of automobile design, with Isotta Fraschini (1900) and, especially, Alfa Romeo. The arrival of the engineer Nicola Romeo in 1915 increased the dynamism of the cars produced by Alfa (Anonima Lombarda Fabbrica Automobili, founded in 1910). And so began the heyday of the red Alfa Romeo race cars with engines by Vittorio Jano. Enzo Ferrari founded his Scuderia Ferrari racing team in 1929, at first using Alfa Romeos but eventually developing his own racers, starting with the *125 S* of 1947.

The studio BBPR was formed in 1932 by Gian Luigi Banfi, Ludovico Barbianao di Belgiojoso, Enrico Peressutti, and Ernesto Nathan Rogers. These architects—urbanists above all—were not averse to working on small-scale projects and organized the *Form of the Useful* exhibition at the Ninth Triennale. Other architect-designers included Giulio Minoletti, who was responsible for the emblematic Italian luxury train, the

la nouvelle
FIAT
la nuova
500
der neue
el nuevo
the new

BELOW
316. Carlo Felice Bianchi Anderloni (1916–2003)
Ferrari 166 MM Touring Barchetta, 1948–53

OPPOSITE, TOP
317. **Franco Albini behind his *Veliero* bookcase**, 1938, manufactured originally by Poggi and now by Cassina

OPPOSITE, BOTTOM
318. Franco Albini (1905–1977)
***LB7* bookcase**, 1957
Made by Poggi (Italy)
Rosewood, lacquered aluminum
Private collection

Settebello electric for Breda (1949); and Franco Albini,[14] who opened his studio in 1930 and, while giving precedence to architecture, was already participating in the Triennales at that time. The bookcase with a guy-rope structure that Albini produced in a single prototype for his home on via De Togni in Milan (1938), and which now appears in Cassina's *I Maestri* catalog, is a veritable manifesto (plate 317). Its glass shelves, suspended by steel cables and ashwood masts, recall the yards and sails of a boat. (The bookcase is in fact called *Veliero*, or "Sailboat.") Albini pursued his work in the 1950s with the same coherence and in the same rationalist vein, although his personal poetics led him to take a greater interest in means of construction, technological innovations, and the relationship between materials of various forms and colors. Three of his creations may serve as examples: the small *Luisa* chair (1950), the *Fiorenza* chair (1955), and the *LB7* bookcase (1957; plate 318), which defined the image of the Poggi firm in Pavia.

As of 1951, Albini was assisted by Franca Helg, perhaps the first woman to work in the field in Italy.[15] For her, "design is constant experimentation" that requires "patient research" and is characterized by a concern for line and detail. She would take on a wide variety of projects, from television sets (the *Orion* for Brionvega, 1963) to lighting (for Sirrah, 1968) to furniture (for Poggi and Bonacina, although it is difficult to distinguish her contribution from Albini's) to dishware (for San Lorenzo in 1971, plate 321; and Richard Ginori in 1986). In 1963, Albini and Helg, along with Bob Noorda, set to work on the visual identity of Milan's first metro line, a project that would win them the Compasso d'Oro the next year. Their guiding principle in this undertaking was modularity, allowing for almost infinite expansion, as would later be demonstrated by Metro Line 2.

Also part of that generation of architects was the Venetian Carlo Scarpa,[16] who often worked with Venini on the design of very delicate glass objects (plate 322). Influenced by the work of the pioneer of organic architecture, Frank Lloyd Wright, Scarpa took a rather classical approach. In the 1960s, with the collaboration of Dino Gavina, he created a heterogeneous series of fine furniture for Simon International, including the *Doge* table (1968), and various light fixtures for Flos.

Finally, there was Carlo Mollino, the enfant terrible of architecture and design.[17] Mollino, from a family of artists, graduated from the University of Turin in 1931. An ingenious dabbler, he was interested in

photography, set design, aeronautical engineering, and furniture, as well as automobile racing. He created two race cars, the *Osca 1100* (1954) and the DaMolNar *Bisiluro* ("Double Torpedo"; 1955). Most of his projects remained at the prototype stage or were produced on a very small scale (although they would be reissued in the 1970s). His eclectic and highly imaginative idiom transformed tables, chairs, sofas, and mirrors by evoking the animal world or the human body (the female body in particular); in his surrealist universe, the familiar was rendered unique (plates 319, 320). His influences included Antoni Gaudí, as well as Charles Rennie Mackintosh, Charles and Ray Eames, Le Corbusier, and Alvar Aalto.

By contrast, two figures in the history of Italian design had their roots in Futurism: Marcello Nizzoli,[18] a painter by training, and Bruno Munari,[19] also an artist-designer. They may be considered the first prominent nonarchitect designers in Italy.

For them, Futurism was a method of design attuned to the needs of their contemporaries. This is attested by the objects they produced, which are always marked by a kind of minimalism: they possess an essential form, to quote Munari, that "emerges naturally from the nature of the service they offer."

Nizzoli very quickly moved toward abstraction, a prelude to his days among the ranks of the rationalists. His 1931 encounter in Milan with the Neapolitan Edoardo Persico was a determining factor. With Persico, a critic and historian of art and architecture, and coeditor of *Casabella*, Nizzoli designed the physical layout for several stores. Although he received his first commission as a graphic designer for Campari billboards, his name is almost synonymous with Olivetti. The latter firm had been founded by Camillo Olivetti in Ivrea (Piedmont) in 1908 to produce calculators and typewriters. Thanks to Adriano Olivetti (1901–1960), who succeeded his father in 1929, the company distinguished itself with a business culture focused equally on fine modern products and on the working environment. It therefore turned to designers for innovative graphic design, public relations, and product design (plates 323, 324). (In that field, the painter Alberto Magnelli's role in creating the *MP1*, Olivetti's first portable typewriter, in 1932 is significant.) But the firm also encouraged them to imagine a sort of ideal city, where employee housing and factory buildings would be harmoniously integrated into the urban space.

Nizzoli was enlisted to work in Olivetti's publicity department in 1935, and he went on to serve as product

OPPOSITE
319. Carlo Mollino (1905–1973)
***Arabesco* table**, 1950
Made by Apelli e Varesio (Italy)
Maple plywood, glass, brass
Musée des Beaux-Arts, Montreal

BELOW
320. **Interior of a house in Turin designed by Carlo Mollino**, c. 1950

OVERLEAF
321. Franca Helg (1929–1989)
***Pannocchia* and *Spirale* bowls**, 1971
Made by San Lorenzo (Italy)
Silver

PAGE 269
322. Carlo Scarpa (1906–1978)
Vases, 1940
Made by Venini (Italy)
Glass
Private collection

LEFT
323. Ettore Sottsass (1917–2007)
***Valentine* portable typewriter**, 1969
Made by Olivetti (Italy)
Museum of Modern Art, New York

BELOW
324. **Advertisement for the Olivetti *Valentine* portable typewriter by Ettore Sottsass**, 1969
Musée National d'Art Moderne, Centre Georges Pompidou, Paris

OPPOSITE
325. **Advertisement for the Olivetti *Lettera 22* portable typewriter by Marcello Nizzoli**, 1950
Museum of Modern Art, New York

manager at the firm until 1960. He has to his credit several typewriters: the *MC Summa 40* (1940); the *Lexicon 80* (1948), "one of the first examples of an object whose sleek casing was not designed as a mere cosmetic envelope but the result of a careful correspondence to the internal structure of the mechanism it protected," according to Gillo Dorfles; and the *Lettera 22* (1950), the portable typewriter par excellence (plate 325).

As for Munari, he was not only a designer and painter but also a sculptor, an illustrator, a graphic artist, and a pedagogue at heart. He took a real interest in the eventual owners of his objects. Behind his "useless machines," his "aerosculptures" of 1933, there is almost always an implicit message for the user: "You try, too!" He was also interested in children, for whom he published an innovative series of books beginning in 1945. Munari was one of the founders of the Movimento per l'Arte Concreta (MAC; Concrete Art Movement)[20] and in 1952 launched his first products: two foam-rubber toys, the cat *Meo* and the little monkey *Zizi* (winners of the first Compasso d'Oro in 1954).[21] He would also design for Danese,[22] which was founded in the mid-1950s by Bruno Danese and Jacqueline Vodoz to produce everyday objects for the home and office and later specialized in the promotion of educational games. Danese's products, designed by Munari as well as Enzo Mari, were distinguished by a forthright functionalism and an immediately recognizable look (plates 327, 328). Munari's designs for the firm include the famous *Cubo* ashtray of 1957 (plate 326) and a series of lamps in geometric shapes, for which he also designed the packaging; notable among these is the *Falkland* pendant lamp of 1964, with a long elastic mesh that covers metal rings of varying diameters, giving the piece a pronounced verticality (hence its nickname *Calza*, or "Stocking"; plate 329). For the firm Robots, Munari designed a structure that combines a child's bed and a play environment: the *Abitacolo* of 1970, awarded the Compasso d'Oro in 1979.

olivetti Lettera 22
GROS MONTI & C.
"V 3 - 131 It."
Visto Questura di Torino del 7 - 7 - 1951

JUIN
23
LUN

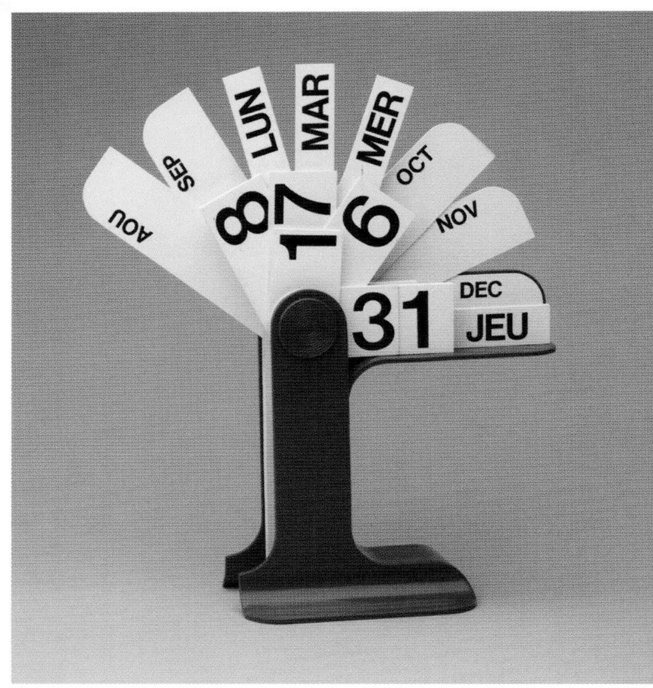
AOU
SEP
LUN
MAR
MER
OCT
NOV
8
17
6
31
DEC
JEU

OPPOSITE, TOP
326. Bruno Munari (1907–1998)
***Cubo* ashtray**, 1957
Made by Danese (Italy)
Anodized aluminum, melamine
Museum of Modern Art, New York

OPPOSITE, BOTTOM
327–28. Enzo Mari (b. 1932)
***Timor* perpetual calendar**, 1966
Made by Danese (Italy)
Plastic
Museum of Modern Art, New York

ABOVE
329. Bruno Munari (1907–1998)
***Falkland* pendant lamp**, 1964
Made by Danese (Italy)
Jersey, aluminum
Centre National des Arts Plastiques/Fonds National d'Art Contemporain, Paris

Vespa
Vespa 125
L. 128.000
Vespa 150
L. 148.000
Vespa G.S.
L. 178.000
PIAGGIO
PRODUZIONE PIAGGIO 1960

OPPOSITE, TOP
330. Corradino d'Ascanio (1891–1981)
***Vespa GS 150* scooter**, 1955
Made by Piaggio (Italy)
Museum of Modern Art, New York

OPPOSITE, BOTTOM
331. **Advertisement for *Vespa* scooters**, 1960

BELOW
332. Cesare Pallavicino and Pierluigi Torre (1902–1989)
***Lambretta* scotter**, 1949–50
Made by Innocenti (Italy)

THE POSTWAR ERA: NEW VENTURES

The reconstruction of Italy began on April 25, 1945, in the liberated city of Milan. It was marked by the express desire to get the country back on its feet as soon as possible. The capital of Lombardy had lost many of its houses: more than 50 percent of the buildings had been destroyed or seriously damaged. The Galleria Vittorio Emanuele was eviscerated, the Teatro alla Scala no longer had a roof, and the Pinacoteca de Brera was collapsing under rubble.

Very quickly, the architects formed a loose association, the Movimento Studi per l'Architettura (MSA), to continue the work of the rationalists. After the war, then, it was rationalism, in parallel with the radical restructuring of industry, that would profoundly mark Italian design.

It was natural that 1947 should be the year of the "Reconstruction Triennale." That eighth Triennale, organized by Piero Bottoni, set out the themes that would provide the guiding thread for subsequent Triennales: urbanism, social architecture, and products and furniture designed for the working- and middle-class homes being built in the suburbs. The Triennale's Housing Salon therefore gave preeminence to the questions of uniformity and standardization, while basic furniture made with inexpensive materials and the relationship between form and function held a prominent place at the *Objects for the Home* exhibition. After the event, one concrete object would remain: the experimental QT8 (Quartiere Triennale Ottava), a neighborhood designed and built on the outskirts of Milan to demonstrate the feasibility of low-cost construction. This may be the only Italian example of the synthesis of the arts advocated by Le Corbusier.

The attention granted to the synthesis of the arts and to an interdisciplinary approach was just as essential for the architects, designers, and artists who formed the Movimento per l'Arte Concreta in December 1948. Its founders included Dorfles and Munari, soon joined by Vittoriano Viganò and other architects, the industrial designer Joe Colombo, graphic designers (Giancarlo Iliprandi, for example), and various artists (notably, the abstract painter Luigi Veronesi).

After the war, industry converted back to civilian production as efficiently as possible. Piaggo and Innocenti began to produce motor scooters that repurposed the now-obsolete components of the aeronautics industry. The *Vespa* (1946–47) was produced in the factories of Pontedera in Tuscany (plate 330). Its name, of course, refers to the lightness of the wasp (*vespa* in

BELOW
333. Marco Zanuso (1916–2001)
***Lady* chair**, 1951
Made by Arflex (Italy)
Metal, latex foam, fabric
Museum of Modern Art, New York

OPPOSITE
334. Marco Zanuso (1916–2001)
***Sleep-O-Matic* convertible sofa**, 1954
Made by Arflex (Italy)
Steel, latex foam, fabric
Private collection

Italian) but is also derived from the acronym VESPA (Veicoli Economici Società per Azioni; Economical Vehicles Joint Stock Company).

The *Vespa* was soon joined by the *Lambretta* (1949–50), built in Milan on the banks of the Lambro River (plate 332).[23] Both scooters were designed by aeronautical engineers: the *Vespa* by Corradino d'Ascanio (who had developed a helicopter prototype in 1930); the *Lambretta*, with a tubular frame, by Cesare Pallavicino and Pierluigi Torre. These models not only enhanced the design of motor scooters but also revolutionized transportation, becoming the symbol of a newly motorized Italy (plate 331).

In the area of furniture, Pirelli[24] began to experiment at this time with elastic fabrics and foam rubber, a material that had been invented to protect airplane fuel tanks from shell fragments.[25] A handful of young entrepreneurs who had been involved with these experiments founded Arflex (1950), a firm that immediately attracted talented designers. Its mass production of chairs helped to supplant traditional furniture making and create a new kind of home environment. For Arflex, the architect Marco Zanuso created the famous *Lady* chair, lightweight and with soft curved lines (1951; plate 333) and the *Sleep-O-Matic*, a sofa that converts into a bed (1954; plate 334). Older, more established designers also collaborated with Arflex, such as Franco Albini and the members of the BBPR group. Over time, Giancarlo De Carlo, Tito Agnoli, Pierluigi Spadolini, Joe Colombo, and Cini Boeri[26]—one of the first Italian women to design houses and consumer products—would work for the agency. Boeri started out as a designer and architect alongside Ponti and Zanuso, her teacher at the Politecnico di Milano. She also created some original furniture, such as the *Serpentone* ("Big Snake," 1971), an infinitely expandable sofa made up of soft polyurethane rings (plate 335); and the *Strips* series of modular sofas (in collaboration with Laura Griziotti), which received the Compasso d'Oro in 1979 (plate 337). Her *Ghost* chair, designed with Tomu Katayanagi for Fiam in 1987, was a great success (plate 336).

Aurelio Zanotta also entered the world of furniture and accessories in 1954, founding a company that would encourage many designers to experiment with new materials, such as steel tubing and polymers like Perspex and Plexiglas. Meanwhile, Marco Zanuso[27] managed to combine crystal and steel in his *Marcuso* pedestal table (1969). Zanuso also designed an interesting fully equipped kitchen for the newly created Elam company in 1966, along with all sorts of other appliances, from television sets to sewing machines to lamps, over the years. But perhaps his greatest achievement was to theorize the dignity of design, of small-scale projects versus large-scale architecture, although he continued to play a leading role in the latter field, and not only in Italy.

Another company that would become a key player in Italian design and manufacturing got its start at this time: Kartell,[28] founded in 1949 by Giulio Castelli, an

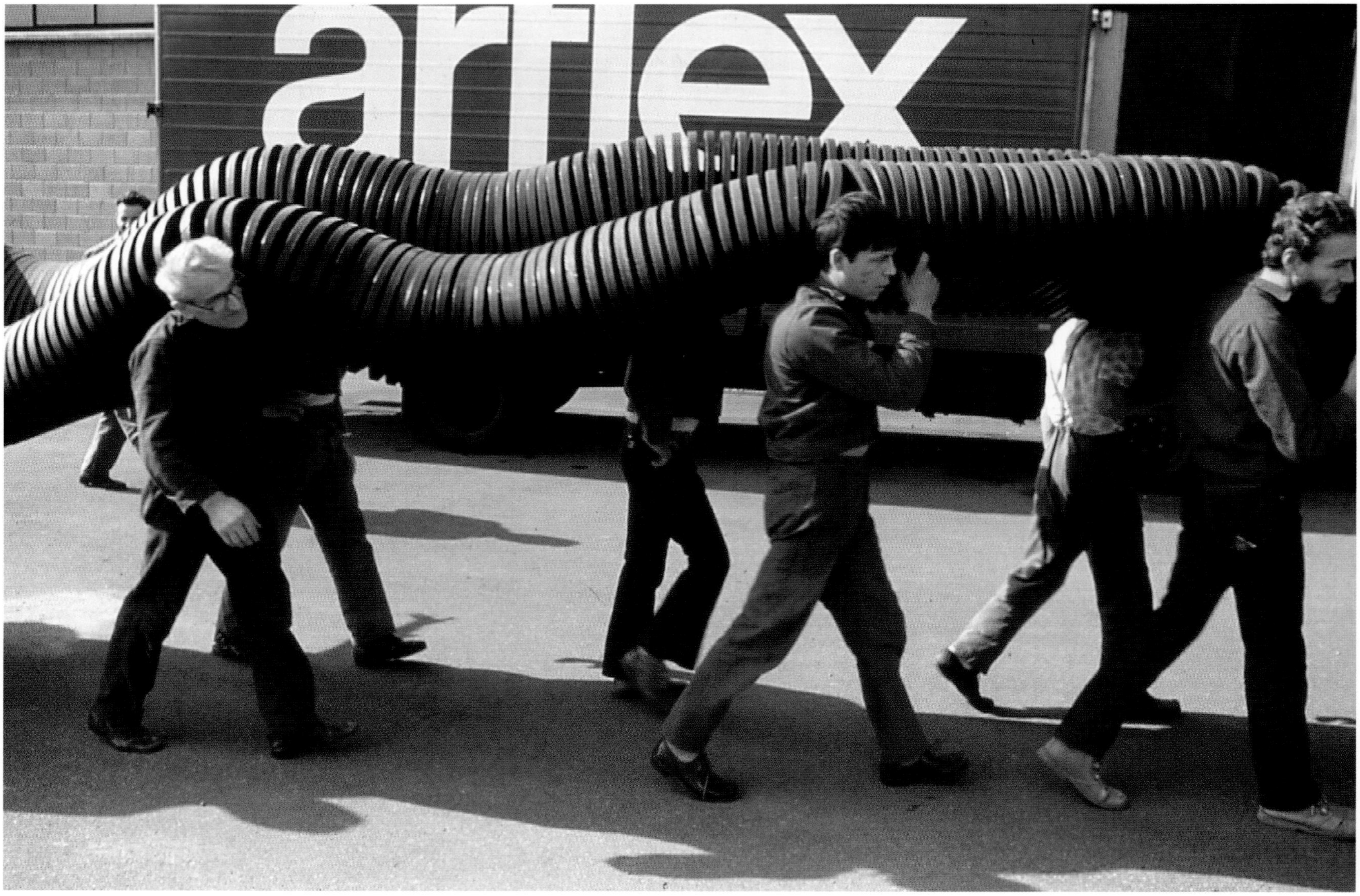

ABOVE
335. **Transporting the *Serpentone* sofa, designed by Cini Boeri and made by Arflex (Italy)**, 1971

OPPOSITE, TOP
336. Cini Boeri (b. 1924) and Tomu Katayanagi (b. 1950)
***Ghost* chair**, 1987
Made by Fiam (Italy)
Glass
Musée des Arts Décoratifs, Paris

OPPOSITE, BOTTOM
337. Cini Boeri (b. 1924)
***Strips* bed**, 1972
Made by Arflex (Italy)
Wood frame, polyurethane foam, heat-sensitive polyester fiber, velvet

engineer and a student of Giulio Natta at the Politecnico di Milano. Natta had long been researching polymers with the support of the Montecatini company, and in 1963 he would receive the Nobel Prize in chemistry for his work on this subject. Castelli was thus aware of the potential of the new materials and technologies, but he also heeded the suggestions of several architects, including Anna Ferrieri, who would become his wife.

Numerous designers worked for Kartell, on the most varied projects. They renewed the home environment with small, inexpensive, and ingenious accessories that replaced the traditional tin buckets, aluminum sieves, glass lemon presses, and rattan carpet beaters. Thanks to plastics, they could offer these items in novel and harmonious shapes and explosive colors. Roberto Menghi designed receptacles, watering cans, and jerry cans (plate 339) with curved lines, rounded corners, and ergonomic handles, as well as various car accessories (such as the innovative *Portasky* ski rack, with Carlo Barassi). Gino Colombini's reimagined household goods stand out for their unusual colors (plate 338), while the Castiglioni brothers created light fixtures of a new kind. Traditional items of furniture such as chairs were also updated: for example, the *4999*, an all-plastic chair for children, designed by Zanuso in 1964 (with Richard Sapper); and Joe Colombo's *Universale* chair (1965–67), made of ABS, with curved lines and streamlined surfaces, the first full-size chair molded entirely from plastic (plate 340). Anna Castelli Ferrieri[29] also created original pieces, such as the *4905/7* stackable tables, the *Componibili* system of square and round storage modules (1967; plate 341), and the *4830* series of stools (1979). All these families of multiuse products, cheerful and colorful objects with a rigorous style and a low selling price, breathed new life into the domestic setting. Everything was extremely functional—not only the design and form of each product, but also its mass production once the mold was made.

Azucena had a very different project in mind. This small business was founded in 1947 by three architects —Ignazio Gardella, Luigi Caccia Dominioni, and Corrado Corradi dell'Acqua—to produce limited editions of furniture and interior decor that they designed with an eye toward historicism. These objects, which could be used in different contexts, were not intended for a large public. Dominioni, for example, created the *Catilina* chair (1958) as well as the *Franz Joseph* bed (whose very name spoke volumes).

Ernesto Gismondi ventured (initially with Sergio Mazza) into the world of lighting: his company Artemide began the mass production of light fixtures in

OPPOSITE, TOP
338. Gino Colombini (b. 1915)
Dustpan, 1957
Made by Kartell (Italy)
Polystyrene
Musée National d'Art Moderne, Centre Georges Pompidou, Paris

OPPOSITE, MIDDLE LEFT
339. Roberto Menghi (1920–2006)
Jerry can, 1958
Made by Pirelli (Italy)
Polyethylene
Museum of Modern Art, New York

OPPOSITE, BOTTOM RIGHT
340. Joe Colombo (1930–1971)
***Universale* chairs**, 1965–67
Made by Kartell (Italy)
ABS plastic
Museum of Modern Art, New York

BELOW
341. Anna Castelli Ferrieri (1920–2006)
Storage modules, 1967–69
Made by Kartell (Italy)
ABS plastic

BELOW
342. Osvaldo Borsani (1911–1985)
***P40* chair with auto-adjusting armrests**, 1955
Made by Tecno (Italy)
Steel painted black, fabric
Vitra Design Museum, Weil am Rhein, Germany

OPPOSITE
343. Achille Castiglioni (1918–2002) and Pier Giacomo Castiglioni (1913–1968)
***Arco* lamp**, 1962
Made by Flos (Italy)
Marble, stainless steel, aluminum, electrical components

1959. Flos, founded in Merano to make indoor lights with a spray-on plastic called "cocoon," would begin producing designs by the Castiglioni brothers in Brescia in 1961, such as the very innovative *Arco* lamp (1962), destined to become a design icon (plate 343). With its long steel arm arcing up from a marble base, it provided a freestanding source of overhead light, unattached to the wall or ceiling.

Historic businesses were also overhauled: at the initiative of Osvaldo Borsani, an architect who received his degree in 1937, his family's Atelier Borsani in Varedo (outside Milan) became Tecno, a name well-suited to its research into new materials and its new range of modular products. The *D70* sofa (1954); the *P40* chair (1955), equipped with flexible armrests and an adjustable back (plate 342); and the articulated *L77* bed (1957), also adjustable, broke new ground. Arteluce[30] reopened its doors with a store-workshop, where Gino Sarfatti,[31] trained in naval and aeronautical engineering, continued the work he had begun in 1939, designing and producing more than four hundred lamps that were innovative not only in their materials and light sources but also in their production techniques and visual impact (plates 344–46).[32]

In the Marches—a region of central Italy that made a significant contribution to Italian design, conceiving and producing all sorts of items, from the smallest household articles to high-tonnage ships—the third generation of the Guzzini of Recanati was ready to set to work. Since 1912, Fratelli Guzzini (which had called on the design talents of Gio Ponti and especially Luigi Massoni) had offered items for the table and home, first in ox horn, then in Plexiglas (it is said to have salvaged the windows of airplanes stockpiled by the Azienda Rilievo Alienazione Residuati),[33] and finally in plastic. The innovative solution of two-tone melded plastic sheets dates to 1953, injection molding to 1955–60. But the Guzzini were not about to stop there: in 1963, they added a second company, iGuzzini Illuminazione, for which personalities such as Piero Castiglioni and Renzo Piano, as well as many foreign designers, would work over the years.

During those same years, Giuseppe Brion was a technician at Phonola, then at Radiomarelli, where his wife, (Ono) Rina Tomasin, one of the first women entrepreneurs, also worked. He laid the foundations for an important though short-lived venture: in their free time, Giuseppe and his wife assembled radios for third parties, and then began to produce their own electronics under the brand names Vega (after the star), Radio

LEFT
344. Gino Sarfatti (1912–1985)
***2097/30* ceiling lamp**, 1958
Made by Arteluce (Italy)
Chrome-plated steel
Musée National d'Art Moderne, Centre Georges Pompidou, Paris

BELOW
345. Gino Sarfatti (1912–1985)
***566* desk lamp**, 1956
Made by Arteluce (Italy)
Brass, lacquered metal, painted steel, electrical components
Musée National d'Art Moderne, Centre Georges Pompidou, Paris

OPPOSITE
346. Gino Sarfatti (1912–1985)
***534* table lamp**, 1951
Made by Arteluce (Italy)
Brass, lacquered metal, painted steel, electrical components
Galerie Kreo, Paris

Vega Television, and finally Brionvega. Initially the firm was very successful,[34] thanks to the work of several designers: Bonetto, Zanuso, Sapper, and the Castiglioni brothers. The Brions had a unique ability to grasp the potential of the different approaches these designers pursued: Rodolfo Bonetto's "technological" look, Zanuso and Sapper's minimalism avant la lettre, the irony of the Castiglioni brothers' Dadaist language (plates 347–50). All were allowed absolute freedom in their choice of materials—plastics, wood, or metal. Despite their different forms, many of the firm's products had at least one quality in common: they occupied a space halfway between the static and the portable.

That was true of Zanuso and Sapper's *TS 505* radio (1964); the Castiglioni brothers' *RR 126* stereo system (1966); Mario Bellini's *RR 130* hi-fi system (1970); and Sapper's *Sound Box* portable radio-recorder (1974). The objective was to create appliances that could be used throughout the house and sometimes even outdoors, on those first excursions that the Italians were coming to enjoy so much.[35] Zanuso's *Black ST 201* television (1969) had, like certain other appliances of its type, the look of an abstract object, anticipating minimalism. Zanuso and Sapper were always up-to-date, almost predicting the future at times, as attested by the *Grillo* (Siemens, 1965), the precursor of all the flip phones we use today, a "phone held in the palm of the hand" (plate 356). More than that, it is a piece of interior decor, both casual and suggestive.

MILAN AND THE SPREAD OF DESIGN

In the early 1950s, the Milan of industry and commerce, of business and culture, began to consolidate. The close relationship between art and industry, which dates back to the 1930s and is one of the city's characteristic features, was fueled by a dense network of artisanal workshops, specialized publications, and, above all, major events—not only the Triennale but also international showcases such as the Milan Trade Fair. During those years, a certain fervor surrounded the culture of "production," in the best sense of the word: in other words, the culture of industrial design, of the project.

In 1951, the Ninth Triennale—with staging by Luciano Baldessari, a large light installation by Lucio Fontana, and Antonia Campi's ceramic panels—welcomed two exhibits fundamental to the history of design. The *Form of the Useful* exhibit, whose name would later become a slogan, addressed the question of industrial design, with the notion of utility replacing that of mass production, while the *Home Salon* provided a first sampling of the furniture and utensils being produced. Ponti wanted the Triennale to be "open to all trends":

347. Achille Castiglioni (1918–2002) and Pier Giacomo Castiglioni (1913–1968)
***RR 126* stereo system**, 1966
Made by Brionvega (Italy)
Musée National d'Art Moderne, Centre Georges Pompidou, Paris

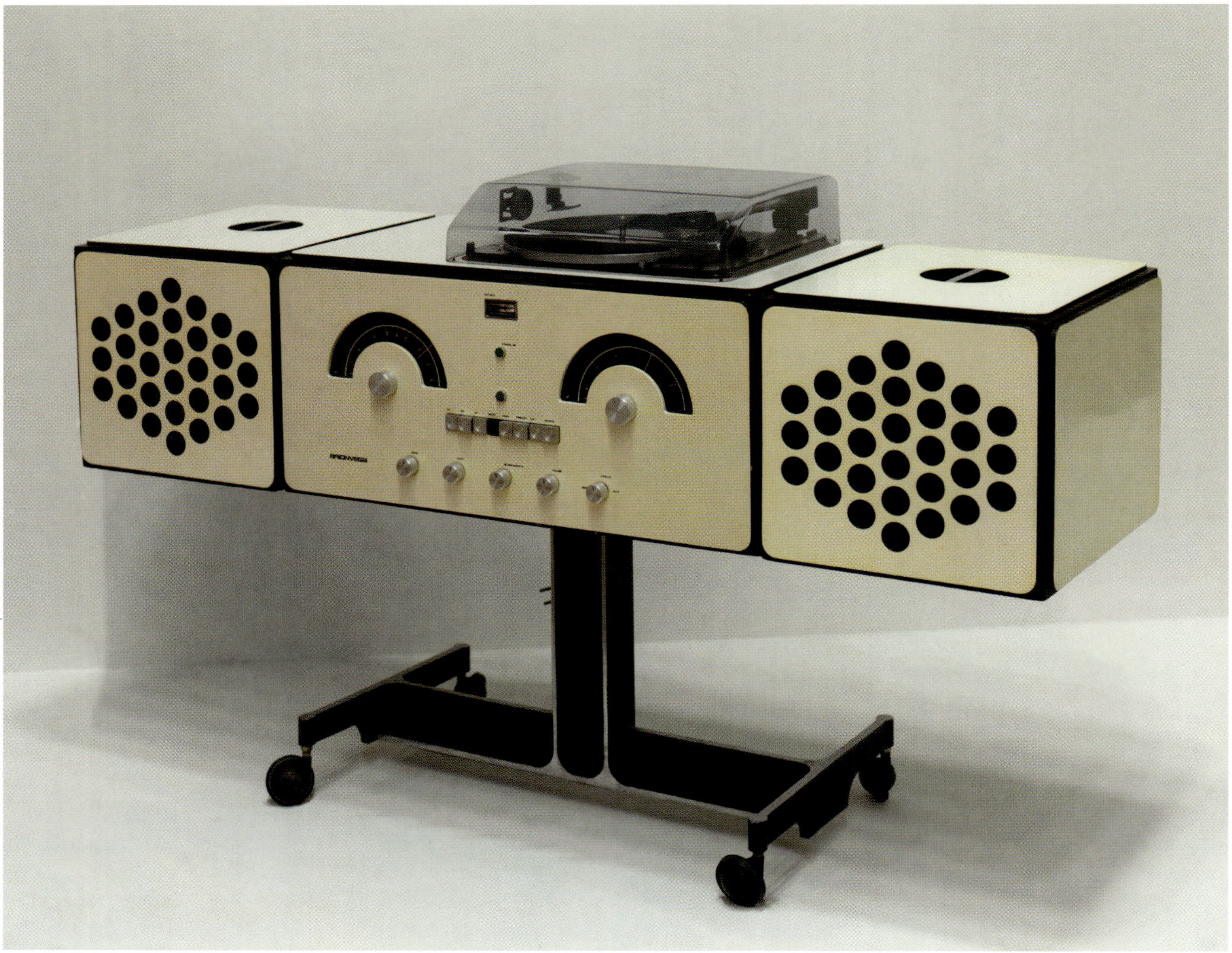

the different commissioners emphasized the modernist style, while Sweden and Finland displayed, in an elegant space, their famous furniture, sought out by all who wanted to modernize their way of life.

In the meantime, the famous department store La Rinascente had reopened its doors, with its own graphic arts and publicity department run by Max Huber, who sought not only to sell Italian taste but also to influence it. From 1952 on, the store held an influential annual exhibition called *The Aesthetics of the Product*, which gave rise to the idea of awarding a prize to the best industrial designers and their creations.

Similarly, the thirtieth Milan Trade Fair included *Art and the Industrial Aesthetic*, an exhibit that would return the following year, in 1953. Launched in 1920, the Milan Trade Fair displayed many remarkable experiments in style and materials, and connected architects, designers, and artists to industry. In 1946, it again became a springboard for industrial design. At that first postwar fair, the prototype of the *Candy* electric washing machine, Italy's first such device for the home, was on view (produced by the Officine Meccaniche Eden Fumaggali of Monza, newly reconverted to civilian use). Also displayed were machine tools, shoes, toys, electrical appliances, and equipment for processing plastics. The following year, the furniture manufacturers, like the automobile manufacturers, were granted an entire pavilion. In 1961, the Milan furniture fair, properly the Salone Internazionale del Mobile (organized by Cosmit, the Comitato per il Salone del Mobile Italiano), became completely autonomous, and in subsequent years it was held in September, replacing the traditional Trade Fair in April. That shift played a large role in drawing the public's attention to furniture made in Italy. The Milan furniture fair was an attentive witness to changing trends, first in the aisles of the Trade Fair and later in the new pavilions in Rho Pero. From the 1980s on, it would become the epicenter for a series of concurrent events, which increasingly mobilized the entire city and were often intended to broaden the official themes of the fair. Every year in April, therefore, Milan became "the city of design."

The review *Domus* (no. 269, 1952) had published a *Manifesto of Industrial Design* that—already!—detailed a few of the paradoxes of the Italian case, and in particular the official nonexistence in Italy of the profession of designer (a situation that persists today), even though the designer's work, for its part, had a concrete existence, and indeed a great aesthetic coherence. It was acknowledged that "the moment had come for industrial design, taste, aesthetics, and products for the home."

BELOW
348. Marco Zanuso (1916–2001)
***TS 502* radio**, 1964
Made by Brionvega (Italy)
Museum of Modern Art, New York

OVERLEAF
349. Richard Sapper (b. 1932) and Marco Zanuso (1916–2001)
***Doney 14* portable television set**, 1962
Made by Brionvega (Italy)
Centre National des Arts Plastiques/Fonds National d'Art Contemporain, Paris

PAGE 289
350. Richard Sapper (b. 1932) and Marco Zanuso (1916–2001)
***Algol 11* television set**, 1964
Made by Brionvega (Italy)
Musée National d'Art Moderne, Centre Georges Pompidou, Paris

BRIONVEGA

RIGHT
351. **View of the tenth Triennale di Milano**, 1954

OPPOSITE, LEFT
352. **View of the eleventh Triennale di Milano**, 1957

OPPOSITE, RIGHT
353. ***Comparative Exhibition of Glass and Steel* organized by Franco Albini and Jay Doblin at the twelfth Triennale di Milano**, 1960

It was in 1954 that industrial design truly gained the starring role. The tenth Triennale played host to the first annual conference of the International Congress of Industrial Design, as well as an accompanying exhibition (plate 357). At the conference, the position taken by Giulio Carlo Argan caused a tremendous stir. As Enzo Frateili wrote, "[Argan] compared design to urban planning, drawing a parallel between housing developments and the distribution of goods and their formal quality. He then identified the presence of two spatial aspects of the object: the space of function and the space of representation."[36] The purely cosmetic role of industrial design, as it was conceived in the U.S., was therefore rejected. Discussions centered on producing an industrial object in accordance with an artistic, even intelligent design. There were even predictions that such a path would not be at all difficult for the Italians to take. The philosopher Enzo Paci observed that "the designer is situated between art and society," that he "invents forms that have never been produced and that, in their organic unity, express new ways of life." The figure of the designer was thus becoming emancipated from that of the architect and the artist. At the close of the congress, a motion expressed the desire that the Triennale might encourage the creation of a serious school of industrial design. This motion was perhaps inspired by Max Bill, who had spoken at length of the Hochschule für Gestaltung (a sort of "New Bauhaus") he had founded in Ulm, where many Italians would go as teachers or students.

In displaying Scandinavian, English, and German creations (from the Ulm school and Braun), and in putting the spotlight on the Italians, the exhibition presented alongside the congress sought to demonstrate that the Bauhaus was outmoded. The 150 objects on view—all in production—provided a snapshot (though a somewhat arbitrary one) of postwar industry.

Stile Industria appeared during the same period. That extraordinary review, which began as a column in *Domus* and was edited by Alberto Rosselli, would, for nine years, be a primary venue for discussing all aspects of design, architecture, production, and style—even the ethics of the designer. It was also at that time (1954) that the first Compasso d'Oro was awarded by La Rinascente. Stemming from an idea shared by Ponti and Rosselli, this prize was intended to recognize manufacturers, artisans, and designers involved in research—both technical and formal—on industrial production, those who had succeeded in synthesizing a product's aesthetic quality and its functional characteristics. Originally, the award included all consumer goods within its purview. The jury was always composed of leading personalities: not only Ponti and Rosselli but also Zanuso, Rogers, Albini, and Pier Giacomo Castiglioni. Very quickly, the Compasso d'Oro became an occasion to take stock of the state of design (production, producers, designers, modes, and means). In that respect, it has the value of a historical document: the list of winners tells the story of Italian design and of the different designers who succeeded one another season after season. The fifth (1959) through eighth (1964) Compasso d'Oro awards were administered by the Associazione per il Disegno Industriale (ADI), which would subsequently organize the competition on its own. The first cycle of the Compasso d'Oro came to an end in 1970, but the prize would be reestablished in 1979 after lengthy debates.[37]

The ADI was founded in 1956 by a group of designers and enlightened manufacturers who wished to promote and institutionalize the still avant-garde

contributions of design. The ADI had its first headquarters at Ponti's studio on via Dezza in Milan. Sixty people attended the initial meeting, convened at the Museo delle Scienza e della Tecnologia,[38] a site that was chosen quite deliberately, since it was there that Ponti had imagined creating a section devoted specifically to industrial design. (He wanted design to be linked not to art but to technology.) Rossi was elected the first ADI president, while two designers—Enrico Peressutti, a member of the BBPR agency, and Albe Steiner—and two manufacturers, the Milanese Giulio Castelli and the Venetian Antonio Pellizzari, served on the steering committee.

From the start, the association's activities were quite varied. It organized congresses and exhibitions to publicize signature products in Italy and abroad, and hosted the congress of the International Council of Societies of Industrial Design (ICSID) in Venice in 1961, and again in Milan in 1983, when the theme was "From the Spoon to the City." This slogan, coined in 1952 by Ernesto Nathan Rogers, echoed a 1917 saying by the Turin Futurist Nicola Galante: "From the *palazzo* to the fork." It had the merit of encapsulating the designer's field of operation, and it affirmed the inseparability of form and function, which Rogers had adopted as a guiding principle of the Italian reconstruction.

The problems facing the field of design are reflected in the proceedings of the ADI over the years:[39] the need to protect the designer's professional status, to establish a school for training designers, and to create a "permanent display of products," in other words, to found a museum of design. That last point would come up often and would in a certain sense be resolved in the first decade of the twenty-first century.[40]

The eleventh Triennale (1957) followed the same line as the previous one: it too hosted an exhibition on industrial design, highlighting, in a deliberately didactic manner, both artisanal products and those manufactured by the most advanced technology (plate 352). To borrow Rosselli's words, "the hypothesis of a European culture of design now appears to be an imminent reality."

From the twelfth to the sixteenth Triennale (in 1982), industrial design would no longer be presented separately. Nonetheless, the twelfth Triennale (1960), whose theme was "Home and School," included an interesting comparative exhibition centered on materials: Albini and Jay Doblin selected products in glass and steel from the four corners of the world and exhibited them up and down the grand staircase (plate 353). The thirteenth Triennale, which adopted the guiding theme of "Free Time" and emphasized the concept of the overall environment, was held in 1964, which would go down in history as the year of Pop art. Not only the Triennale but also the Venice Biennale of that year would instill a Pop spirit in many everyday objects.

The fourteenth Triennale opened in May 1968, within a context of heightened criticism (and the protest movement did not spare the Triennale). Its theme, "The Masses," anticipated a debate that would soon be of great relevance. Giancarlo De Carlo, Rosselli, Steiner, Vigano, and Zanuso were on the council that set out the main lines of the event, which was divided into different sections ("Errors," "Information," "Perspectives"). A choice position was given to a group of young Italians—M. Platania, A. Barrese, S. Delfino, A. Grassi, G. Laminarca, A. Marangoni, and J. Gardella—who examined the ecological challenges presented by the lack of potable water, a subject little known at the time. Visitors reached the *Italian Expressions and Products* space

354. Vico Magistretti (1920–2006)
***Guione* floor lamp**, 1970
Made by Artemide (Italy)
Lacquered metal, electrical components
Museum of Modern Art, New York

though a tire tunnel, which had the effect of emphasizing the strangeness of that exhibit. Presented side by side were various proposals for a different living environment, such as the "programmable, flexible, and adjustable housing system" designed by Joe Colombo.

In the 1950s, Milan was also home to a few specialized showrooms that would become true reference points for design. In 1953, Fernando and Maddalena De Padova had opened a space on via Montenapoleone (later moved to corso Venezia), with the initial aim of raising public awareness about Scandinavian design, which was enjoying a certain success thanks to the Triennales. Then their space itself became a site of production, with the support of several designers, especially Vico Magistretti.[41] An architect and the son of an architect, Magistretti paid particular attention to the home and to interior design, a subject he approached as a designer not only of buildings but also of furniture and objects, many of which are now considered contemporary classics (plate 354). He designed the *Selene* chair (1967) and the archetypal *Eclisse* lamp (1967) for Artemide (plate 355); the *Maralunga* sofa (1973) for Cassina; and a pioneering textile bed, first in a long series, the *Nathalie* (1978) for Flou, which presented itself as a "bed stylist," the distributor of a single model.

In 1957, also in Milan, Arform opened at the initiative of Paolo Tilche, to market and later produce modern furniture and accessories. So too did Centro Fly, a store that impressed customers with its escalators, background music, and labels (which clearly displayed the names of the designers—a real innovation at the time). For the most part, Centro Fly showcased kitchenware: "A million new things for a million beautiful things."

The capital of Lombardy was, moveover, a center of specialty publishing, which contributed significantly to the development and diffusion of Italian industrial design, particularly furniture and decorative accessories. In 1989, Eugenio Battisti counted fifty-six magazines in Italy devoted to architecture, fourteen to urban planning, twenty-six to furniture, and eighteen to design in the narrow sense. Of all these titles dealing with *progetto* (design), forty-seven were published in Milan.

While under the direction of Ernesto Nathan Rogers (from 1954 on), *Casabella* would be less attentive to design. It was not until 1970–76, with Alessandro Mendini as editor, that it would become an organ for publicizing experiments in radical design.[42] *Domus*, which returned to Ponti's hands in 1949, regularly took an interest in furniture and decoration throughout the 1950s. The review, which Cesare Casati coordinated between 1961 and 1978, also organized the Eurodomus exhibitions, which took place in Genoa (May 1966), Turin (spring 1968 and May 1972), and Milan (May 1970, at the Palazzo dell'Arte). The aim of these events was to present experimental solutions for the home, by placing furniture within the contemporary world and bringing it face-to-face with products from other countries. New materials were also prominently featured.

In addition to the short but intense run of *Stile Industria*, already alluded to, a few other reviews are worth mentioning. *Interni* first appeared in 1954 as a "Magazine of Interior Decoration" and continues to be on the cutting edge of furniture and accessories. In the last few decades, it has also organized many satellite events during the Milan furniture fair. In 1961, Piera Peroni founded *Abitare* (initially titled *Casa Novità*), a review that has always been careful to address the relationship between human beings and their living environment, although it has undergone various changes in direction and editorial policy.[43] The first issue of *Casa Vogue*,

BELOW
355. Vico Magistretti (1920–2006)
***Eclisse* lamps**, 1967
Made by Artemide (Italy)
Lacquered aluminum, electrical components
Museum of Modern Art, New York

BOTTOM
356. Richard Sapper (b. 1932) and Marco Zanuso (1916–2001)
***Grillo* telephone**, 1965
Made by Sit-Siemens (Italy)
Museum of Modern Art, New York

RIGHT
357. Achille Castiglioni (1918–2002) and Pier Giacomo Castiglioni (1913–1968)
***Snoopy* lamp**, 1967
Made by Flos (Italy)
Enameled metal, marble
Museum of Modern Art, New York

OPPOSITE
358. Gae Aulenti (1927–2012)
***Sgarsul* rocking chair**, 1962
Made by Poltronova (Italy)
Lacquered wood, leather
Photograph by Carol-Marc Lavrillier, 1972
Musée National d'Art Moderne, Centre Georges Pompidou, Paris

initially conceived as a supplement to *Vogue*, appeared in 1968. That review, under the direction of Isa Vercelloni until 1992, showcased homes as they were lived in and paid constant attention to social phenomena. Also noteworthy are *Forme* ("of the useful" is implied), launched by Luigi Massoni in 1963, and *Formaluce*, edited by A. M. Prina from 1968 on and dedicated to lighting.

While these magazines are all brought out by conventional publishers, manufacturers have also created their own periodicals. For example, from 1964 to 1991 the Busnelli Industrial Group published the semiannual *Caleidoscopio*, whose columns welcomed lively discussions of the culture of contemporary design. In 1967, eight businesses working in the areas of furniture and lighting (Arflex, Artemide, Bernini, Boffi, Cassina, Flos, Icf, and Tecno) partnered to publish a joint review, *Ottagono*.[44] The state-owned giant Finmeccanica published *Civiltà delle Macchine*; Olivetti, *Zodiac*; and Pirelli-Linoleum, *Edilizia moderna*. Issue 85 of *Edilizia moderna*, which appeared in 1964 and was devoted to industrial design, won the Compasso d'Oro in 1967. Finally, from 1956 to 1969, *Qualità*—the publication of Kartell, launched by Giulio Castelli and designed by Michele Provinciali—explored, in eleven installments, the question of synthetic materials in the culture of design.

THE TURN OF THE 1960S

In the 1960s, some designers yielded to the temptations of the Neoliberty current, whose manifesto was the exhibition *New Models for Italian Furniture*, held in Milan in March 1960. They revisited, in a critical way, an experiment that had been very successful in Italy at the beginning of the twentieth century, namely, the adoption of American-inspired "styling." Among the major representatives of Neoliberty were Vittorio Gregotti, Ludovico Meneghetti, and Giotto Stoppino, who together created the *Cavour* chair (1959, SIM-Novara, reissued by Frau); Gae Aulenti, who designed the *Sgarsul* rocking chair in curved beechwood (1962, Poltronova; plate 358); and the Castiglioni brothers, who conceived the *Sanluca* chair (1960, Gavina), with a plastic frame entirely covered in leather and fabric, as well as the *Spinamatic* beer tap.

Achille[45] and Pier Giacomo Castiglioni can be credited with a particular approach to design, with a clearly recognizable style: the ready-made, an obvious reference to Dadaism. The one constant in all their work is a hint of lightheartedness, combined with a form of irony that seeks to demystify and minimize the designer's role. The Castiglioni brothers had a unique capacity to assemble humble objects (or parts of objects) and give them a new expressiveness, an autonomy, and a true dignity (plate 357). For example, the *Mezzador* (Sharecropper) stool is composed of a tractor seat mounted on a steel blade, with a beechwood base (plate 366). (The third Castiglioni brother, Livio, would work independently in the areas of lighting and sound, becoming the leading Italian lighting designer.)

The various objects labeled "Pop design" are distinguished by their whimsical shapes and their inventive use of materials (including unusual ones, especially

tires), and often by their ironic bombast and large dimensions. The *Blow* chair, designed for Zanotta by DDL (De Pas, D'Urbino, and Lomazzi, 1967)[46] is *the* symbol of Pop design (plate 359): it nearly realizes the utopian objective of Marcel Breuer, who dreamed of being able to sit on "a transparent and elastic column of air," as Rainer Wick remarked in the pages of *Kunstforum International* in 1968. Made of molded polyurethane foam, DDL's *Joe* chair (distributed by Poltronova in 1970), whose name alludes to baseball legend Joe DiMaggio, can also be called Pop (plate 361). So too can the *Sacco*, the free-form, self-adjusting beanbag chair by Gatti, Teodoro, and Paolini (1968, Zanotta), filled with thousands of polystyrene pellets (plate 360). The Pop creations of the Gufram Company[47] (founded 1968) include the *Pratone* chair by Ceretti, Derossi, and Rosso (1971), made up of tall blades of foam-rubber "grass" that create the effect of an artificial "lawn" to stretch out on (plate 362), and Drocco and Mello's *Cactus* coatrack (1970). These objects, which at first had little success, achieved cult status during the 1980s, with the vogue for retro furniture.

In the 1960s, other experiments readily attested to the relationship between art and design. Dino Gavina,[48] for example, commissioned artists like Man Ray to design everyday objects. A rule-breaking pioneer, Gavina began to produce furniture at a store-workshop in Bologna in 1953 in collaboration with Mollino, the Castiglioni brothers, and Scarpa. His *Bastiano*, the archetype for a new kind of a sofa, has a very simple structure: it is a basket filled with cushions. He also produced furniture designed by Marcel Breuer, whom he met in New York in 1962.

The furniture produced by Poltronova (founded in Tuscany in 1956 by Sergio Cammilli, after a decisive meeting with Ettore Sottsass)[49] is also quite poetic. An architect and designer who graduated from the University of Turin but was active in Milan from 1947 on, Sottsass took over the computer design department of Olivetti in 1957. In that capacity, he designed the *Elea 9000*, the first large computer produced in Italy (1960). He advocated a new kind of design that went beyond functionalism, and constantly endeavored to find more sensuous ways to use form and color. Beginning in the early 1970s, he would play a prominent role in the adventure of radical design. He was also behind a

OPPOSITE
359. Jonathan De Pas (1932–1991), Donato D'Urbino (b. 1935), and Paolo Lomazzi (b. 1936)
***Blow* armchair**, 1967
Made by Zanotta (Italy)
Inflatable PVC
Museum of Modern Art, New York

BELOW, LEFT
360. Piero Gatti (b. 1940), Francesco Teodoro (1939–2005), and Cesare Paolini (1937–1983)
***Sacco* chair**, 1968
Made by Zanotta (Italy)
PVC-coated fabric, polystyrene pellets
Musée des Arts Décoratifs, Paris

BELOW, RIGHT
361. Jonathan De Pas (1932–1991), Donato D'Urbino (b. 1935), and Paolo Lomazzi (b. 1936)
***Joe* chair**, 1970
Made by Poltronova (Italy)
Polyurethane foam, leather
Museum of Modern Art, New York

BOTTOM
362. Giorgio Ceretti (b. 1932), Pietro Derossi (b. 1933), and Riccardo Rosso
***Pratone* chair**, 1971
Made by Gufram (Italy)
Polyurethane foam
Vitra Design Museum, Weil am Rhein, Germany

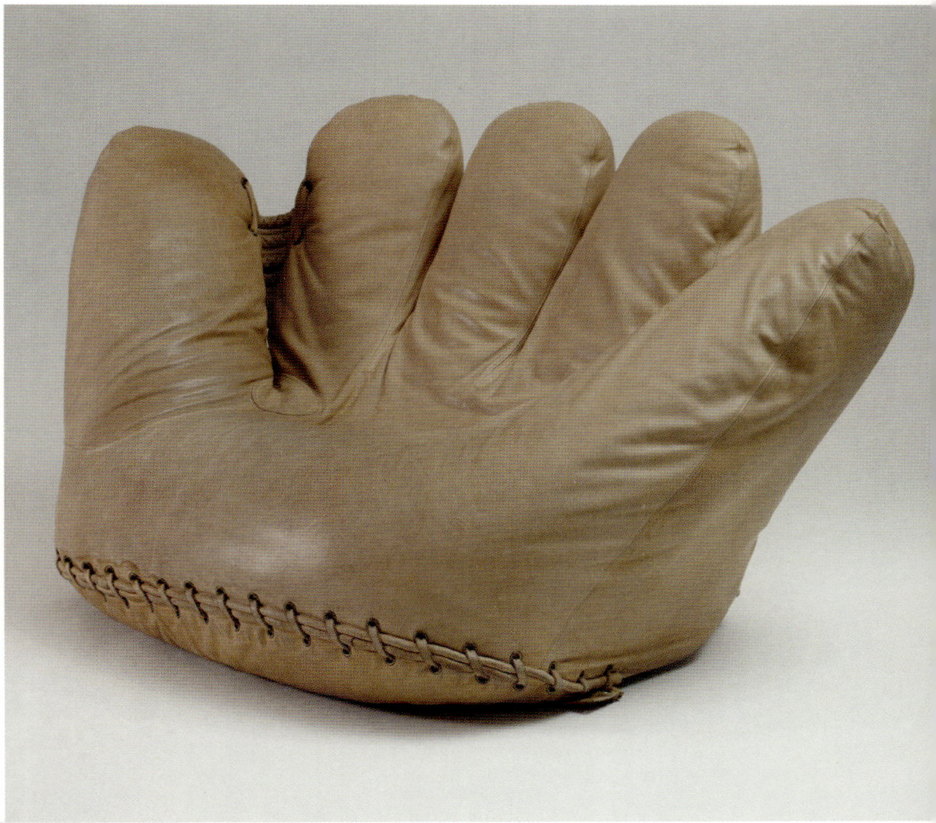

BELOW
363. Joe Colombo (1930–1971)
Sketch of the *Tube* chair, 1969
Pencil on paper
Musée des Arts Décoratifs, Paris

OPPOSITE, TOP
364. Joe Colombo (1930–1971)
***Tube* chair**, 1970
Made by Flexform (Italy)
PVC, polyurethane, metal, burlap
Musée des Arts Décoratifs, Paris

OPPOSITE, BOTTOM LEFT
365. Joe Colombo (1930–1971)
***Boby* storage unit on castors**, 1969
Made by Bieffeplast (Italy)
ABS plastic
Museum of Modern Art, New York

OPPOSITE, BOTTOM RIGHT
366. Achille Castiglioni (1918–2002)
***Mezzadro* stool**, 1970 (prototype designed for the eleventh Triennale di Milano, 1957)
Made by Zanotta (Italy)
Tractor seat mounted on steel blade and beechwood base, secured by a bicycle wing nut
Museum of Modern Art, New York

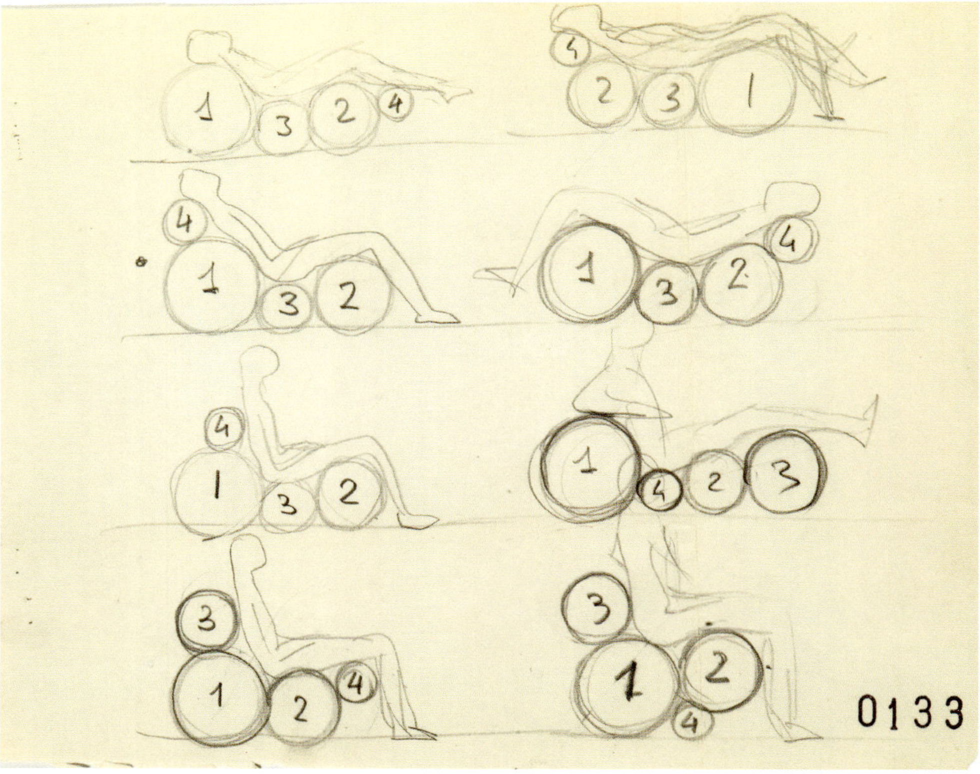

cutting-edge housing project, created for *The New Domestic Landscape: Achievements and Problems of Italian Design*, a major exhibition organized by Emilio Ambasz for the Museum of Modern Art in New York. This exhibition, which surveyed the different avant-garde currents of the 1960s, was an important watershed in the culture of Italian design. Sottsass championed plastic in the Environments section, with movable cabinets and wheeled containers (designed for Kartell) that could be used to mark off the different areas of the house.

For a few years, in fact, domestic interiors had been undergoing a transformation, which was reflected in the objects they contained. The first room to change was the kitchen. The tiny space reserved for food preparation expanded to become a multipurpose environment, the true heart of the home, as it had been in earlier times. Its furnishings also became more diverse, adapting themselves to the new dimensions and then the new functions of the kitchen. A gradual shift occurred from the modular kitchen to a new, more comfortable arrangement, and from craftsmanship to standardized industrial production. First there were modules made of metal, laminates, and wood, such as the *T12* (Boffi), which Gian Casé displayed at the Triennale of 1960. Then there were the audacious and visionary (for the time) "monoblocs," inspired by the principles of the total living environment. For the most part, these units remained at the prototype stage. The *Mini Kitchen* on castors (Boffi, 1963–66, still in production) initially met with little success. And yet, in a space measuring only eighteen cubic feet (0.5 m^3), it contained all the indispensable elements. It was designed by Joe Colombo.[50]

A painter by training, Colombo created many products during his short career, in addition to his avant-garde plans for housing modules. New materials (fiberglass, ABS, polyethylene, PVC, methacrylate) prompted him to imagine unusual combinations and to introduce new types of object (plates 363–65). The *Acrilica* lamp, for example, had a C-shaped Plexiglas diffuser and a metal base housing a little fluorescent tube (1962). The *Smoke* drinking glass abandoned the traditional symmetry (1964), while the *Universale* chair was made of a single piece of molded plastic (Kartell, 1965). The *Alogena* lamp became an archetype: it was the first to use a halogen bulb (1970, O-Luce). But in particular, Colombo thought that "all the objects needed in a house ought to be integrated into the usable spaces . . . and as such ought to be called 'equipment'"—like the *Kitchen-Box* unit, both kitchen and dining room; the *Central-Living* block; the *Night-Cell*, which served as a bedroom, wardrobe, and bathroom (for Bayer's *Visiona* exhibit); and the *Cabriolet*, a "bed–living system" (1970; plate 367).

The bathroom expanded as well, increasingly becoming a place devoted to the care of the body. Fixtures were still made from ceramic, but they were no longer limited to white: Antonia Campi,[51] first for SCI, then for Richard Ginori–Pozzi Ginori, introduced unexpected colors and color combinations, more sculptural forms, and better ergonomics. A few decades later, the unit and monobloc solutions imagined by Giulio Minoletti

367. Joe Colombo (1930–1971)
Experimental interior *Visiona 1* for Bayer AG, exhibited at the 1969 Cologne furniture fair
Joe Colombo Archives, Milan

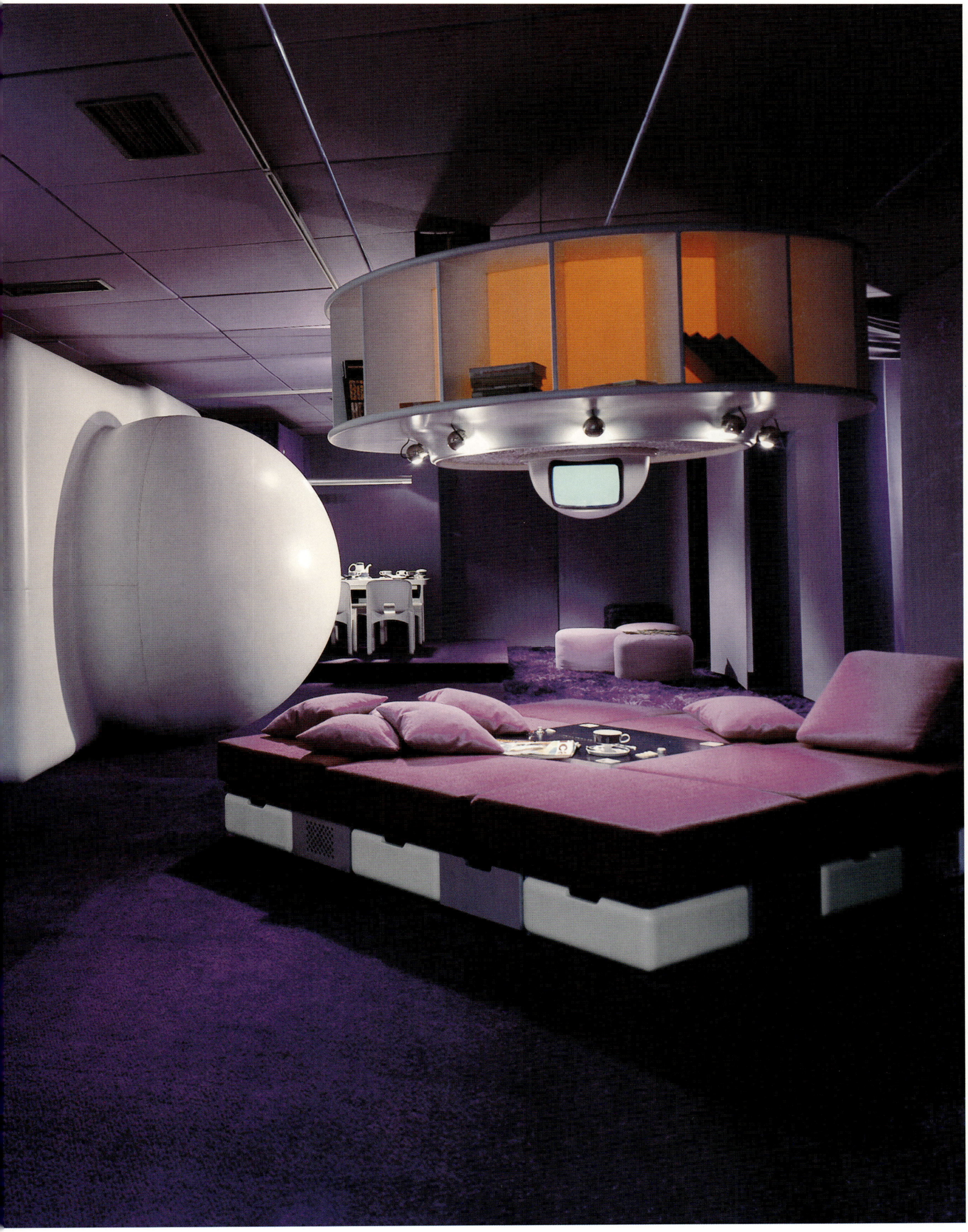

368–70. **Gaetano Pesce's *UP5* chair expands after being removed from its vacuum-sealed packaging**

and Rosselli, among others, had gone out of style. Working for Teuco, Giovanna Talocci—initially with the collaboration of Fabio Lenci—chose to use plastics, which, thanks to their malleability, allowed for a multitude of possibilities.

Ideas proliferated, and some companies opened research centers. At C&B, between 1965 and 1973, Cesare Cassina and Piero Busnelli offered designers the opportunity to experiment with cutting-edge technologies. Gaetano Pesce used vacuum compression to create the *Up* series of lounge chairs in polyurethane foam (1969; plate 371), which rapidly became icons of Italian and international industrial design. The form of the *UP5* chair was inspired by votive statues of fertility goddesses, and is virtually a manifesto on women's condition (plates 368–70). And though Pesce still preferred plastics, he put them to alternative uses: in the *Dalila* chair (Cassina, 1980), the *Tramonto a New York* sofa (Cassina, 1980; plate 374), the *Feltri* chair (Cassina, 1987; plate 373), and the *Vesuvio* coffeemaker (Zani & Zani, 1993), for example.

A designer, architect, and artist, Pesce formulated his *First Manifesto for a Flexible Architecture*[52] at a conference in Jyväskylä (Finland) in 1965. In Padua in 1959, he had joined Gruppo N with nine other artists. This was one of many groups that, under the label "programmed art," then "kinetic art" and "gestalt art," sought to exploit all the possibilities of a work (mechanical, electro-

371. Gaetano Pesce (b. 1939)
***UP5* chair, *UP6* ottoman, and four *UP3* armchairs**, 1969
Made by C&B (Italy)
Polyurethane foam, jersey
Courtesy Gaetano Pesce

magnetic, and so on) by "programming" its movement in an almost scientific fashion. After 1961, this international current in art came to be known as New Tendencies, and its adherents included not only individual artists and designers, such as Munari, Enzo Mari,[53] and Dadamaino, but also groups that theorized the anonymity of products and the disappearance of the artist's signature. After Gruppo N in Padua, Gruppo T formed in Milan; one of its members was Gabriele De Vecchi, who would become a noted designer of silver.

In Verrucchio (Emilia-Romagna), Argan surveyed the international situation, presented the problem of art's social function, and called for reducing the gap between artistic production and collective use. In the years that followed, the Milanese Gruppo Mid (Antonio Barrese, Alfonso Grassi, Gianfranco Laminarca, and Alberto Marangoni) also tackled design (plate 378). In 1966, the question of the multiple in art and, even more, new investigations in the field of visual perception, led the group in the direction of large-scale industrial production.[54] They created, among other things, the Bialcol bottle (1975, still in production) and the *Black* razor for Bic (1985).

Enzo Mari took the same route from art to design. He created many products for Driade, a company that had its roots in the prefab sector, distributing Antonia Astori's *Programma 1* modules from 1968 on. In 1970, Mari, the "theorist of design," published *The Function of Esthetic Research*, which announced the revalorization of the figure of the craftsman (a theme he would take up again in 1981, for his exhibition *Where is the Craftsman?*). For Mari, the craftsman is one who invents, possesses a trade, and is able to adapt his own work to fit the needs of the moment.

OPPOSITE
372. **Advertisement for the *Moloch* lamp by Gaetano Pesce (1971), made by Bracciodiferro (Italy)**
Courtesy of Gaetano Pesce

LEFT
373. Gaetano Pesce (b. 1939)
***Feltri* chair**, 1987
Made by Cassina (Italy)
Wool felt, epoxy resin, hemp, polyester wadding, quilting
Musée d'Art Moderne, Saint-Étienne Métropole

BELOW
374. Gaetano Pesce (b. 1939)
***Tramonto a New York* sofa**, 1980
Made by Cassina (Italy)
Laminated wood, polyurethane foam, polyester wadding, fabric
Musée d'Art Moderne, Saint-Étienne Métropole

ABOVE
375. Bruno Munari (1907–1998)
***Abitacolo* bed and play environment**, 1970
Made by Robots (now Rexite) (Italy)

OPPOSITE, TOP
376. Marco Zanuso (1916–2001) and Richard Sapper (b. 1932)
***4999* stackable children's chairs**, 1964
Made by Kartell (Italy)
Injection-molded polyethylene

OPPOSITE, BOTTOM LEFT
377. Pio Manzù (1939–1969)
***Cronotime* table clock**, 1966
Made by Riz Italoro (Alessi since 2010) (Italy)

OPPOSITE, BOTTOM RIGHT
378. Gruppo Mid
***Girondella* (kinetic object)**, 1965
Made by Danese (Italy)
Plastic, polystyrene, corundum granules
Museum of Modern Art, New York

12
9
3
6

BELOW
379. Superstudio
***Passiflora* lamp**, 1966
Made by Poltronova (Italy)
Perspex
Musée National d'Art Moderne, Centre Georges Pompidou, Paris

OPPOSITE
380. Superstudio
***Bazaar* sofa**, 1969
Made by Giovannetti (Italy)
Polyester, polyurethane, fabric
Musée National d'Art Moderne, Centre Georges Pompidou, Paris

THE 1970S AND '80S: EXPERIMENTS AND ECLECTICISM

The 1960s, a time of dogmatism as much as irreverence, witnessed the coexistence of design, architecture, and fashion, and laid the foundations for the eclecticism of the following decade. Elsewhere in Europe and across the Atlantic, the radical experiment found its guiding principles in the American writers of the Beat Generation (from Jack Kerouac to Allen Ginsberg), in a militancy inspired by Marxism, and in the hippies. In Italy, that movement was characterized by a greater attention to content (rather than form) and to ideology (rather than image). Tomás Maldonado defined the Italian version of the radical experiment as a "romantic aestheticization of the political act of protest." The task assigned to objects (and to design) was to unsettle the user and raise consciousness. The movement, active between 1968 and 1972, attracted several groups of designers. Most were from Florence, either because utopianism (as an instrument of critical analysis) looked like the most concrete path in that city affected by underdevelopment or because two major representatives of organic architecture, Leonardo Ricci[55] and Leonardo Savioli,[56] were teaching at the time in the Department of Architecture at the University of Florence. Ricci argued that "form is a consequence of the life potential inherent in the object that is preparing to come into being."

Archizoom was formed in 1966 by Andrea Branzi, Gilberto Corretti, Paolo Deganello, and Massimo Morozzi. This group, which would dissolve in Milan in 1973,[57] worked primarily in the field of architecture but also held exhibitions, conceived installations, and created various products, such as the *Superonda*, a modular polyurethane sofa (Poltronova, 1966), and the very innovative *A&O* series of chairs, designed in 1973. The *A&O*, produced by Cassina[58] in 1975, after the group had dissolved, represents a new kind of object, composed of apparently cheap materials that nevertheless bear a strong expressive charge.

The year 1966 also saw the advent of Superstudio, whose members, Adolfo Natalini, Cristiano Toraldo di Francia, Roberto and Alessandro Magris, Giampiero Frassinelli, and Alessandro Poli, engaged in theoretical research around the *progetto* in architecture and design. Superstudio, which disbanded in 1978, designed the *Sofo* modular chairs (Poltronova, 1966) and the *Quaderna* table (Zanotta, 1971), among other pieces (plates 379–80). Gruppo UFO entered the arena in 1967, in the wake of student protests in the Department of Architecture at the University of Florence. It continued until 1978, coordinated by Lapo Binazzi,[59] who sought

BELOW
381. Alessandro Mendini (b. 1931)
Terra chairs, 1974
Made by Bracciodiferro (Italy)
Plexiglas filled with earth and cork
Courtesy Fragile, Milan-Paris

OPPOSITE
382. Giancarlo Piretti (b. 1940)
Plia chairs, 1968
Made by Castelli (Italy)
Aluminum frame, methacrylate, rubber
Museum of Modern Art, New York

to make architecture "spectacular"—to turn it into an event, an urban and environmental guerrilla action.

Another Florentine group was active in the radical movement from 1969 on: Gruppo 9999,[60] which engaged in photomontage and graphic art, video, film, and happenings. Turin was not to be outdone: there, in 1971, Giorgio Ceretti, Pietro Derossi, Carlo Gianmarco, Riccardo Rosso, and Maurizio Vogliazzo created Gruppo Strum (an abbreviation of "For an Instrumental Architecture"). Their activities took on an educational, sociopolitical, conceptual, and cultural character, and they particularly distinguished themselves in the radical movement with their seminars and contributions to critical theory. Among their creations in the field of design, in addition to the aforementioned *Pratone* (Gufram, 1971), is the *Torneraj* chair (1972).

Most of these groups participated in the famous New York exhibition *The New Domestic Landscape* (1972). In addition, Archizoom, Superstudio, Gruppo 9999, UFO, Ziggurat,[61] Remo Buti, Riccardo Dalisi, Ugo La Pietra, Gaetano Pesce, Gianni Pettena, and Ettore Sottsass joined a cooperative that was founded in January 1973 by the editorial staff of *Casabella*. This anti-school of design, called Global Tools, a "free school for individual creativity," was, however, never really operational.

From 1970 to 1976, *Casabella* was edited by Alessandro Mendini, an architect, designer, and theorist, and a partner in the Nizzoli atelier. He reenergized the review by linking it to the new tendencies, before going on to create a new magazine, *Modo*, in 1977. Subtitled "A Journal of the Culture of Design," *Modo* opened its columns to everything related to the world of design and manufacturing. It also took an interest in the social phenomena that were gradually transforming industrial production into commodification, which resulted in kitsch. And because this new, resolutely "trendy" magazine carried not only news but criticism, it is an indispensable tool for studying its era. *Modo* continued until 2006 under the editorship of Franco Raggi, Andrea Branzi, Cristina Morozzi, and Almerico de Angelis. Mendini edited *Domus* from 1980 to 1985 and continued to develop his philosophy and methodology through exhibitions and conferences. A "cultural provocateur," he considered every project an essential tool and believed that art and design were closely linked in a permanent dialogue—one that, in Mendini's case, was founded on entertainment and utopianism.

Mendini also collaborated with Pesce on the Bracciodiferro project for Cassina. In 1971, the company began to test the market by distributing several limited editions to the gallery circuit. Assembled by Aldo Cichero into a collection that would be manufactured in Genoa, this experimental furniture perfectly embodied the spirit of the times and the provocations of the antiestablishment designers. For Bracciodiferro, Pesce designed the *Moloch* lamp (1971), which he hoped would be "useful for large spaces." A true icon of design, it is a blown-up replica of the desk lamp used by so many designers (plate 372). He also created the various versions of the *Golgotha* chair, made of natural resin, as well as the table and desk of the same name; the *Guanto* ("Glove") coatrack; and the *Pugno* ("Fist") chair. Mendini's designs for Bracciodiferro included the *Terra* chair in Plexiglas, cork, and earth (plate 381), the *Voragine*

table in aluminum and brass, the cast-bronze *Letargo* and *Senza Luce* lamps, and the *Monumentino da Casa*, a ziggurat chair in charred or lacquered wood. These pieces were exhibited in Berlin in 1975, next to furniture designed by masters of rationalism, to highlight the dialectical and ideological relationship between these two approaches. This first experiment with one-offs and limited editions led some designers to consider standardization obsolete and to advance the hypothesis that, with technology, it was easy make unique pieces.

Gavina,[62] completely at odds with the reigning ambience, launched a new collection, Ultramobile (1971), with the aim of introducing functional works of art into interior decoration. To that end, surrealist objects were adapted for daily use. Result: the *Magritta* armchair (1971), half upside-down bowler hat, half green apple, a tribute by Sebastian Matta. During that time, Cammilli also dreamed up "inhabitable furniture" for Poltronova, and Mario Ceroli designed his "livable sculptures," the *Furniture of the Valley*.

The collections offered by Alchimia since 1976 are better known and have met with more critical success, especially in the press. Such was the case for the *Bauhaus I* collection, launched in 1979, and for *Bauhaus II*, introduced the following year. In Milan, Alessandro and Adriana Guerriero founded the Group for the Design of Images for the Twentieth Century to produce nonindustrial furniture. This group, which planned to hold exhibitions, initially included Sottsass, Branzi, and Mendini, but ultimately came to represent Mendini's offerings almost exclusively: the *Kandinsky* sofa (1980; plate 383), the *Proust* armchair (1980, with the collaboration of Prospero Rasulo; plate 384), and the *Mobile Infinito* ("Infinite Furniture") series of 1981. The proposed transgressions were almost a return to the grammar of Futurism. The group's products, marking the transition between the radical and the neo-modern, also included clothing, decor, household items, and books, as well as seminars and performances.

The Memphis group, too, orchestrated a "cultural operation": the International Movement for the New Applied Arts, officially launched in September 1981—in the midst of the Milan furniture fair—with a provocative exhibition in a showroom on corso Europa.

OPPOSITE
383. Alessandro Mendini (b. 1931)
***Kandinsky* couch**, 1980
Made by Studio Alchimia (Italy)
Wood, fabric painted in acrylic
Musée National d'Art Moderne, Centre Georges Pompidou, Paris

BELOW
384. Alessandro Mendini (b. 1931) and Prospero Rasulo (b. 1953)
***Proust* armchair**, 1980
Made by Studio Alchimia (Italy)
Wood, fabric painted in acrylic
Centre National des Arts Plastiques/Fonds National d'Art Contemporain, Paris, on loan to the Musée des Arts Décoratifs, Paris

BELOW
385–87. **Cover and interior pages of the *Memphis Furniture Milano* catalog**, 1981

OPPOSITE, TOP LEFT
388. Ettore Sottsass (1917–2007)
***Casablanca* sideboard**, 1981
Made by Memphis (Italy)
Wood, plastic laminate
Victoria and Albert Museum, London

OPPOSITE, TOP RIGHT
389. Michele De Lucchi (b. 1951)
***Kristall* table**, 1981
Made by Memphis (Italy)
Plastic laminate, lacquered wood, metal
Private collection

BOTTOM
390. Michele De Lucchi (b. 1951)
***First* chair**, 1983
Made by Memphis (Italy)
Wood, tubular steel, rubber
Vitra Design Museum, Weil am Rhein, Germany

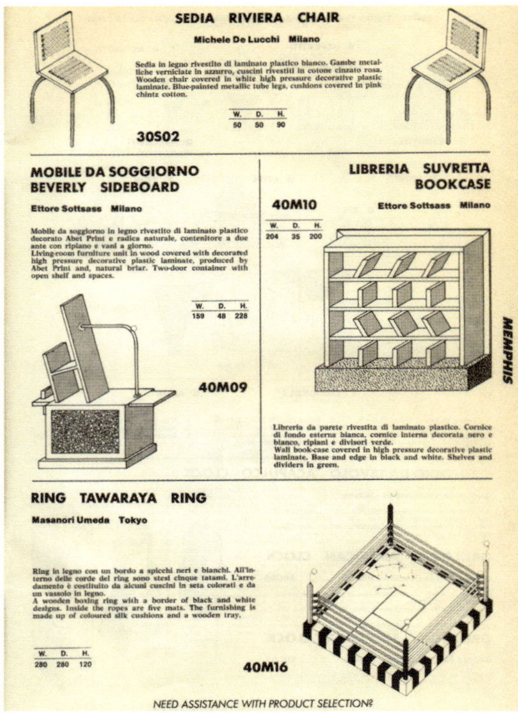

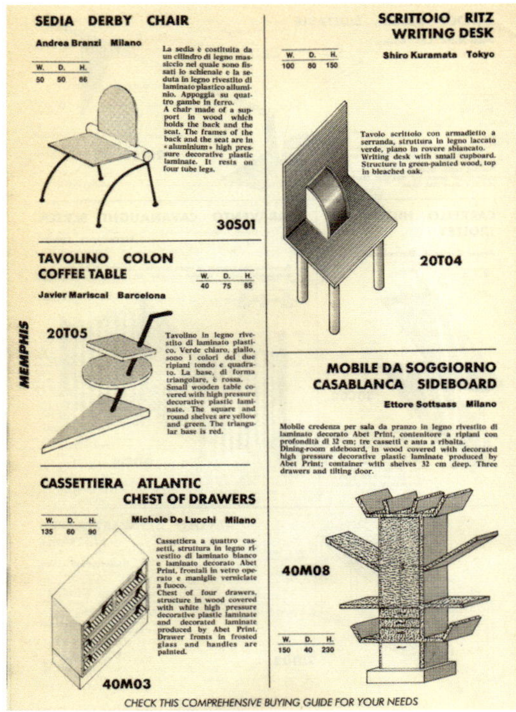

That first exhibition was conceived by Ettore Sottsass, Barbara Radice,[63] and Michele De Lucchi and realized with the assistance of a handful of young architects. It featured intentionally nonfunctional objects that quoted historical idioms, drawing inspiration from the pioneers of the modern movement in particular: Sottsass's *Carlton* bookcase (1981), itself an icon of design, is one example. Very quickly, the explosion of colors and the light-heartedness of the pieces, all one of a kind, found a host of imitators (plates 385–92).

The experiment came to an end in 1987. Two years later, Memphis launched Meta Memphis, which combined art and design to produce the *Ad Usum Dimorae* line of furniture and accessories, conceived primarily by artists.[64] Then, in 1991, it founded Memphis Extra, which offered glass objects. Sottsass and De Lucchi always chose a joyful design for the limited editions and a rigorous functionality for the objects to be mass-produced. And not by chance, perhaps: both had begun their careers at Olivetti, with the pioneering Nizzoli. In 1990, De Lucchi would begin to produce his own work, which he divided among different ateliers, depending on the materials used. Ultimately, he created Produzione Privata, "which he saw as an oasis of freedom—the possibility of realizing projects without taking the market into account."

It was also through the use of distinctive materials (plastic laminate, for example) that the image of contemporary furniture underwent a transformation. Abet Laminati was founded in Bra (Piedmont) in 1946 to produce synthetic finishes, "the first not to be taken from organic plants but developed by man," as Ponti boasted. That business supported the Guerrieros' group and Memphis in their endeavor to create new surfaces and decors.

Alberto Alessi was also interested in limited editions, and in 1971 he established a research center, Arts and Crafts, whose efforts would be focused on design, rather than marketing. His family's business, which had been founded in Omegna, on Lake Orta, in 1921, originally specialized in the production of coffeepots and trays in nickel-plated brass and nickel silver.[65] After he joined the firm, Alessi would revolutionize the domestic landscape with products concerned as much with form as with function. With the playful objects created by the Castiglioni brothers, Sapper, Sottsass, and then younger designers such as Stefano Giovannoni, the Alessi company now worked on two fronts:

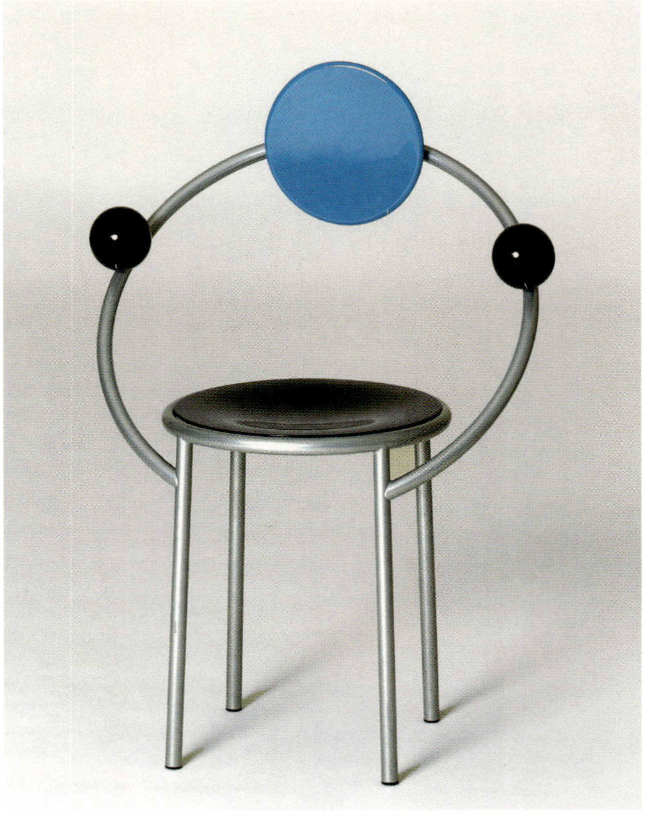

OPPOSITE
391. Ettore Sottsass (1917–2007)
***Carlton* bookcase**, 1981
Made by Memphis (Italy)
Wood, plastic laminate
Metropolitan Museum of Art, New York

BELOW
392. Ettore Sottsass (1917–2007)
***Attide* cup**, 1986
Made by Memphis (Italy)
Blown and cut glass, metal fasteners
Musée des Arts Décoratifs, Paris

LEFT
393. Richard Sapper (b. 1932)
***9090* coffeepot**, 1978
Made by Alessi (Italy)
Stainless steel, handle in steel with black oxide finish
Museum of Modern Art, New York

BELOW
394. Luigi Caccia Dominioni (b. 1913), with Achille Castiglioni (1918–2002) and Pier Giacomo Castiglioni (1913–1968)
***Dry* cutlery**, 1938
Made by Ditta Spoggi (Italy)
Silver
Museum of Modern Art, New York

OPPOSITE
395. Achille Castiglioni (1918–2002) and Pier Giacomo Castiglioni (1913–1968)
Condiment set, 1980–84
Made by Alessi (Italy)
Glass, stainless steel
Victoria and Albert Museum, London

mass-produced but refined objects made of plastics, and experimental collections, the result of research into traditional forms such as the coffeepot.[66] Riccardo Dalisi's efforts in this category (1987) were rewarded with a Compasso d'Oro, as was Sapper's *9090* coffeepot (1978; plate 393) and the *Dry* cutlery by the Castiglioni brothers and Dominioni (plate 394). The Officina Alessi has produced experimental objects and reissues since 1983, whereas the A di Alessi brand offers more democratic, "poppier" products in a more accessible price range.

In the early 1980s, a broad debate got under way about "postmodernism," a term that has enjoyed great success, thanks to the writings of Jean-François Lyotard and Gianni Vattimo. The "crisis of reason" called into question all the models of thought on which the modern age had been founded. In the field of design, this corresponded to a challenge to modernism and the international style, which were both accused of neglecting the formal values of objects and their historical meanings, privileging only a functional aesthetic. Greater emphasis was now placed on the expressiveness of the image. The styles of the past were quoted and combined with one another, and greater importance was given to tradition and local memory.

In 1985, the term "neo-eclecticism" also came into use. The magazine *AreA*, published since 1981, became the mouthpiece for that current in its early days. It gave pride of place to classic furniture and accessories, and adapted for its own use the principles of eclecticism, which at the turn of the twentieth century had extolled the mixing of historical styles. *AreA* did embrace certain theories of postmodernism, especially the view that the aesthetic function of the object ought to prevail over its practical use, and that designers should draw from the stylistic repertoire of the past, from classicism to romanticism. Craftsmanship and one-of-a-kind pieces were encouraged, and designs that were stimulating at an emotional level began to be produced once more.

Certain boundary breakers, practitioners of both art and design, like Ugo La Pietra, Nando Vigo, and Prospero Rasulo, paid particular attention to the decorative elements of the object and developed the field of "artistic design." Others invented the concept of "ethnic design," that is, design that remained in contact with the local culture. The term was coined by the same La Pietra, a designer, architect, theorist, and researcher

BELOW, LEFT
396. Mario Bellini (b. 1935)
***Cupola* soup tureen**, 1985
Made by Rosenthal (Germany)
Enameled porcelain
Musée des Arts Décoratifs, Paris

BELOW, RIGHT
397. Mario Bellini (b. 1935)
***Divisumma 18* calculator**, 1973
Made by Olivetti (Italy)
Musée National d'Art Moderne, Centre Georges Pompidou, Paris

OPPOSITE
398. Mario Bellini (b. 1935)
***Teneride* armchair**, 1970
Made by Cassina (Italy)
Polyurethane foam, metal
Musée National d'Art Moderne, Centre Georges Pompidou, Paris

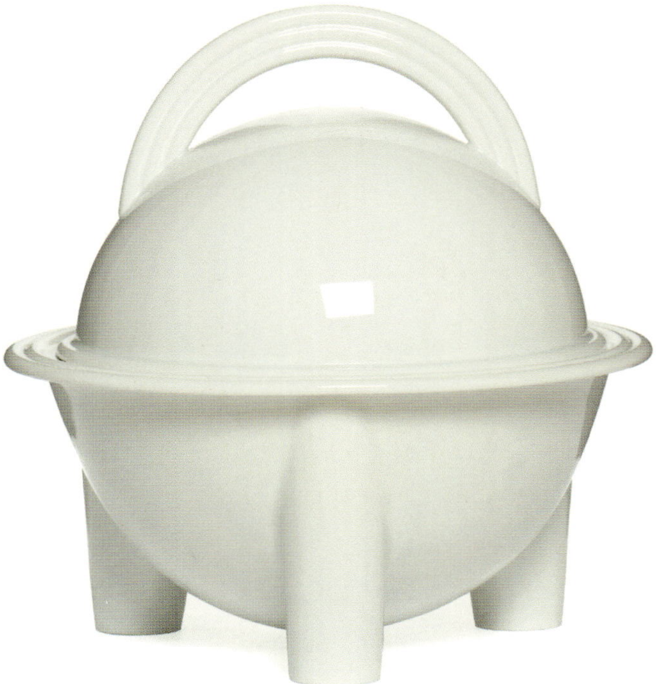

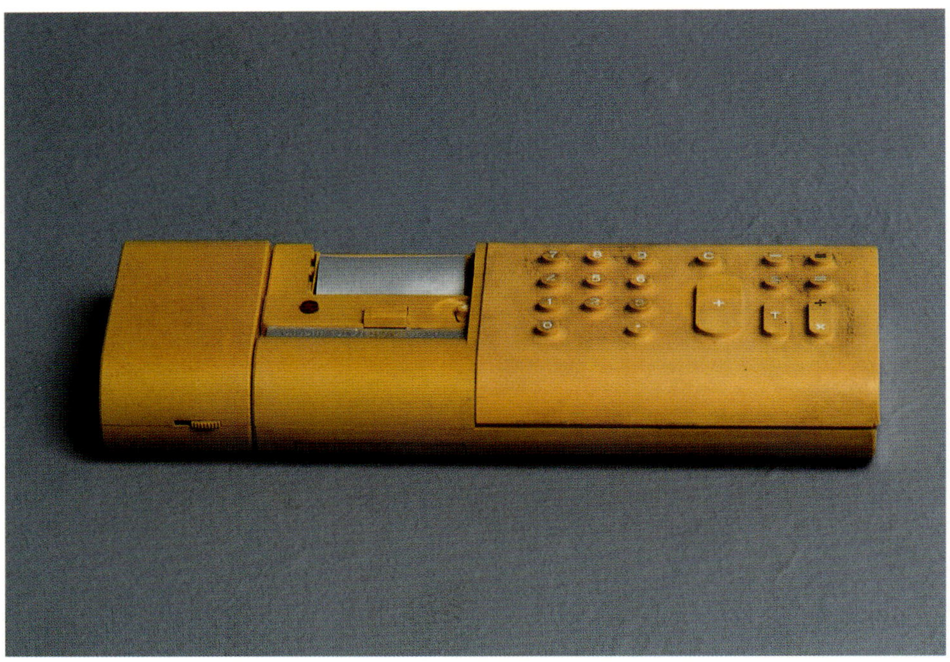

who produced the Telematic House for the 1983 Milan furniture fair, in an attempt to anticipate possible future scenarios. François Burkhardt used the phrase "critical regionalism" to characterize a type of design that emphasized its relationship to artisanal craft, while Branzi invented the label "New Wave Design" to encompass a heterogeneous series of experiments associated with decorative art and artisanal crafts. Paolo Navone[67] gave concrete form to that notion: his objects, halfway between design and craft, attest to the value of the unique object, in contrast to the monotony of mass production.

During those same years, many established companies created brands not only to diversify their offerings and present collections by individual designers, but also to test new markets, as Cassina had previously done with Bracciodiferro. In 1983, Zanotta[68] launched Zabro to distribute artists' pieces, and Acerbis International founded Morphos, intended for a select clientele. In 1984, Driade created Aleph (named after a surrealist short story by Borges), a multilingual laboratory where several non-Italian designers worked.[69] The same year, the famous Bolognese firm Castelli introduced the Castilia brand, for which Giancarlo Piretti designed the *Dilungo* table (1985–86), equipped with a pull-out shelf—in accordance with a philosophy of showing all the possible extensions of the work. He also created the *Plia* chair (plate 382) and the *Diletto* bed, which mechanically adjusted to different positions (with the footrest serving as a multifunctional support). This bed was conceived by Mario Bellini, a designer and architect long employed by Olivetti, where he was responsible for the *Divisumma 18* calculator (1973; plate 397), the *Praxis 35* typewriter, and the *Quaderno* portable computer (1992). Deeply interested in the culture of design, Bellini served as editor of *Domus* from 1985 to 1991 and as exhibition commissioner of the Triennale di Milano, and taught in Italy and abroad. He also created many other classics of Italian design, such as the *Bambole* series of sofas for BBB Italia, a successful example of modular furniture (1971–72); television sets for Brionvega; and lamps for Artemide and Flos. His designs earned him several Compasso d'Oro awards. At MoMA in 1972, he exhibited *Kar-a-Sutra*, a curious prototype of a mobile living unit, innovative but now lost. Bellini was a presence in the field of automotive design as well, designing for Renault, Fiat, and Lancia.

LEFT
399. Bertone, with Pininfarina
***Giulietta Spider* convertible**, 1957
Made by Alfa Romeo (Italy)

BELOW
400. **Advertisement for the Alfa Romeo *2500 Berlina Gran Turismo* and *1900 Berlina Monoscocca***, 1951
Private collection

OPPOSITE
401. Pinin Farina (1893–1966)
***Flaminia* coupé**, 1957
Made by Lancia (Italy)

FROM THE COACHBUILDER TO THE AUTOMOTIVE DESIGNER

Italian cars are objects on a human scale, especially compared to their large American cousins. They have their roots in the country's creative spirit and expert craftsmanship. Many Italian automakers have enjoyed continued success over the decades,[70] particularly the luxury brands, and Italian automotive designers have often been recruited by foreign marques. Take the case of Walter de Silva: After beginning his career with Fiat, he moved to Alfa Romeo in 1986, where he developed a number of innovative designs, such as the *156*, a 1998 model, and the *147*, a 2000 model. In 1999, he was recruited by the Volkswagen Group, and since 2007 he has been head of design for all seven of the company's brands.

We have already discussed Fiat, the Turin firm that made the automobile an everyday means of transportation in Italy. Another pioneering Italian marque from the same city is Lancia; founded in 1906 by Vincenzo Lancia, a race car driver and entrepreneur, it has always distinguished itself by its technological innovations.[71] The *Lambda* (1923) was the very first car to have a unibody construction, which the *Aprilia* (1937) would also have. Even after World War II, "modern" Lancias enjoyed great success, especially in motorsports. The *Delta Integrale* (1979), designed by Giorgetto Giugiaro, was a leading rally car, while the *Stratos* (1974), designed by Nuccio Bertone, was a standout on the track.

Bertone inherited the coachworks founded by his father in 1912 and built it into a leading automotive research center after World War II. The Bertone company designed production cars such as the Giulietta *Sprint* of 1954 (plate 399), as well as concept cars that represent most advanced expression of Italian style and technological innovation: among others, the Chevrolet *Corvair Testudo* (1963), the Alfa Romeo *Carabo* (1968), and the *ZER* (1994), which set the speed record for an electric car, reaching almost 189 miles per hour (303 km/h).

One legendary enterprise of the past should also be mentioned, that of Cesare Isotta and the Fraschini brothers. At the dawn of the twentieth century, they began to produce luxury automobiles in Milan under the Isotta Fraschini brand, including some of the most prestigious vehicles of their day, famous for being entirely handcrafted. The business had its highs and lows,

running into serious financial difficulties after each of the world wars, as many carmakers did. It later began to produce buses, whose bodies were designed by Ugo Zagato, the owner of a coachworks that had pioneered a lightweight, all-metal mode of construction.

However, not all Italian coachbuilders had adopted the latest technological innovations: in 1963, they produced only 43,000 of the million cars sold in Italy. Many could not withstand the competition, but others were able to reinvent themselves as consultants to major automakers, Italian or foreign. By the 1970s, the traditional figure of the coachbuilder had largely disappeared, to be replaced by that of the automotive designer, who conceived and built a prototype and also considered methods for manufacturing it, before entrusting its mass production to a larger firm.

Pinin Farina's coachworks was among those that were able to adapt with the times. His son Sergio took over the day-to-day operation of the firm in the 1950s.[72] Early in that decade, Pininfarina was still making only two car bodies per day, but it began to augment its expert craftsmanship with new technologies. An increase in production quickly resulted, and Pinin and Sergio grasped the full potential of the shift. They took a decisive step, building a large new factory in Grugliasco, on the outskirts of Turin. In this new facility, which opened in 1959, Pininfarina focused its efforts on a form of automotive design that took both aesthetics and aerodynamics into account. Particularly through its close collaboration with Ferrari, the firm would set the standard for sports cars and race cars.

As we have seen, Pininfarina already had the Cisitalia 202 (1947) to its credit. This small coupé, the first automobile to be exhibited at a museum (MoMA, in 1951), was considered the Trojan horse of the Italian style, because of the attention it attracted across Europe and in the U.S. With the Lancia *Flaminia* (1957), Pininfarina created a body with sharp ridges and angles, and vast expanses of glass, all on coil springs (plate 401). This same line of development led to the *Spider Duetto*, designed for Alfa Romeo in 1966 and kept in production with slight variations until the early 1990s (plate 402). The firm continues to design production models, such as the Ferrari *575M Maranello* and the Ferrari *Enzo*, as well as one-of-a-kinds like the Ferrari *P4/5*.

But a broader range of products now bears the Pininfarina logo, from the torch for the Winter Olympic Games in Turin (2006) to the *Sirio* tram (2002, for AnsaldoBreda), a practical and functional mode of transportation, now in operation in many cities in Italy and elsewhere. Pininfarina also worked on the styling and aerodynamics of the *ETR 500* (1985), the first nontilting high-speed train made in Italy, developed by the TREVI consortium (TREno Veloce Italiano).[73]

Another automotive design firm that has worked on trains—specifically, their passenger compartments—is Italdesign, founded in Turin in 1968 by Giorgetto Giugiaro, with Luciano Bosio and Aldo Mantovani. Italdesign had a very specific objective: to provide carmakers with an integrated service that included not only the design of the vehicle, the production of the prototype, and the complete engineering of the project, but also a cost analysis and production schedule. Their first project was the Alfa Romeo *Alfasud*, which was produced in the factories of Pomigliana d'Arco (in the province of Naples), in order to aid the economic development of southern Italy. Immediately after, Italdesign began its long collaboration with the Volkswagen Group.

Since the 1970s, the studio has designed countless popular models for a number of brands, from the

Volkswagen *Golf* (1981) to the Fiat *Uno* (1983) and *Punto* (1993). One particular success was the Fiat *Panda* (1981), a versatile, inexpensive, and roomy city car that won the Compasso d'Oro (plate 404). Italdesign's extensive research efforts have also led to innovative concept cars, such as the Alfa Romeo *New York Taxi*, in which particular attention was paid to the relationship between the exterior dimensions and the interior capacity, and the *Capsula* (1982). The latter's fully equipped and autonomous chassis (which included the engine, transmission, fuel tank, spare tire, trunk, power brakes, radiator, and lights) could be adapted to a sedan, a pickup, a commercial van, an ambulance, or a tow truck. The sleek aerodynamic body of the *Brivido* touring car (2012), constructed in aluminum, carbon fiber, and glass, is painted in red Xirillac pearl to accentuate its sportiness. A special department, Giugiaro Design, was created in 1981 to extend the same integrated concept—"from the idea to the actual object"—to other areas of industrial design. Among this department's most recent projects, both large and small, are the interiors for the new *Alstom* train in Italy, the *Ultrac Vortic* and *Sportrac 5* tires for Vredestein, and the *Arco* line of ceramic cookware for TVS.[74]

Giugiaro was trained at Fiat's Centro Stile (established in 1958 under the direction of Dante Giacosa), as were many other designers, including Pio Manzù, who managed to carry out several influential projects before his untimely death in 1969, at the age of thirty.[75] A graduate of the Ulm HfG, where he also taught, Manzù came to prominence in 1962, when he won an international competition for a coupé body on an Austin Healey *100* chassis. He created the Fiat *127*, one of the first cars to come in two body styles, and worked on different aspects of Fiat's *City Taxi* concept car of 1968 (plate 403). He also organized the exhibition *The Lines of the Italian Car* (1967, San Marino), which traced the history of automobile design in Italy for the first time. Finally, he conceived of the famous *Parentesi* pendant lamp (Flos, 1970; plate 407), whose design was completed by Achille Castiglioni. This innovative lamp features a spotlight that slides and pivots on a steel cord between the floor and ceiling.

Alongside the entry-level models developed by established automakers, there is the unusual case of the very economical *Isetta*, which anticipated the *Smart* car. A tiny yet comfortable two-seater for city driving, it had an unusual configuration: riders entered the passenger compartment through a single door in front. Designed by Ermengildo Preti, an aeronautical engineer, for Iso Rivolta in Milan in 1953, the *Isetta* was not properly appreciated in Italy, where only twenty per day were built. In Germany, however, it was a remarkable success: it was produced by BMW at the rate of three hundred per day from 1955 to 1962 and came in a range of unusual colors, from apple green to fluorescent pink.

OPPOSITE
402. Pinin Farina (1893–1966)
***Spider Duetto* convertible**, 1966
Made by Alfa Romeo (Italy)

BELOW
403. Pio Manzù (1939–1969)
***City Taxi* concept car**, 1968
Made by Fiat (Italy)

BOTTOM
404. Giorgetto Giugiaro (b. 1938)
***Panda* city car**, 1981
Made by Fiat (Italy)

OPPOSITE
405. Gae Aulenti (1927–2012)
***Pipistrello* lamp**, 1965
Made by Luce Martinelli (Italy)
Lacquered aluminum, methacrylate, electrical components
Musée National d'Art Moderne, Centre Georges Pompidou, Paris

BELOW
406. Ennio Lucini (1934–1997)
***Cespuglio* lamps**, 1969
Made by iGuzzini Illuminazione (Italy)
Plexiglas, electrical components

OTHER FIELDS OF DESIGN

Besides furniture and accessories, another well-known field of Italian design is lighting. After World War II, new materials such as methacrylate and plastics made it possible to develop lamps with novel forms, while advances in the technology of light sources impacted not only the aesthetics but also the ethos of lighting design, prompting Italian designers and entrepreneurs to "design for the future." The LED was about to cause a revolution in energy efficiency, and would also affect the style of many lamps, such as those created by C. De Bevilacqua for Artemide and Danese.

Fluorescent lights date to 1936 but did not enter the home until 1949, when Achille and Pier Giacomo Castiglioni used them in their *Tubino* lamp for Arredoluce (reissued by Flos in 1974), which extended the bulb's cylindrical shape to the entire metal frame. Forms grew increasingly diverse, encompassing pendant, wall-mounted, floor, and desk lamps, and even chair lamps, such as the *Libra-Lux* by Roberto Menghi (1948, Lamperti), which can balance on a soft armrest. Gino Sarfatti continued to design sculptural lamps for Arteluce,[76] such as the *Piedemolle* (1961). Then there was Artemide, founded by E. Gismondi and later joined by C. De Bevilacqua; and Flos, founded in Merano in 1962 to produce lamps in a plastic material called "cocoon."[77] The Castigliani brothers brought Flos to prominence (plate 410), but it was Sergio Gandini (and later his son Piero) who would become its soul. Based in Brescia, they sought out other famous names in Italy and abroad; for example, Philippe Starck designed the *Miss Sissi* lamp for Flos in 1991.

The Castiglioni brothers were more interested in the light source than its container, and made use of unusual components. Their *Tojo* (1962), for example, used a car headlight, a transformer, and a fishing pole, whose guides held the electrical wire. The aforementioned *Arco* (1962) is no mere lamp, but an archetype. Equipped with a marble base that provided a counterweight to the large stainless steel arc, and with an adjustable reflector in polished aluminum, it constituted an "independent light source." Under the name iGuzzini, the Guzzini brothers,[78] Adolfo and Giannunzio in particular, began producing indoor and outdoor lights—initially decorative, and later increasingly technical (plate 406). They often collaborated with architects on site-specific solutions, which they then adapted for industrial production. Examples include Renzo Piano's *Lingotto* (1993), designed for the exhibition center in Turin (in a former industrial space), and Gae Aulenti's lamps for the Palazzo Grassi in Venice, which were developed (with Piero Castiglioni) into the *Cestello* series of lighting arrays (1986).

Aulenti also designed the resolutely expressive *Pipistrello* ("Bat") (Martinelli, 1965; plate 405), while Mario Bellini created the anthropomorphic *Chiaro* lamp (Flos, 1967). Magistretti set to work as well, devising such lamps as the *Guione* (1970; plate 354) and the

LEFT
407. Pio Manzù (1939–1969) and Achille Castiglioni (1918–2002)
***Parentesi* light**, 1969
Made by Flos (Italy)
Stainless steel wire, rubber housing, metal base

BELOW
408. Paolo Rizzatto (b. 1941)
***Costanza* lamps**, 1986
Made by Luceplan (Italy)
Aluminum, polycarbonate

very successful *Eclisse*, a fixed spherical structure with a half-sphere pivoting inside it, allowing one to direct the light and regulate its intensity (1967, Artemide; plate 355). The designer of the first low-voltage halogen lamp (1970) was Joe Colombo. But the record for number produced must perhaps go to the *Tizio*, an articulated desk lamp with an innovative zoomorphic form, designed by Sapper (Artemide, 1972). Livio Castiglioni (assisted by his son Piero, who would carry on his father's legacy, thanks to his own creativity) tended increasingly to minimize the supporting structure of his lamps, to improve the performance of the light source. With Gianfranco Frattini, he created the entirely flexible *Boalum*, using a PVC tube (Artemide, 1971). Ettore Sottsass designed "fun" lamps for Memphis (the *Tahiti* and *Ashoka*, 1981), as did De Lucchi (the *Oceanic*, 1981). By contrast, Alberto Meda and Paolo Rizzatto's lamps are distinguished by their studied elegance and a constant attention to the light source, materials, and production techniques; in collaboration with Sandra Severi, they created many pieces for Luceplan (founded in 1979 by Gino Sarfatti's son Riccardo).[79] The *Costanza* (1986; plate 408), which revisited the traditional lamp shade, is a Rizzatto design, while the *Lola* lamp (1986; plate 409), whose name underscores its feminine lightness, was designed jointly by Meda and Rizzatto. Made of aluminum and carbon fiber, the *Lola* diffuses an intense and functional light, and its height can be adjusted by means of an extendable rod.[80] Meanwhile, Fontana Arte[81] followed its own path, working with a large number of designers, including Aulenti, Castiglioni, Sottsass, and Piano.

In the field of transport, Italy has made important contributions not only to automotive design but also to naval architecture.[82] Italy has a number of shipyards, and several of its architects have also designed watercraft, from the largest to the smallest. Many cruise ships produced in the shipyards of Genoa and Trieste are distinguished by their luxurious accommodations. The Riva motorboat, called "the Rolls-Royce of the sea," was famous on the French Riviera in the mid-1950s and is still built on Lake Iseo. Its production is industrial in scale (between 1950 and 1970, four thousand Rivas were launched), but also of high quality: even into the postwar era, these luxury items were crafted by hand. Giorgio Barilanni designed one of Riva's most iconic models, the *Aquarama*, an eight-person boat measuring 26 feet 4 inches by 8 feet 7 inches (8.02 by 2.62 m). It has been in production since 1962, and is still made of varnished mahogany. Carlo Pagani, who breathed new life into the Milan department store La Rinascente after World War II, was responsible for its interior. Wood eventually gave way to fiberglass: Carlo Riva, who represented the third generation of the firm, was the first boatbuilder in Italy to use that material, in 1970.

In certain respects, Riva's boats came as a response to those of Baglietto (established in Varazze, Liguria), which were built to order, barely a few hundred per year. Baglietto's first shipyard opened its doors in 1854, in a garden shed a hundred yards from the sea. There the founder, Pietro Baglietto, started out building small

LEFT
409. Alberto Meda (b. 1945) and Paolo Rizzatto (b. 1941)
***Lola* lamp**, 1986
Made by Luceplan (Italy)
Carbon fiber, steel sheet
Musée des Arts Décoratifs, Paris

OVERLEAF
410. Achille Castiglioni (1918–2002)
***Taraxacum 88* pendant lamp**, 1988
Made by Flos (Italy)
Aluminum, 60 bulbs
Centre National des Arts Plastiques/Fonds National d'Art Contemporain, Paris

PAGE 331
411. Paolo Ulian (b. 1961) and Giuseppe Ulian (b. 1959)
***Anemone* lamp**, 1998
Bic pens, bulb

fishing boats, but he quickly turned to pleasure craft and yachts. And though Baglietto is certainly a historic shipyard,[83] today the Italian naval architect par excellence is Andrea Vallicelli, whose name remains inextricably linked to that of the *Azzurra*, the very first Italian boat to participate in the America's Cup, in 1983.

A few significant achievements in the field of industrial design should also be mentioned. We have already alluded to the case of Olivetti, whose history encompasses not only that of the firm itself, founded in 1908, but also that of the Piedmontese city of Ivrea, and of the enlightened family that gave the firm its name. In 1932, Olivetti introduced its first portable typewriter, whose case was designed by the engineer Aldo Magnelli with advice from his painter brother Alberto. In 1935, Adriano Olivetti, son of founder Camillo, reshaped almost the entire city of Ivrea in the name of modernity, to respond to new needs inside and outside the factory. Even today, the urban center of Ivrea is an open-air museum of architecture.[84] In seeking out artists and architects with a social conscience, he opened the way for a new relationship between technology and humanist culture. He was responsible for the very thoughtful review published by the Olivetti company and for the founding of the Comunità movement (1946), whose objective was to disseminate a distinctive industrial culture. The architects Figini and Pollini were joined by Xanti Schawinsky, a Swiss visual designer and photographer trained at the Bauhaus. While working on the *Studio 42* typewriter (1935), he enlisted the services of Nizzoli, who created his first product for Olivetti in 1940, the *Multisumma*, which was followed by the *Lettera 22* (1950), the quintessential portable typewriter.

In the following years, Olivetti remained the leader in its sector, while also entering the promising new field of electronics. Sottsass's and Hans von Klier's designs for the company were the emblem of future possibilities, to which, oddly enough, most of Italian industry paid little attention. Nevertheless, in terms of its design, the *Elea 9003* computer anticipated concepts that only the great science fiction films were depicting at the time (Stanley Kubrick's *2001: A Space Odyssey* of 1968, for example). As Sottsass himself recalled in 1993: "I imagined . . . black metal structures enclosed in sheets of aluminum plate . . . then, ultimately, these cases, once assembled, would become a sort of absolutely mysterious silver architecture, because you would see nothing there but a monolith of shining, metaphysical metal, a sort of maleficent, or at least impenetrable, god." An utterly cosmic image, worthy of a great visionary—as that artist-designer-architect has always been.

The landmark, albeit brief, history of Brionvega and its electronics has already been mentioned. But Brionvega was also important in another sector, one that has recently been undermined by outsourcing but was once a real focus of innovation in Italy, namely, that of white goods. This term[85] encompasses all kitchen appliances—from the refrigerator to the dishwasher, stove, and exhaust hood, as well as the washing machine. Ventures such as Candy, Indesit, Ariston, Merloni, and

RIGHT
412. Marco Zanuso (1916–2001)
***Ariante* portable table fan,**
1973
Made by Vortice (Italy)
Unbreakable plastic, electrical components
Museum of Modern Art, New York

OPPOSITE
413. Roberto Pezzetta (b. 1946)
***OZ* refrigerator,** 1998
Made by Zanussi (Italy)
Musée National d'Art Moderne, Centre Georges Pompidou, Paris

Zerowatt made design the leading edge of their production. By 1995, Italy had become virtually the appliance factory for the European continent: nearly half of all white goods were made in the country. This great undertaking, which began after World War II, grew up among established machine shops that understood the new possibilities of industrial design. In 1947, the Fumagalli brothers exhibited the first washing machine made in Italy at the Milan Trade Fair. (It was called "Candy," after the American tune by that name).

Several other young entrepreneurs, with the assistance of the Centri Stile Interni, also developed products that went on to conquer the European market. The business daily *Il Sole 24 Ore* wrote in the 1960s that "household appliances are to Italy what watches are to Switzerland." And the *Financial Times* would even go so far as to compare the perfection of these two characteristic national products.

Thus arose an industry that responded to the need to mechanize the home, to provide it with "household helpers" that were durable, functional, aesthetically pleasing, and less expensive than those of foreign brands. The refrigerator, for example, was entirely reinvented, thanks to the use of plastics: all its internal surfaces were molded in ABS (plate 413). Zanuso (who also designed small household appliances), Gino Valle, and Luigi Spadolino were among the first designers to enter this sector.

In the Marches, Elica, a manufacturer of exhaust hoods, used design to add a new élan to its products in the early twenty-first century: Gaetano Pesce contributed a "hood sculpture" decorated with fruits and vegetables. Elica found a precedent in the Milanese company Vortice, established by Attilio Pagani in the 1950s, which produced air purifiers and took its name from its first product. (Zanuso would later work with Vortice; plate 412.)

The contributions of Italian design to the sporting equipment sector[86]—where performance is always a high priority—are perhaps less well-known, but significant nevertheless. They include bicycles, from Bianchi to Cinelli, and also motorcycles. Some of the most notable recent bicycles have been designed by Alessandra Cusatelli, but Rinaldo Donzelli's *Graziella* is also worthy of mention. Produced by Carnielli from 1963 on, it solved the problem of the folding bike for city use and, indeed, became the quintessential example of that type. To mention only two motorcycle companies: Ducati, founded in Bologna in 1926, is "the builder of sports motorcycles par excellence," and Gilera, founded in Milan in 1909, is the oldest brand of Italian motorcycles still in production. The races these firms organized were an effective public relations instrument, of course, and a valuable tool for developing new production models.[87]

Some designers have created sporting equipment for people with limited mobility. For example, the Turin designer-entrepreneur Danilo Ragona drew on his personal history to improve the lives of the disabled. Under the Able to Enjoy brand, he has produced innovative ultralight wheelchairs made of light and versatile carbon fiber, which allow their users to engage in sports from tennis to handball. With the support of Pininfarina Extra, he also developed the *Monoski* (2008) for disabled skiers.

Technogym, founded by Nerio Alessandri in Cesena (Romagna) in 1987, manufactures equipment for athletes in training, as well as people who simply want to stay in shape. Since 1996, it has been the official supplier of the training centers for the Olympic Games. A mechanical engineer who loves motors, Alessandri began by designing machines for his own workouts, but now relies on medical, scientific, and technological research—and on design. In the company's research center, the designer Antonio Citterio has created equipment for both fitness clubs and individuals. His *Kinesis* (2008), for example, accommodates more than two hundred fitness exercises.

The quest for physical well-being, and the popularity of the Mediterranean diet, has also given rise to "food design." This trend has manifested itself in the many products for the kitchen designed by Fratelli Guzzini and Alessi. In this connection, we may recall that, back in 1933, the very first espresso machine was invented on the banks of Lake Orta (Piedmont), with an aluminum boiler and a one-of-a-kind multifaceted shape. It would adopt an octagonal silhouette in the 1950s but retain the same icon: Signore Bialetti, a little man in a mustache. In Italy, it would replace the traditional Neopolitan coffeemaker.

RIGHT
414. **Display of prototypes and models at the Alessi Museum in Crusinallo, Italy**

OPPOSITE
415. **Alessi factory and museum in Crusinallo, Italy**

A NEW CENTURY

And so we arrive at the dawn of the twenty-first century. During the last years of the twentieth, individuality assumed an increasingly important place, and traditional design, which respects the technical, functional, and ergonomic values of the product, enjoyed a resurgence. But while products have become increasingly diversified, no real stylistic revolution seems to have occurred. Some have spoken of *Design povero*, linking design to experiments in conceptual art. In practice, this term is primarily used to refer to a type of design that privileges humble materials, such as recycled paper, and combines objects at hand to create others with different functions. This current has an illustrious precedent in Mari's *Metamobile* (Gavina, 1976).[88] In 1974, others began to use the expression "neo-minimalism" to refer to a nascent trend toward simplicity, in reaction to the exuberance of form in the previous decade. Naturalness, lightness, and sobriety mark that approach, which targets formal more than functional aspects. Consider the approach of the designer-manufacturer (a combination that is not unusual in Italy) Giulio Cappellini, who revived the old family business by acquainting the Italians with the creations of young international designers, from Jasper Morrison (his *Thinking Man's Chair* of 1988, in lacquered metal) to Tom Dixon (his *S Chair* of 1989, with its lithe silhouette).

In the late 1980s, a number of other Italian manufacturers also opened their doors to foreign designers from throughout Europe and across the Atlantic. This is another distinctive feature of Italian design culture and of the country's industrial sector, which is composed of small firms concentrated in particular areas. These flexible businesses are able to express the very essence of Italian taste and the art of living, which is highly attractive to foreign designers—as is the network of relationships provided by a city such as Milan.

Just as Bertone, Giugiaro, Pininfarina, and others were engaged by foreign companies, many foreign designers came to work in Italy, where they discovered their chosen homeland. Andries Van Onck from the Netherlands, for example, worked for Olivetti beginning in 1957, and later became an independent designer in Milan. The Japanese designer Isao Hosoe has lived in Italy since 1967, working first with Ponti and Rosselli;[89] and Hosoe's countryman Makio Hasuike ran his own studio in Italy beginning in 1968, and in 1982 founded Mh Way, which produces bags and desk accessories.

Italian entrepreneurs are always attentive to new discoveries in design and to the emergence of new design "stars," while realizing that it takes time for a designer's image to attach itself to his or her works. Moreover, they are careful to preserve the cultural identity of the products that foreign designers create for them. Alessi, Driade, Flos, and Magis (a northeast Italian company founded by Eugenio Perazza in 1976, which func-

tions as a sort of international laboratory that anticipates future trends) were among the first Italian companies to seek out foreign designers. Philippe Starck,[90] for example, has a strong presence in Italy, thanks to his playful products: he designed the *Costes* chair for Aleph (1981); the *Arà* table lamp, with its distinctive horn shape, for Flos (1988); and the more provocative *Gun* table lamp, with its shaft in the shape of a gold pistol, also for Flos (2005). In the late 1980s, Kartell commissioned the *Dr. Glob* chair (1988–89), among other projects, from Starck, and in 1990 Alessi invited him to reconceive one of its classic pieces, a stainless steel tray. And, if we are to believe the legend, shortly thereafter Starck sent in a paper napkin containing a sketch for the *Juicy Salif* citrus squeezer, one of the company's best sellers, which, even beyond its practical utility, is a real conversation piece. Recently, Alessi has also turned to Asian design, commissioning eight architects from China to design trays and containers (the firm's most typical products), in an explicit confrontation between East and West.

In 1994, the Israeli Ron Arad, also attracted by the presence of artisans and manufacturers who know the value of design, opened a studio along the lakefront of Como. For Kartell, he designed the *Bookworm*, a flexible bookcase that unrolls along the wall (1995). Alessi, Cappellini, Magis, and Driade assigned him several design projects, which he always realized with his signature spirals and curves. More recently, after working with Castiglioni and Magistretti, the Spaniard Patricia Urquiola settled in Milan. In 1998, with Patrizia Moroso—who worked in her family's furniture factory, founded in the early 1950s—Urquiola undertook a new approach to the chair: casual, flexible, convertible. The Moroso factory is located in Friuli, a border region where Manzano chairs have been produced for more than a hundred years. In that productive triangle, little known to the general public, anonymous skilled craftsmen have produced millions of chairs in their shops over the generations. These chairs are exported worldwide under the brand names of illustrious American, German, or French enterprises. Today, the Friulian

furniture makers are facing a full-blown crisis, because of the outsourcing of production to low-cost factories in Asia. Only those that have invested in design are still hanging on: Moroso[91] of course, but also Gervasoni,[92] which is in its third generation.

As for the other characteristics of Italian design and production in recent years, we lack the historical distance needed to assess them critically. However, at least one of the many paradoxes of Italian design seems to have been resolved, namely, the absence of schools of design. Before the 1990s, there were no schools or departments devoted specifically to design in the Italian universities, nor even courses in design. Until then, only the ambiguously named ISIAs (Higher Institutes for Artistic Industry), founded in the 1970s in Rome, Florence, Urbino, and Faenza, offered public education in design. Many who are now recognized designers came out of these institutions: for example, Paolo and Giuseppe Ulian, who, through a skillful recycling of rubbish, an intelligent approach to construction, and an ethical view of the product, were able to propose a new type of design inspired by consumer behavior. Examples of their work include the *Mat Walk* bath mat (2001) or the *Anemone* lamp, made of Bic pens (1998; plate 411).

During the entire second half of the twentieth century, the training of designers took place exclusively in private schools (with the exception of certain courses offered in schools of architecture and, later, academies of fine art). In Milan, the Società Umanitaria, a nineteenth-century institution, created a school of design in 1953 with the help of the visual designer Albe Steiner, following the model of the Ulm HfG. In 1954, Nino Di Salvatore opened the Scuola Politecnica di Design, first in Novare, then in Milan, employing both Munari and Huber as instructors. There are also two international institutions on Italian soil: the Istituto Europeo di Design (IED), active since 1966 and with branches throughout Europe and Latin America; and the Nuova Accademia di Belle Arti (NABA), which was founded in 1980 in response to the deficiencies in Italian academies of fine arts. NABA has become a training center for careers in design, fashion, and communications. And, finally, one postgraduate course: the Domus Academy, created in 1983 at the initiative of Maria Grazia Mazzocchi, with the support of Mendini, who was editor of *Domus* at the time. The establishment of the Domus Academy attested to a growing awareness of the importance of Italian design, and to a desire to teach the Italian mode of industrial creativity.

In the 1990s, as private schools continued to multiply, and not only in Milan, drawing in a flood of foreign students, Italian design was finally institutionalized in the universities. Various departments began to offer courses that would develop, over the years, into schools of design with complete degree programs (for the bachelor's, the master's, and the doctorate). In 1993, the Politecnico di Milano underwent such a transformation, with the school of design taking its place beside

OPPOSITE
416. Antonio Citterio (b. 1950) and Toan Nguyen (b. 1969)
***Kelvin LED* desk lamp**, 2009
Made by Flos (Italy)
Aluminum, LEDs

BELOW
417. Claudia Raimondo (b. 1957)
***Joy no. 1* centerpiece bowl**, 2013
Made by Alessi (Italy)
Stainless steel

those of engineering and architecture. Meanwhile, in Venice, the Istituto Universitario di Architettura di Venezia (IUAV) created a school of arts and design. In Turin, Genoa, Florence, Camerino (Macerata), Rome, Naples, and Palermo, courses in design are offered in the architecture schools. And in the last ten years, the Libera Università di Bolzano has established a trilingual school of arts and design.

Italy also lacked a museum that could display the history of industrial design. In the early 1950s, Ponti developed a plan for such an institution, which he wanted to install within the Museo della Scienza e della Tecnologia in Milan. In the end, a large space designed by Michele De Lucchi in the Palazzo dell'Arte was devoted to industrial design. The roots of this endeavor lay with the Triennale, although it must be said that this event has changed profoundly in the last two decades. Ever since it broadened its range of activities to include research and documentation (1990), it has hosted exhibitions covering everything from architecture to fashion. The themes of the Triennales have also become much more heterogeneous since the 1970s, and design no longer plays the starring role.

The Triennale Design Museum, which opened in December 2007, is characterized by its dynamism and its capacity for innovation. The first exhibitions were conceived by Andrea Branzi. To respond to the question "What is Italian design?" he displayed *The Seven Obsessions of Italian Design* (2007–8), attempting to outline the shape of a culture that has taken on the contemporary world while cultivating a close relationship with ancient history and tradition. In *The Dream Factories: Men, Ideas, Ventures, and Paradoxes of Italian Design*, Alberto Alessi told the stories of businesses large and small, which in Italy are closely linked to the biographies of their founders.

For the last fifteen years or so, the theme of memory—a valuable tool for consciously projecting oneself into the future—has been addressed by many design firms. They have made an effort to organize their archives and collections, knowing that their history and identity represent an important aspect of Italian design and entrepreneurial culture (plates 414, 415).[93]

The "culture of design" has become a focus of growing interest, and design has become a social phenomenon even outside Milan, although that city remains the historic center of gravity for its development. The Association for Industrial Design, which has its headquarters in Milan, has been overhauled and is now subdivided into thirteen territorial delegations. It has assembled a collection that includes almost all the three hundred design products awarded the Compasso d'Oro since the creation of the prize fifty years ago, as well as the two thousand that received an honorable mention. The Ministry of Culture has declared this collection, managed by a foundation created in 2001, to be of "extraordinary historical and artistic interest."

For Milan (if not Italy as a whole), the coming years will no doubt confirm the city's innate capacity to influence taste and to cast its spell, as already suggested by its system of international *saloni*. The traditional Milan furniture fair now coexists with events dedicated to different types of design, for the office, lighting, construction, the kitchen, and the bathroom. These exhibitions are a landmark of urban culture in its broad sense, and now the model has been expanded to Turin, which the International Council of Societies of Industrial Design designated the "world capital of design" in 2008.

GREAT BRITAIN

PENNY SPARKE

PRECEDING PAGE
418. Ron Arad (b. 1951)
***This Mortal Coil* bookcase**, 1993
Made by One Off Ltd (U.K.)
See plate 459.

BELOW
419. Wells Coates (1895–1958)
***AD-65* Radio**, 1932
Made by E. K. Cole Ltd (U.K.)
Bakelite, stainless steel, fabric
Victoria and Albert Museum, London

OPPOSITE
420. **Exhibition of furniture from the Utility project at the London Building Centre**, 1942
Imperial War Museum, London

Industrial design was born in the United Kingdom in the second half of the eighteenth century. The spinning jenny and the Jacquard loom, for example, revolutionized the ways in which textiles were designed and made. In addition, the economic expansion of those years enabled people who previously had relatively few possessions to own more goods. To meet their demands, manufacturers had to ensure that those goods were visually appealing.

Through the nineteenth century, the middle classes continued to expand, and their capacity to consume increased. However, by the second half of the nineteenth century, an energetic campaign for design reform, championed most famously by William Morris and the adherents of the Arts and Crafts movement, had emerged to address what some saw as the adverse effects of design's alliance with industry. In the early twentieth century, the Design and Industries Association (DIA) was formed on the model of the earlier Deutscher Werkbund.

In the middle years of the twentieth century, Great Britain took an ambivalent approach to modernist European design. When it represented itself at exhibitions, Britain frequently called on its indigenous traditions, invented or otherwise. However, a number of British designers did rise to the challenge of modernism in the interwar years, creating many exciting new forms that quickly became icons of the age. Some took up the Bauhaus baton, designing objects for the home that focused on function and displayed a minimal, geometric aesthetic. Walter Gropius and Marcel Breuer both spent time in Britain during their exile from Nazi Germany and on their way to permanent residence in the U.S. While in London, Breuer worked with Jack Pritchard's company, Isokon, creating a collection of furniture in bent plywood. The aesthetically radical plastic radio cabinets designed by Serge Chermayeff and Wells Coates, for example, also drew on Bauhaus principles (plate 419).

WARTIME DESIGN

During World War II, design transferred its focus from the commercial marketplace to the national war effort. Many British designers who had been trained in the interwar years were now called on to apply their talents to new challenges. Several architects and designers—among them the theater designer Oliver Messel, the graphic designer Richard Guyatt, the architects Basil Spence and Hugh Casson, and the industrial designers Robert Goodden, James Gardner, and Ashley Havinden—were asked to help create camouflage. The British Ministry of Information depended especially on the skills of graphic and advertising artists. Of the many who were involved in these activities, Austin Cooper, Tom Eckersley, Abram Games, James Gardner, Milner Gray, Ashley Havinden, F. H. K. Henrion, E. McKnight Kauffer, Tom Purvis, and Hans Schleger (known as Zero) stand out.

The Utility scheme was an important wartime project that brought a limited range of consumer goods—furniture, household items, and clothing in particular—to the people who needed them most. Utility products were designed to operate outside the fashion and style systems and to have a universal appeal (plate 420). The idea was first conceived by Hugh Dalton, the president of the Board of Trade, who brought in

BELOW
421. **Cover of the first issue of *Design*, the journal of the Council of Industrial Design,** January 1949
Design Council Archive, London

OPPOSITE
422. **Cover of the catalog to the exhibition *Britain Can Make It*, organized in London by the Council of Industrial Design,** 1946

the furniture designer Gordon Russell to help him develop it. The aim was to produce a limited set of simple, standardized items at a controlled price. Utility furniture consisted of basic wooden pieces that recalled traditional, vernacular designs, albeit with a modern twist.

The British excelled at military aircraft design, and their wartime successes included the *Spitfire* and the *Hurricane*. Steel and aluminum were needed for airplane production, and many existing civilian objects made from those materials were redeployed to that end. Countless stories could be told of wartime technologies and designs that were transformed into postwar consumer goods. For example, the object that now rules all of our lives—the computer—got a wartime technological boost at Bletchley Park, Britain's codebreaking establishment in Buckinghamshire. The *Colossus*, in operation from 1944, was a breakthrough in the development of the programmable, digital electronic computer.

DESIGN FOR PEACETIME

The years between 1939 and 1951 had seen a transformation of the modern world, and by the latter year the democratization of material culture that had been promised before the war was beginning to become a reality. While the war had prioritized the basic needs of human existence and technological innovation for military purposes, peacetime brought an inevitable swing back to a focus on individuals and their psychological needs. To meet these needs, new demands would constantly be made of the visual, material, and spatial culture of the postwar world.

An official watchdog, the Council of Industrial Design, was set up to promote "good design." Born of a coalition government, it remained in place under the Labor government that soon came to power. The first director of the council, Gordon Russell, had been in charge of the Utility scheme. Russell lived in Broadway in the Cotswolds and had links with the legacy of the nineteenth-century Arts and Crafts movement. His appointment consolidated the importance of that movement, its values, and its approach to design in the postwar years. *Design* magazine, the mouthpiece of the council, was launched in 1949; it promoted a strategy of following the U.S. and Scandinavia in developing design into a true economic and cultural force (plate 421). The first step was to create a taste for "good design" among consumers, and with it a strong home market that, in turn, could be transformed into an international market for well-designed British goods. The Design Centre was opened in 1956 in the Haymarket as one means of presenting good design to the British public.

Other developments of the 1940s included an effort by manufacturers themselves to employ professionally trained designers to create stunningly modern items—furniture, textiles, glass, ceramics, and metal goods, for the most part—for the progressive home. The Council of Industrial Design was an important interface in that context, putting designers in touch with manufacturers and setting up the Duke of Edinburgh awards scheme to recognize successful designs. A reorganization of art and design education also took place at that time, in order to supply a new generation of designers to work with industry. London's Central School of Art and Design was especially prominent in the early postwar years, while the Royal College of Art (RCA), initially established in 1837 and connected with the Victoria and Albert Museum, reorganized itself in the 1950s to address the needs of the day. A number of key British designers of those years—David Queensberry, David Mellor, Robert Welch, and Robert Goodden among them—were RCA graduates. A professionalization of design, especially interior design, also took place in

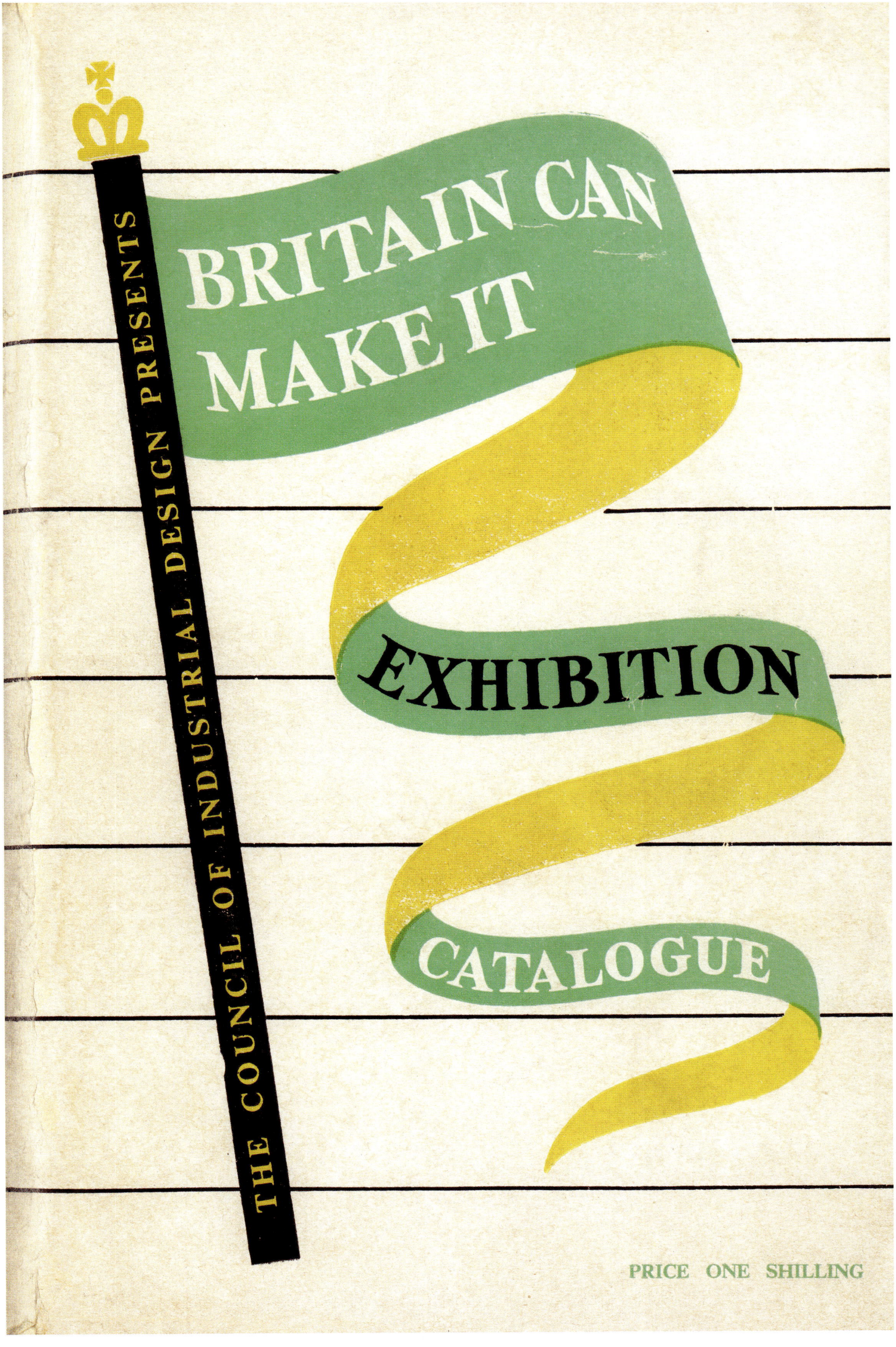
THE COUNCIL OF INDUSTRIAL DESIGN PRESENTS
BRITAIN CAN MAKE IT
EXHIBITION
CATALOGUE
PRICE ONE SHILLING

BELOW
423. Ernest Race (1913–1964)
BA side chair, 1945
Made by Race Furniture Ltd (U.K.)
Aluminum, plywood, fabric
Museum of Modern Art, New York

RIGHT
424. Ernest Race (1913–1964)
***Springbok* chairs**, 1951
Made by Race Furniture Ltd (U.K.)
Steel, plastic
Private collection

OPPOSITE
425. Ernest Race (1913–1964)
***DA 1* armchair**, 1946
Made by Race Furniture Ltd (U.K.)
Steel, aluminum, foam, fabric
Victoria and Albert Museum, London

those years, making it increasingly hard for amateurs to achieve adequate results by themselves.

Among the first acts of the Council of Industrial Design was to sponsor the *Britain Can Make It* (BCMI) exhibition of 1946 (plate 422). In the wartime spirit of public education, the exhibition was intended to open the British public's eyes to the importance of standards in design. Conceived by Stafford Cripps, BCMI displayed mostly prewar goods, as industry had not yet had a chance convert back to peacetime production. There was a strong emphasis on the role of the designer in the postwar context, however, and the Design Research Unit (a consultancy cofounded by the influential critic Herbert Read) created a display—"What Industrial Design Means"—that charted the development of the design of an egg cup from the chicken through the final product. Many of the camouflage and Ministry of Information designers were involved, transferring their skills to peacetime to create didactic settings that engaged the public in a debate about the meaning of design. A design quiz was included so that visitors could register their views on the subject. One of the most controversial sections of the exhibition presented a series of furnished rooms intended for a range of different social and income groups. From a "Kitchen in a Cottage in a Modern Mining Village" to the "House of the Managing Director of an Engineering Works," these specimen rooms spanned the social spectrum, revealing a strong interest in the link between design and social class.

One of the stars of the BCMI exhibition was designer-entrepreneur Ernest Race's little *BA* chair, which had an aluminum frame and elegant tapering legs (plate 423). Race had been in the auxiliary fire service during the war, and afterward he had set up a furniture company. He was quick to realize the potential of redundant wartime materials and set about utilizing scrap aluminum and ex-RAF upholstery materials (plate 425). The *BA* side chair benefited from the Spitfires to Saucepans campaign that was promoted at BCMI. A handful of other postwar designs—including a spaceship by Wells Coates, an aluminum-bodied sewing machine by Henrion, a streamlined bicycle by B.G. Bowden, and an air-conditioned bed by F.C. Ashford—were also offered to whet the appetite of a nation hungry for new products. Race went on to create some of the iconic seating designs displayed at the 1951 Festival of Britain, in particular the *Antelope* and *Springbok* metal-rod chairs that were scattered across the South Bank (plate 424).

The Festival of Britain's two signature buildings—the Dome of Discovery and the Skylon—visually echoed the Perisphere and Trylon of the 1939 New York World's Fair and expressed a similar technological utopianism. With rationing nearly at an end, it was time once again to look forward to a prosperous future, made possible by technology. The festival proved hugely popular, and exhibits such as the Homes and Gardens and the Lion and Unicorn pavilions crossed a spectrum from the traditional to the ultramodern. The work undertaken by the Festival Pattern Group, who used the organic patterns seen under microscopes to decorate brightly colored textiles and ceramics, helped create a new aesthetic for the postwar era.

BELOW
426. **Advertisement in *Plate* magazine**, October 1955

OPPOSITE
427. **Cover of *Do It Yourself* magazine**, June 1959

THE CONTEMPORARY HOME

The two decades following the end of World War II saw a dramatic shift in the British public's engagement with modernity in the realm of everyday life, particularly in the home. The interwar years had seen a rather cautious response to new influences and a general reluctance to move beyond the comfort of "Tudorbethan" suburban homes and heavily upholstered three-piece suites. While the new middle-class and lower middle-class consumers of that time had found it hard to resist the nostalgic pull of Victorianism, the post-1945 years witnessed a greater willingness to embrace the new, and to express that forward-looking approach at home. This transition was greatly facilitated by new housing developments such as Harlow and Cumbernauld New Towns.

The ideal of the "modern" or, as it was referred to at this time, the "contemporary" home was widely disseminated through the mass media. To a considerable extent, it displaced the concept of the "traditional home" as an aspiration, especially for people who had not previously had the opportunity to be homeowners. Indeed, for a large sector of the population, inhabiting a home furnished in the contemporary style represented an important sign of the arrival of modernity in their lives.

The American version of the contemporary home was disseminated in Britain through numerous popular television programs, including *I Love Lucy*. The contemporary style also featured in the sets developed for several British TV series. But most of all, it filled the pages of popular women's magazines on both sides of the Atlantic (plate 426). Mary Grieve, the editor of *Woman* magazine, understood the close link between consumption and women's entrance into modern life, and geared her magazine to exploit that association. For new consumers, buying their first house (often in suburbia) and equipping it with contemporary furniture, kitchen gadgets, and a large refrigerator (in one of a range of novel colors with exotic names) represented their primary material means of proclaiming their membership in the affluent postwar world.

The Do-It-Yourself movement was also imported to Britain from the U.S. in those years, offering people the opportunity to modernize their own homes economically. The consumer-oriented magazine *Do-It-Yourself* succeeded the much more technically focused *Practical Mechanics* of the prewar era, and its front covers showed couples working together to build cupboards, strip walls, and improve the backyard (plate 427). The myth of "togetherness" was widely disseminated, helping to confirm the importance of the close-knit nuclear family in the postwar era. On another level, it served to consolidate the isolation of the family unit and the loss of a traditional sense of community as, increasingly, identity and social status were formed through consumption and participation in the fashion cycle.

It was clearly felt that consumption, rather than production, was the key to Britain's future, and that consumption needed to operate within a context of educated taste, which at that time meant a preference for

COMPLETE GUIDE TO CEILING REPAIRS
JUNE 1959
Do it yourself
FOR THE PRACTICAL MAN ABOUT THE HOUSE
1/-
FEATURES INCLUDE
BUILD A QUARRY TILED VERANDAH
MAKE A SUN LEISURE-BENCH
ALL ABOUT BUYING A HOUSE
SIX EASY-TO-GROW HOUSE PLANTS
WORKSHOP TOOLS YOU CAN MAKE
BUILDING FIREPLACES & FLUES
INSTALLING A WATER CLOSET
USING PEGBOARD IN THE HOME

BELOW
428. Carl Jacobs (b. 1925)
***Jason* chair**, c. 1950
Made by Kandya Ltd (U.K.)
Beech, beech plywood
Victoria and Albert Museum, London

OPPOSITE, TOP LEFT
429. David Harman-Powell (b. 1931)
Stackable cups, c. 1968
Made by Ekco Plastics Ltd (U.K.)
Plastic
Victoria and Albert Museum, London

OPPOSITE, TOP RIGHT
430. Ronald E. Brookes
Plates, c. 1960
Made by Brookes & Adams Ltd (U.K.)
Melamine
Victoria and Albert Museum, London

OPPOSITE, BOTTOM
431. Kenneth Grange (b. 1929)
***Chef* mixer**, 1966
Made by Kenwood (U.K.)
Victoria and Albert Museum, London

the modern. Research on the emergence of the concept of the "teenager" in these years has also demonstrated that the consumption of certain essential items provided entry to a range of youthful male subcultures—in particular, those of the teddy boys and, later, the mods and rockers.

ANTI-AMERICAN SENTIMENT

In his 1958 book *Culture and Society*, the critic Raymond Williams wrote of the dangers inherent in emulating the American experience. Along with Richard Hoggart (cofounder with Williams of "cultural studies"), Williams feared that society's wholesale embrace of the mass media, including the products of the mass-production industries, would mean a loss of traditional culture and an erosion of standards. He believed that advertising in particular played a leading role in this pernicious transformation.

Immediately after the war, many British designers were hugely optimistic about the potential of plastics to create a brave new world. But others, including Williams and Hoggart, were more cautious, insisting that designers should have an intimate knowledge of these new materials' physical properties and capabilities. Moreover, following an initial enthusiasm for plastics when they entered the commercial marketplace after the war, consumers quickly became disenchanted with them, and doubts began to emerge about plastics' relationship to "good design." As the number of plastic products grew, and quantity inevitably took over from quality, they began to earn a reputation for being "cheap and nasty." That is, their very popularity began to backfire on them.

In Britain, this strong sense of disenchantment was aggravated by the association of plastics with American popular culture. Members of the Council of Industrial Design were among those who became increasingly concerned about the incursion of American commercial values on British soil, and in 1952 Alec Davis, the editor of the council's mouthpiece, *Design* magazine, declared that "American mass-production methods are hardly appropriate to the makers of, say, Staffordshire bone china, Yorkshire woollen cloth, Walsall leather goods, or London-tailored men's wear." In class-conscious Britain, memories of quality handmade goods still ran deep, and many people were suspicious of the brash new merchandise, including Hollywood films and pulp novels, that was being imported from across the Atlantic. In the years before an acceptance of mass culture emerged, these imports were seen as a threat to traditional British values. The Council of Industrial Design's chief information officer Paul Reilly (later to rise to the post of director) was especially anxious about the mass production of plastic products. "The temptation," he wrote, "to adorn a moulding, to imitate carving or to impress a stylistic cliché, such as the 3 parallel lines is certainly hard to resist when one considers the extreme reproduction of plastics."

Views such as these encouraged certain British designers, including Ronald E. Brookes, Gaby Schreiber,

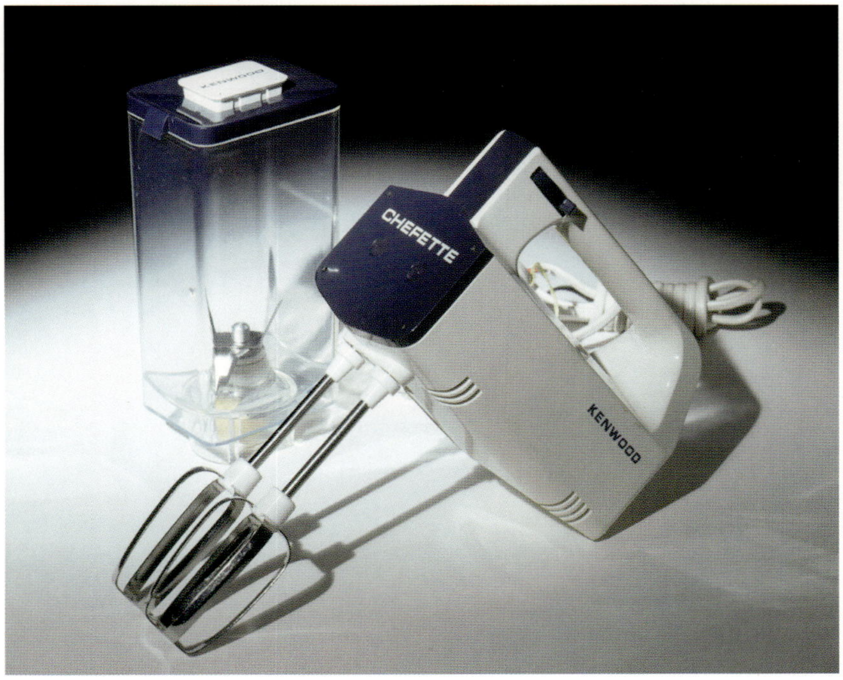

and David Harman-Powell, to rise to the challenge of applying the arts-and-crafts principles of "truth to materials" and "form follows function" to plastic products. The results of their efforts included some elegant pieces that have stood the test of time, among them Brookes's Fiestaware pieces for Brookes & Adams (plate 430), Schreiber's stylish objects for Runcolite, and Harman-Powell's stackable cups for Ekco Plastics (plate 429).

Through the 1950s, the mass media—film, television, magazines, advertising, and mass-produced goods and images themselves—was instrumental in communicating a number of lifestyle models that could be appropriated through consumption. While those models had different impacts in different places, many of them encouraged aspirations that had implications for patterns of consumption and, significantly, for design. A new generation of furniture designers, among them Ernest Race and Robin Day, extended the modernist program by incorporating new materials—including extruded aluminum, bent and molded plywood, and steel rods—into their designs (plate 428). Drawing on both American and Scandinavian examples (the work of Charles and Ray Eames and Arne Jacobsen, among others), Race and Day—the former working with his own manufacturing company and the latter with Hille Ltd.—both produced a number of innovative pieces that stood out in their era.

Day, who had been educated at the Royal College of Art before the war, worked with Clive Latimer on a

BELOW
432. Robin Day (1915–2010), with an engraving by Geoffrey Clarke (b. 1924)
Wall storage system, c. 1951
Made by Hille Ltd. (U.K.)
Steel, ebony, mahogany, plastic
Victoria and Albert Museum, London

OPPOSITE
433. Lucienne Day (1917–2010)
***Calyx* fabric**, 1951
Made by Heal's (U.K.)
Victoria and Albert Museum, London

steel-mounted buffet that won a prize in a competition organized by the Museum of Modern Art in New York in 1948. The publicity surrounding the prize brought Day to the notice of the British furniture manufacturer Hille Ltd., with which he established a long relationship. Working closely with Ray Hille, and subsequently with her daughter Rosamund, Day created a number of classic modern designs for the firm, notably a polypropylene stacking chair (1963) that quickly became ubiquitous and confirmed his status as one of Britain's leading designers (plate 436).

Both Robin Day and his wife, the eminent textile designer Lucienne Day, came to the notice of the design press, and the public, through their work for the Festival of Britain. While Robin designed the bent-plywood seating for the Royal Festival Hall (plate 434), Lucienne created her most famous textile, *Calyx*, whose brightly colored abstract pattern owed much to the work of the painter Joan Miró (plate 433). The couple also worked on a strikingly modern interior exhibited at the Triennale di Milano in 1954. While Lucienne went on to create countless textiles for clients including the Heal's store and, in the 1970s, began making large silk tapestries, Robin became well known as the designer of numerous lines of furniture (plate 432), as well as television sets and radios for the British firm Pye (plate 435), aircraft interiors for BOAC, and carpet designs for Woodward Grosvenor. Working independently for the most part (although sharing a studio for many years), together the Days became a fashionable designer-couple whose lifestyle was admired and widely emulated.

Heals "Calyx" by Lucienne Day
Made in United Kingdom

ABOVE
434. Robin Day (1915–2010)
700 chair for the Royal Festival Hall, 1956
Made by Hille Ltd. (U.K.)
Lacquered metal tubing, laminated wood
Galerie de Casson, Paris

RIGHT
435. Robin Day (1915–2010)
Television set, 1957
Made by Pye Ltd (U.K.)
Victoria and Albert Museum, London

OPPOSITE
436. Robin Day (1915–2010)
***Polyprop* stackable chairs**, 1963
Made by Hille Ltd. (U.K.)
Polypropylene, steel, fabric
Victoria and Albert Museum, London

HOT
SAVELOYS
SIX RARE STEAK KNIVES
SPECIAL ORIGAMI PAPER
PLAYER

OPPOSITE
437. John Maltby (1910–1980)
Poster for Habitat products, 1965
Royal Institute of British Architects, London

BELOW
438. **Cover of the Habitat catalog**, 1974

LIFESTYLE RETAILING

By the early 1960s, it was clear that a mass market for lifestyle retailing and journalism had emerged, and the furniture and textile designer Terence Conran was quick to exploit this trend, opening his first Habitat store on London's Fulham Road in 1964 (plate 438). Although some other progressive retailers—including Dunn's of Bromley, Heal's, and Liberty's—had already branded themselves as "design-led" in the 1950s, they had retained an exclusive appeal and had not played a direct part in the expansion of consumption. Habitat targeted a social group just below the customers of those elite stores, and a younger, more fashion-conscious audience. As well as offering a range of new objects imported from overseas (Vico Magistretti's bright red *Carimate* chair being a special favorite), the key message Conran communicated to his customers was that, as interior decorators already knew well, it was possible to mix old and new furniture. Rather than empower his customers to do that for themselves, however, he offered them a ready-made set of choices. In effect, they were asked to trust his taste and purchase whole interiors from his shop. That might mean combining a Magistretti item with a reproduction Thonet chair or Chesterfield sofa, both of which could also be purchased at Habitat. Only a designer could be trusted to make such courageous decisions! Habitat thrived on this model, subsequently opening a number of other stores around the country (plate 437).

ABOVE
439. **Double-decker bus, *AEC Regent III* model (the predecessor of the *Routemaster*)**, 1952

OPPOSITE, TOP
440. **Advertisement for the Austin *Mini Cooper***, 1969

OPPOSITE, BOTTOM
441. **Sir Alexander Issigonis (1906–1988), designer of the Morris *Minor* and the Austin *Mini***, 1965

TRANSPORT DESIGN

Alec Issigonis's Morris *Minor* was a British automotive icon of the 1950s, but his *Mini* made an even bigger impact in the 1960s (plate 441). He began working on it for the British Motor Corporation (formed by the merger of Morris with Austin) in 1956. The stimulus to create the *Mini* came from the fuel shortages caused by the Suez crisis and the competition from microcars. Issigonis's challenge was to create a very small car for four people, plus luggage, that would use as little fuel as possible. The solution lay in placing the engine in a transverse position and the wheels at the extreme four corners, to allow the maximum amount of interior space. The exterior was a direct result of these decisions, and Issigonis produced a timeless design with little additional ornament or styling. The streamlined curves of the *Minor* had disappeared, giving way to what was essentially a box on wheels. The *Mini* quickly became the car of choice for pop stars, fashion models, and designers and was celebrated as one of Swinging London's cult objects (plate 440).

While Britain acquired a strong reputation for car design in the postwar decades, with models like the Jaguar *E Type* (plates 442, 443) and the *Range Rover* becoming known internationally, other areas of British transport design also flourished in that period,

BELOW
442. ***E Type Roadster***, 1961
Made by Jaguar

BOTTOM
443. **Publicity postcard for the Jaguar *XK-E***, c. 1960

OPPOSITE
444. ***Concorde* 102 supersonic airplane**, 1973
Made by Aérospatiale (France) and British Aerospace (U.K.)

counting such successes as the classic *Routemaster* bus (plate 439) and culminating in the *Concorde* supersonic airplane (plate 444), conceived in collaboration with French engineers. The *Concorde*'s first flight in 1969 was the result of nearly twenty years of development that had led to new advances in aerodynamics, materials, and structures. Like the design of the *Spitfire* before it, the *Concorde*'s strikingly elegant and futuristic nose and wing forms were the result of functional and performance-based criteria. The materials used in its construction—including titanium, stainless steel, and a variety of plastics—were selected after rigorous testing.

OPPOSITE
445. Terence Conran (b. 1931)
Salad set, 1955
Made by W. R. Midwinter Ltd (U.K.)
Glazed earthenware
Victoria and Albert Museum, London

BELOW
446. Robert Heritage (b. 1927)
Sideboard, 1958
Made by Archie Shine Ltd (U.K.)
Rosewood, marble
Private collection

POP DESIGN AND THE RETURN OF MODERNISM

Pop design set in motion a wholesale rejection of the interwar modernism that had insisted an object's outer form should reflect its means of production. Pop objects aimed, instead, to have an expressive relationship with the fashion-conscious society that produced and consumed them. They were intended to have short life cycles, based on the idea that was better throw away an object once it had outlived its ability to express the moment than to keep it. Pop design represented an acceptance on the part of both designers and consumers that design should be led by the needs of the marketplace rather than those of manufacturing. Real disposability—possible, for example, with paper clothing—was one way forward, although it was accompanied by other objects that could not actually be thrown away but resisted permanence through the ability to change their forms and appearance. Although disposability was clearly a problem for plastic products, many of the newer plastics were inherently soft and flexible, and ideally suited to a culture of accelerated change. Such was the chameleon-like nature of plastics that they could don the mantle of good design at one moment and reject it at another. Indeed, without plastics the Pop design movement could not have produced some of its most lasting icons. Alongside the loud music, colorful clothes, and other "fun" lifestyle accessories that characterized youth culture in 1960s Britain, a number of plastic products played a role in overturning what had seemed like the universal values of "good design."

The 1960s saw the English architectural group Archigram introduce flow and movement into their idealized urban visions; they believed the chair was on its way out and pneumatic seating that formed to the body would take its place. The English architectural critic Reyner Banham proposed a plastic bubble as a new housing solution. A number of inflatable plastic chairs emerged, including the *Pumpadinc* armchair designed by Arthur Quarmby and a version covered in fake fur, created by the album cover artist Roger Dean and produced by Hille. Not only was inflatable furniture here one minute and gone the next, but it also suggested a whole new way of living that eroded the distinction between indoors and out.

This new vision for furniture was linked to the general shift in design thinking anticipated by the graphic designer Michael Wolff, who had written in the *Journal of the Society of Industrial Artists* in 1965 that "it will be a great day when furniture and cutlery design, to name but two, swing like the Supremes." By the end of the decade, a new, increasingly youth-focused lifestyle and the prioritization, in the context of mass consumption, of image over object had led in two distinct directions: a flexible form of seating that abandoned the long-standing dependence on chairs in favor of "seating systems," and a new emphasis on surface pattern that encouraged (especially young) people to take back control from designers by adding their own images of flags, targets, and brightly colored stripes to the surfaces of their furniture, both old and new.

The 1950s and early '60s saw the appearance of a considerable amount of elegant neomodernist furniture created by Terence Conran, Alan Turville, Robert Heritage, John and Sylvia Reid, and other British designers and manufactured by firms such as Hille Ltd., Ercol, G-Plan, HK Furniture, Race Ltd., and Archie Shine (plate 446). It was also in the 1950s that a handful of adventurous manufacturers, such as G-Plan, broke away from the norm, offering sets of contemporary-looking furniture to young couples setting up house for the first

BELOW
447. Max Clendinning (b. 1924)
Cupboard with stand, 1965
Made by Liberty (U.K.)
Painted plywood
Victoria and Albert Museum, London

OPPOSITE, TOP
448. Max Clendinning (b. 1924)
***Maxima* armchairs**, 1965
Made by Race Furniture Ltd. (U.K.)
Lacquered plywood, fabric upholstery
Private collection

OPPOSITE, BOTTOM
449. Bernard Holdaway (1934–2009)
***Tom-tom* side table**, c. 1967
Made by Hull Traders Ltd (U.K.)
Cardboard
Victoria and Albert Museum, London

time. These companies took their lead from Scandinavian furniture makers, emulating their commitment to wood (teak and rosewood in particular), to light, elegant forms, and to a minimal use of upholstery. By the end of the 1950s and the early '60s, British firms were also looking to Italy for inspiration. With its debt to tradition, combined with its strikingly tapered wooden legs and modern elegance, Gio Ponti's *Superleggera* chair, manufactured by Cassina in 1957, was one of the most influential pieces to be exhibited at the Triennale di Milano and covered in the burgeoning international design press. Its features were quickly replicated in some of Britain's most successful modern furniture of the period.

By the early 1960s, there were a number of signs that, on one level at least, the British public was finally becoming aware of modern design. And by the middle of that decade, the Pop revolution had infiltrated the everyday environment, bringing with it a new awareness of color and innovative materials. One piece of Italian furniture in particular played an important role in that transformation: Vico Magistretti's *Carimate* chair, designed in 1959 for the restaurant of the Carimate Golf Club in Lombardy and distributed by Cassina three years later. Its rush seat and red-lacquered wood frame embellished many fashionable London restaurants and came to epitomize the new Pop environment. As already noted, it also caught the eye of the young English designer Terence Conran—soon to be one of the most important retailers of modern design in Britain—who produced his own version of it. Magistretti himself, who worked as an adviser to Cassina for over thirty years, until his death in 2006, developed strong ties with London from the early 1960s. He kept a flat in Onslow Square in Kensington and was a visiting professor at the Royal College of Art for more than a decade.

Britain's modern-furniture renaissance of the 1960s was also strongly influenced by a number of other pieces manufactured by Cassina. The Italian company's products were frequently featured in the British design press and were sold to a discerning clientele through the new furniture retailers that appeared in London and elsewhere, among them Aram Designs, Geoffrey Drayton, Oscar Woollens, and Interspace, which was opened in 1968 by the Danes Nanna Ditzel and Kurt Heide. The June 1965 edition of *House and Garden* magazine reported that Italian design was attractive to many young lodgers in Chelsea and Bloomsbury and selected Gianfranco Frattini's *Poltrona Chair 877*, manufactured by Cassina, as one example of that country's achievements in modern design. In February 1968, Tobia Scarpa's knock-down armchair for Cassina was depicted in *Interior Design* magazine, which proclaimed that "this year the Italians have produced the most lively, most controversial, as well as the most likely starters of the prestige race," and praised Italian design for its "delight in new materials and an approach to colour that British designers and manufacturers would do well to examine." In December 1970, the same journal featured a swiveling office chair made of PVC that Mario Bellini designed for Cassina.

A new generation of British designers—including Max Clendinning, Jon Bannenberg, Peter Murdoch, William Plunkett, Bernard Holdaway, Nicholas Frewing, and the team of Jean Schofield and John Wright—embraced the novel concepts of the "knock-down," "inflatable," and "throwaway" (plate 447). They produced such lasting classics as Clendenning's *Maxima* range for Race—a set of chairs in lacquered plywood, sold flat for home assembly (plate 448)—and his side chair for Liberty's, which had a painted plywood frame; Murdoch's so-called "paper" chair, which was actually made of a robust cardboard strengthened with resin (a spotted version appeared in 1963, and one with a more complex pattern five years later; plates 450, 451); and Holdaway's *Tom-tom* range for Hull Traders (plate 449).

LEFT
450. Peter Murdoch (b. 1940)
***Spotty* chair**, 1963
Made by International Paper (U.K.)
Cardboard coated with polyethylene
Musée des Arts Décoratifs, Paris

BELOW
451. Peter Murdoch (b. 1940)
***Thing* chair**, 1968
Made by Perspective Designs Ltd (U.K.)
Cardboard coated with polyethylene
Victoria and Albert Museum, London

OPPOSITE
452. Geoffrey Baxter (1922–1995)
Vases, c. 1965
Made by Whitefriars (U.K.)
Molded glass
Victoria and Albert Museum, London

BELOW
453. David Mellor (1930–2009)
***Thrift* flatware**, 1965
Made by Oneida (U.K.)
Stainless steel

BOTTOM
454. Gerald Benney (1930–2008)
***Saddleback* teapot**, 1960
Silver, wood
Victoria and Albert Museum, London

OPPOSITE, TOP
455. David Queensberry (b. 1929)
Cup and saucer, 1960–62
Made by W. R. Midwinter Ltd (U.K.)
Glazed earthenware
Victoria and Albert Museum, London

OPPOSITE, BOTTOM
456. Tom Arnold (d. 2002) and Enid Seeney (1931–2011)
***Homemaker* table service**, 1957
Made by Ridgway Potteries (U.K.)
Glazed earthenware
Victoria and Albert Museum, London

The decorative arts of the 1960s also succumbed to the pull of modernity, although for the most part they stopped short of going completely Pop. Rather, a neo-modern decorative arts movement emerged. In the field of metalwork, some of the most successful pieces of the mid-1950s onward were designed by the small group of distinguished RCA graduates mentioned earlier: David Mellor, who designed the elegant *Thrift* cutlery for the Ministry of Public Building in 1965 (plate 453); Robert Welch, who created a highly sculptural three-legged candelabra called *Campden* for J and J Wiggin Ltd. in 1957, the Scandinavian-looking *Alveston* tea set for Old Hall Tableware Ltd. in 1962, and a simple and graceful cutlery set for the same company three years later; and the silversmith Gerald Benney, who was responsible for many elegant and highly regarded designs (plate 454), including the pitchers for the tables of the RCA's senior common room. Another RCA alumnus (and professor of ceramics and glass there from 1959), David Queensberry, worked for W. R. Midwinter Ltd. (plate 455) and Webb Corbett Glass, among others, and received the Design Council's Duke of Edinburgh Award for Elegant Design in 1964.

The ceramic designs of those years included the enduringly popular *Homemaker* tea service conceived by Enid Seeney (who created the pattern) and Tom Arnold (who designed the form) for Ridgway Potteries in 1957 (plate 456). Jessie Tait's pieces for Midwinter, among them the *Cuban Fantasy* of 1957, also received much praise and were hugely influential. Glass design responded to the spirit of the age as well: for instance, Geoffrey Baxter's *Unica* and *Brick* vases (1965 and 1966, respectively) abandoned the delicacy of the 1950s for the solid forms and bright colors of British Pop, a shift that pushed glass to its technical limits (plate 452).

BELOW
457. **Ron Arad in his London studio**, 1991

OPPOSITE
458. Ron Arad (b. 1951)
***Big Soft Easy* armchair**, 1991
Made by Moroso (Italy)
Steel, expanded polyurethane foam, fabric
Musée National d'Art Moderne, Centre Georges Pompidou, Paris

NOSTALGIA AND HERITAGE

Through the 1950s and '60s, the modern design movement had created a new image of elite domestic living that, in turn, influenced mass taste. However, by the early 1970s, this movement had all but vanished from view, as a result of the downturn in the economy and the decline of British manufacturing. Suddenly British society began to feel less at home with modernity than it had in the more optimistic climate of the preceding decades. A Victorian revival—all stripped pine and Laura Ashley fabrics—engulfed the popular British home, while, for the social elite, the country house replaced the sophisticated townhouse as a domestic idyll. In essence, the modern style, created by a generation of highly skilled designers and interior decorators, was replaced by an escape to the past and to the country. In the process, the subtle balance between tradition and modernity that had manifested itself in a number of different ways in postwar British homes was irreparably disturbed. Many ceramic, glass, and metal objects were also produced by hand in those years, as part of the burgeoning British craft movement. The primitive aesthetic cultivated by the potters Ruth Duckworth, Lucie Rie, and Hans Coper, for example, provided a perfect counterpoint to the slick look of industrially manufactured pieces.

DESIGN CULTURE

With the newfound affluence of the 1980s, the designer came to the fore once again in Britain, and names such as Ron Arad, Tom Dixon, Nigel Coates, Daniel Weil, Eva Jiřičná, and Ben Kelly filled the pages of popular new design publications like *Blueprint* magazine (plate 457). This was, in fact, the decade of the designer as celebrity, and terms like "hair designer" began to replace more familiar ones, like "hairdresser."

Arad, one of Britain's most prominent and inventive designers, had come from Israel to study at London's prestigious Architectural Association in the mid-1970s. In 1983, he opened a shop called One Off in Covent Garden, where he sold his own designs (plate 459). His best-known creations of that era include the

BELOW
459. Ron Arad (b. 1951)
***This Mortal Coil* bookcase**, 1993
Made by One Off Ltd (U.K.)
Mild steel
Victoria and Albert Museum, London

OPPOSITE
460. Ron Arad (b. 1951)
***Bookworm* bookshelf**, 1993
Made by Kartell (Italy)
PVC plastic
Victoria and Albert Museum, London

Willem
de Kooning
DE KOONING
A BOOK BY ANSELM KIEFER
BERLIN МОСКВА
MOSKAU БЕРЛИН
PUBLIC INFORMATION
ALTROVE
SHARK INFESTED WATERS
Möbel des Jugendstils
MENDINI
paul feiler
Andy Goldsworthy STONE
STONE

BELOW
461. Ron Arad (b. 1951)
***Pappardelle* chair**, 1992
Made by One Off Ltd (U.K.)
Stainless steel
Fonds National d'Art Contemporain, Paris, on loan to the Musée des Arts Décoratifs, Paris

OPPOSITE
462. Ron Arad (b. 1951)
***Ge-Off Sphere* lamp**, 2000
Made by Ron Arad Associates (U.K.)
Stainless steel, mild steel, plastics, stereolithographed polyamide
Musée National d'Art Moderne, Centre Georges Pompidou, Paris

Rover chair, consisting of a leather seat from a Rover car mounted on a tubular steel frame, and a concrete record player. The distressed look of the shop and its products constructed from ready-made parts were described as "postapocalyptic" and linked with the punk movement. The rough edges of Arad's early one-off designs were soon replaced, however, by the much more sophisticated and sculptural forms of the pieces he created for a wide variety of international manufacturers (plate 462). Notable among these were his *Big Soft Easy* chair of 1991 for Moroso (plate 458), his *Bookworm* shelf of 1993 for Kartell (plate 460), and his *Tom Vac* dining-room chair of 1997. His designs experiment with materials and expressive forms. The latter are best exemplified by his *Pappardelle* chair of 1992, which resembles a waterfall made of stainless steel (plate 461). Arad has also earned a reputation in architecture with his 2008 design for the Bauhaus Museum in Tel Aviv, and has recently ventured into new areas such as eyewear.

Tom Dixon also established himself in Britain and went on to become an international name (plate 464). Like Arad, he was self-taught in design and came from abroad, in his case Tunisia, although he moved to England at the age of four. His breakthrough came when he learned to weld, which led to his first notably successful design, the *S Chair* of 1991, manufactured by Cappellini (plate 463). Combining a metal frame with a one-piece wicker seat and back, it possessed a sculptural quality that became the hallmark of Dixon's work. In 1997, he became head of design for the Habitat stores, a role he would also take on at the Finnish firm Artek.

One of the most successful design-related branding exercises of the late twentieth century was undertaken by James Dyson. Among postwar British designers, Dyson stands out as an engineer and inventor who embraced design in order to create products that were both technologically and aesthetically distinctive. Before entering engineering, Dyson had studied furniture and interior design at the Royal College

OPPOSITE
463. Tom Dixon (b. 1959)
S Chair, 1991
Made by Cappellini (Italy)
Steel, rush
Museum of Modern Art, New York

BELOW
464. Tom Dixon (b. 1959)
***Jack* lamp**, 1994
Made by Eurolounge Ltd (U.K.)
Polyethylene

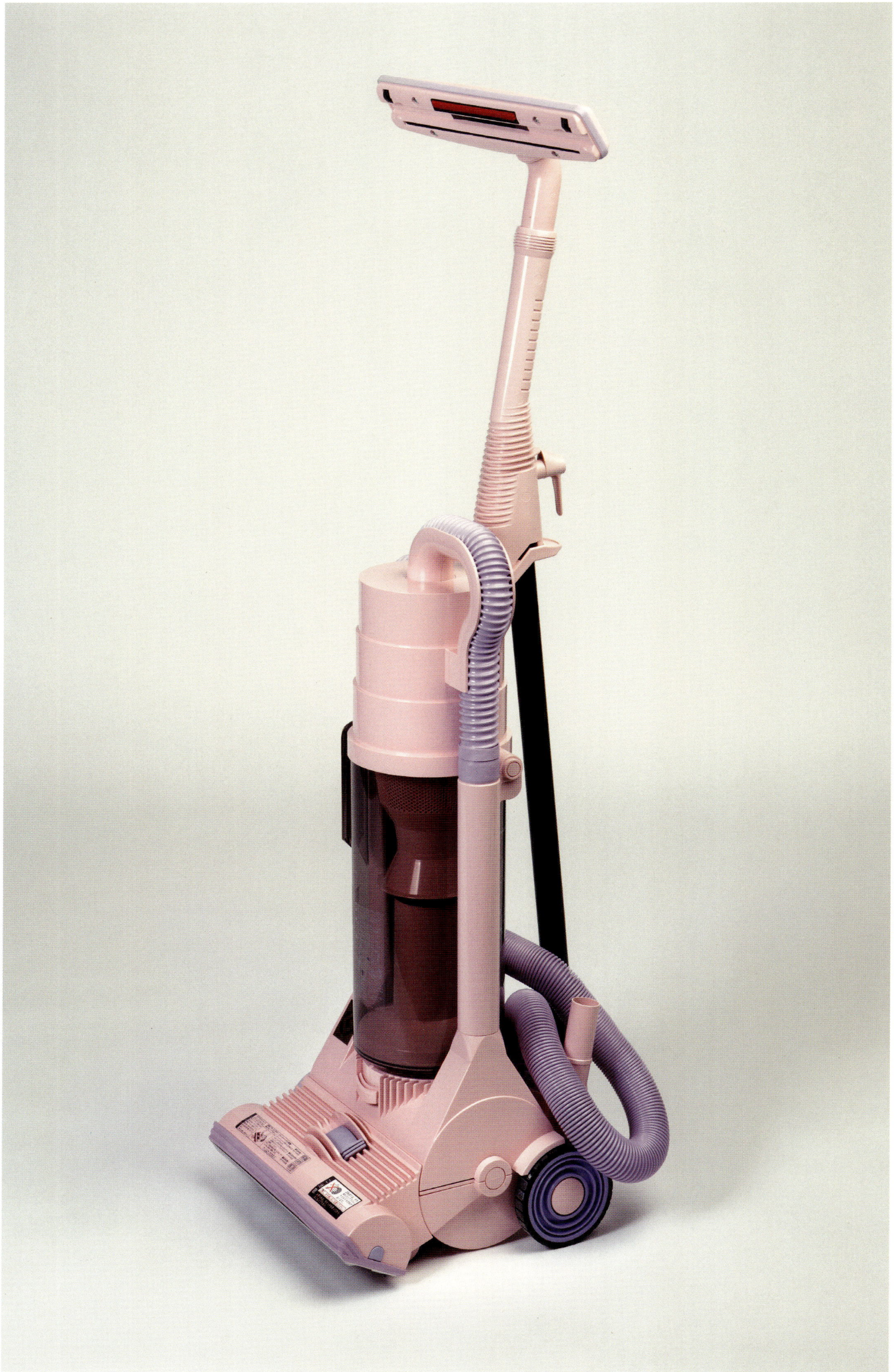

OPPOSITE
465. James Dyson (b. 1947)
***G-Force* vacuum cleaner**, 1986
Made by Dyson Ltd (U.K.)
ABS plastic, polycarbonate, electrical components
Musée National d'Art Moderne, Centre Georges Pompidou, Paris

BELOW
466. James Dyson (b. 1947)
***Air Multiplier* fan**, 2009
Made by Dyson Ltd (U.K.)
ABS plastic, metal, electrical components
Musée National d'Art Moderne, Centre Georges Pompidou, Paris

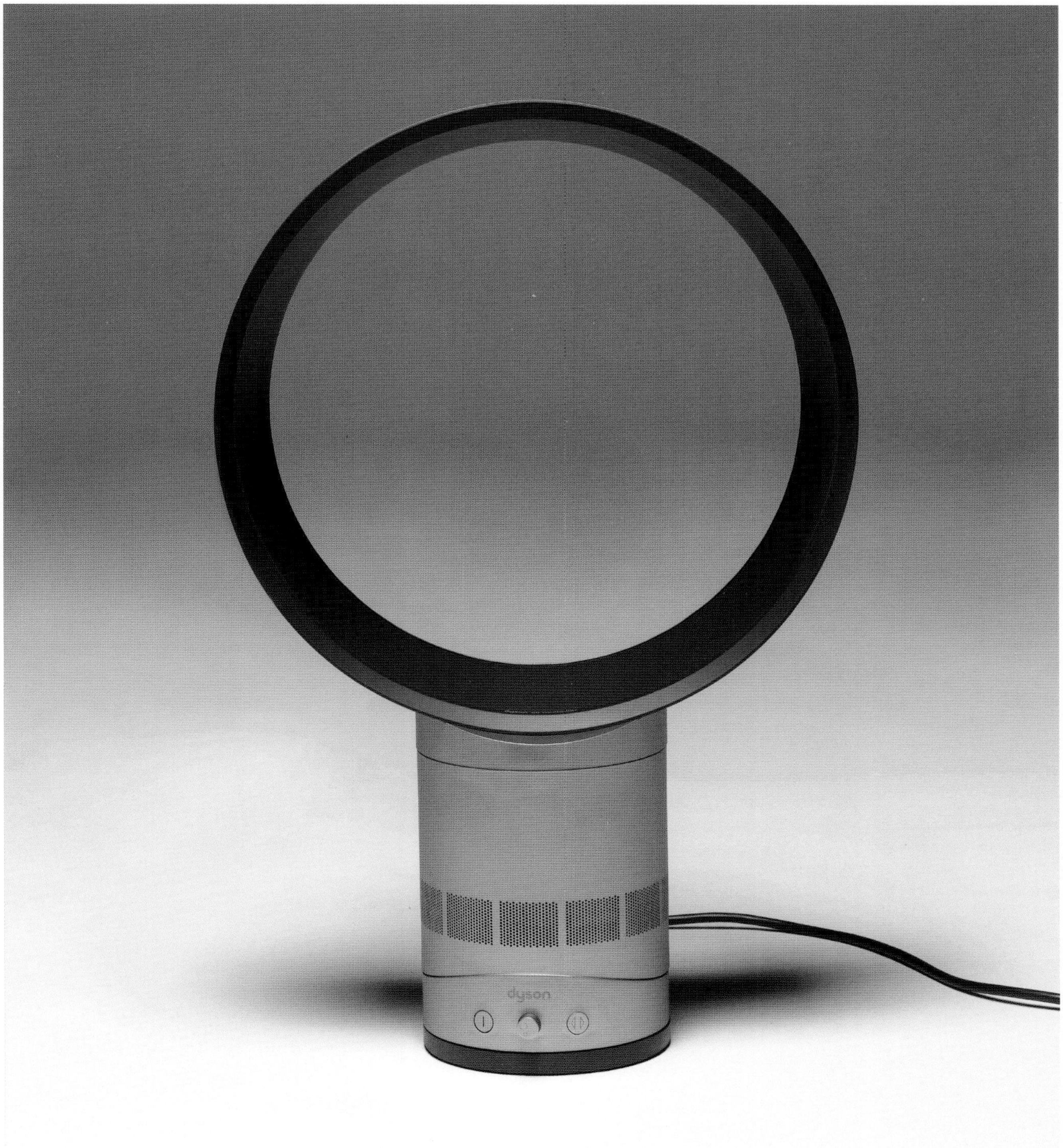

of Art. He conceived of his first product, a boat called the *Sea Truck*, in 1970, while he was still a student. He went on to design the *Ballbarrow*, a wheelbarrow with a ball rather than a wheel. His key invention, however, was the bagless vacuum cleaner. Known first as the *G-Force* and later as the *Cyclone*, it went through numerous prototypes and took several years to get to market (plate 465). Unable to convince British manufacturers to adopt his vision, he launched his vacuum with a Japanese company before eventually deciding to manufacture it himself. In June 1993, he opened his own research center and factory in Wiltshire. Two years later, he employed the ball concept from his wheelbarrow in a vacuum cleaner. Dyson understood the need to make his vacuum cleaner look friendly, so he introduced new colors—pink when it was being made in Japan, and later a whole spectrum of other bright hues, from purple to yellow. Building on the success of what became a wide range of vacuums (which were referred to as "Dysons," displacing the hitherto ubiquitous "Hoover"), he went on to launch a novel washing machine (2000), a hand dryer called the *Airblade* (2006), and, most recently, a fan without external blades, known as the *Air Multiplier* (plate 466).

BELOW
467. Jasper Morrison (b. 1959)
***Hal* chairs**, 2010–11
Made by Vitra (Switzerland)
Polypropylene, wood, metal, various coverings

OPPOSITE
468. Jasper Morrison (b. 1959)
***Low Pad* chair**, 1999
Made by Cappellini (Italy)
Stainless steel, polyurethane foam, brushed steel, fabric
Musée des Arts Décoratifs, Paris

THE NEW CENTURY

In yet another pendulum swing, the 1990s and early 2000s saw a shift away from the dominance of the individual designer and a renewed interest in design teams and cooperative design; in the implications of design for the environment, society, and well-being; and in the "intangible" areas of design, such as service, software, and interface design. By the early 1990s, designers such as Jasper Morrison had already begun to develop a more self-effacing approach to their work, and a new minimalism emerged in the area of interiors, with design groups like Barber Osgerby, founded in 1997, leading the way. Along with Arad and Dixon, Morrison has become one of Britain's best-known designers. Unlike them, however, he has rejected the often extravagant art-based approach to furniture and product design in favor of a much more low-key and style-free one. Another product of the Royal College of Art (1985), Morrison also studied in Berlin. He founded his own design studio in London in 1986 and set off on a career that was to define a new movement in British design, sometimes called the New Simplicity. Morrison introduced a neo-functionalist approach characterized by simple, modest forms applied to a wide spectrum of objects, from furniture (plates 467, 468) to a streetcar for the city of Hanover (plate 477) to a bus shelter (for the Vitra campus in Weil am Rhein). He worked and exhibited alongside the revered Japanese designer Naoto Fukasawa, with whom he also collaborated on objects for the Japanese retail chain Muji. Recent designs by Morrison include a family of outdoor furniture for the Spanish company Kettal, an alarm clock for the Swiss brand Punkt, timepieces for

OPPOSITE, TOP
469. Marc Newson (b. 1963)
***Dish Doctor* dish rack**, 1998
Made by Magis (Italy)
Injection-molded polypropylene
Centre National des Arts Plastiques/Fonds National d'Art Contemporain, Paris

OPPOSITE, BOTTOM
470. Jerszy Seymour (b. 1968)
***Pipe Dreams* watering can**, 2000
Made by Magis (Italy)
Blow-molded polyethylene
Centre National des Arts Plastiques/Fonds National d'Art Contemporain, Paris

LEFT
471. Ross Lovegrove (b. 1958)
***Eye* digital camera**, 1990
Made by Olympus (Japan)

BELOW
472. Sam Hecht (b. 1969)
***Two Timer* clock**, 2009
Made by Established & Sons (U.K.)

LEFT
473. Paul Cocksedge (b. 1978)
Neon lights, 2003
Made by Paul Cocksedge Studio (U.K.)
Hand-molded glass and suspension wires

BELOW
474. Marc Newson (b. 1963)
***Atmos 566* clock**, 2010
Made by Jaeger-LeCoultre (Switzerland)

OPPOSITE
475. Marc Newson (b. 1963)
Pod of Drawers, designed 1987, made 1999
Aluminum, wood, fiberglass
Musée des Arts Décoratifs, Paris

the Swiss watchmaker Rado, footwear for the Spanish firm Camper, and a number of furniture items for the Swiss manufacturer Vitra.

Other British designers who flourished in the 1990s and early 2000s include Marc Newson (plates 469, 474, 475) and the RCA graduate Sam Hecht, who was head of design at IDEO in the mid-1990s and then worked in Japan for three years (plate 472). Later he was retained by Muji and was responsible for many of their products, including the *Second Phone* (2004). Hecht founded the design studio Industrial Facility with his partner Kim Colin in 2002, and they have collaborated with a number of international manufacturers, including Muji, Yamaha, and Magis. More recently, he became design adviser to Herman Miller. Several more RCA alumni, including Sebastian Bergne, Paul Cocksedge (plate 473), Doshi Levien, and Ross Lovegrove (plate 471), also began to make an international impact in these years.

This new wave of British design was encouraged by a group of influential tastemakers, such as Sheridan Coakley, who set up the furniture store SCP in 1985 and played an important role in the successes of Jasper Morrison, Matthew Hilton, and Terence Woodgate, among others. Two decades later, in 2005, the retailer Established & Sons began working with a stable of designers, promoting them both at home and abroad. The arrival of the annual London Design Week and the London Design Festival also contributed to the view that the British capital was at the center of innovative design.

In the early years of the twenty-first century, the impact of British design on the world of high tech was underscored by the work of two London natives in California: Jonathan Ive at Apple Inc. and Bill Moggridge at the design firm IDEO. Britain also continues to lead the way at the experimental, gallery-oriented end of the design spectrum, with Tony Dunne and Fiona Raby's contribution to the field of interaction design, and the presence of the avant-garde Spanish design team El Ultimo Grito in London.

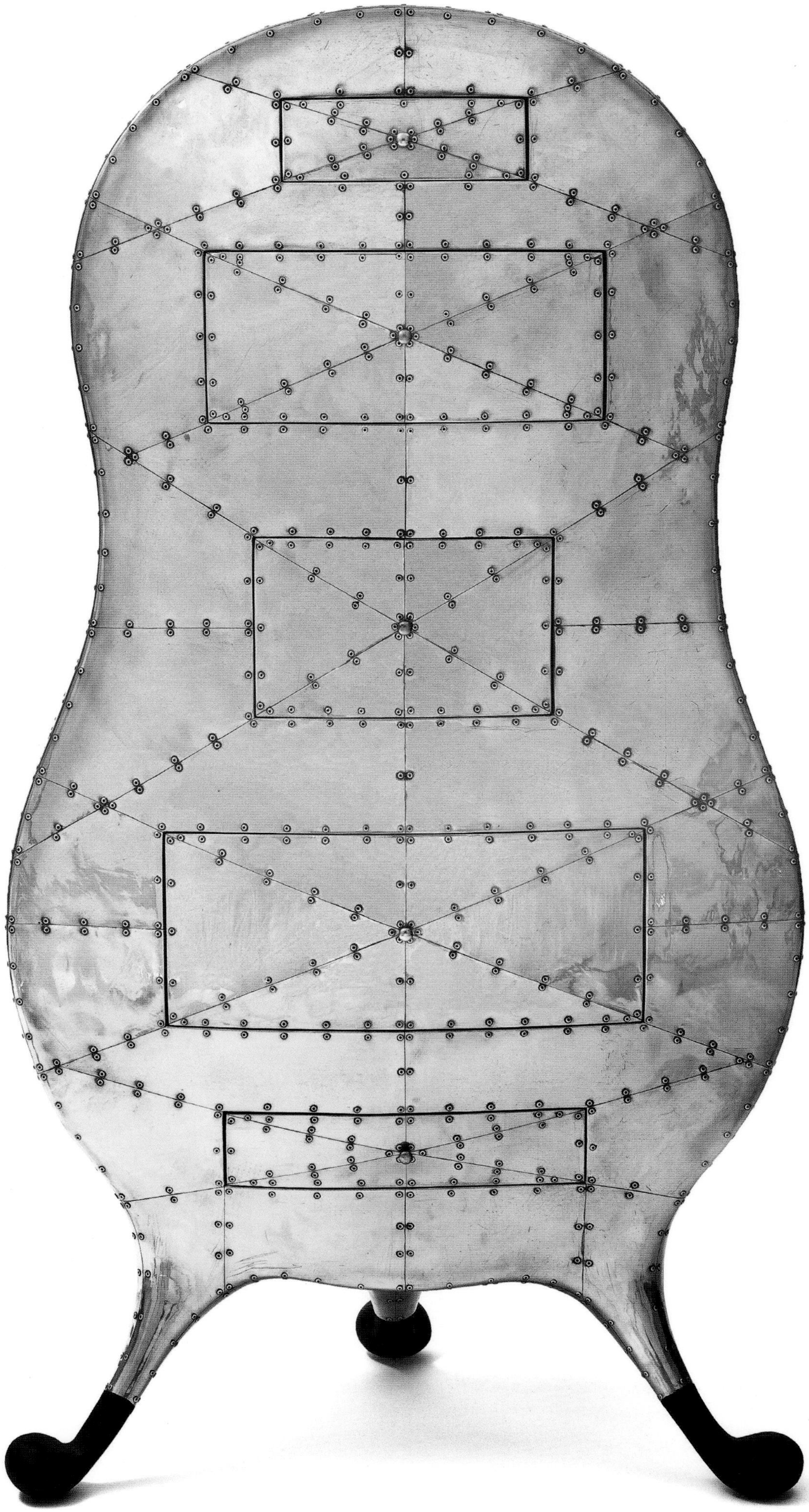

Hackney Central 38
ARRIVA
LT61 BHT

OPPOSITE, TOP
476. Thomas Heatherwick (b. 1970)
New *Routemaster* bus, 2012
Made by Wrightbus (U.K.)

OPPOSITE, BOTTOM
477. Jasper Morrison (b. 1959)
Hanover streetcar, 1997
Made by Linke Hoffman Busch (Germany)

BELOW
478. Kenneth Grange (b. 1929)
***Anglepoise Type 1228* lamps**, 2004
Made by Anglepoise (U.K.)
Aluminum, stainless steel, polycarbonate

KENNETH GRANGE REVISITED

A decade into the twenty-first century, Britain began to look back at its design history. In 2011, London's Design Museum devoted its exhibition space to the work of the most eminent British product designer of the postwar era, Kenneth Grange. The show marked a moment of reflection on Britain's contribution to the world of design over the previous half-century, as well as the search for native designer-heroes to position alongside the United States' Charles and Ray Eames, Italy's Ettore Sottsass, Germany's Dieter Rams, and France's Philippe Starck. Moreover, it consolidated the role of the Design Museum—itself a result of the design culture of the 1980s, and of the generosity of Terence Conran.

Grange had come to the world of industrial design through architecture. His earliest professional experience was working for Bronek Katz, a modernist architect of Polish origin, in the area of exhibition design. He worked in the office of a second architect, Gordon Bowyer, between 1948 and 1951, and then in that of a third, Jack Howe, where he stayed on as a drafting assistant until 1958, when he set up his own office as an industrial designer, building on freelance work he had undertaken for Kodak.

In that same year, the Council of Industrial Design introduced Grange to the appliance maker Kenwood. Grange's redesign of the Kenwood *Chef* was launched at the Ideal Home Exhibition of 1960 (plate 431). With its ability to perform multiple tasks, including chopping, mincing, mixing, and kneading, this device was an immediate success with housewives. Grange's ongoing collaboration with Kenwood generated more than one hundred products, including not only the various models of the *Chef*, the handheld *Chefette*, and more recent food processors like the *Cuisine* and *Gourmet* but also kettles, irons, a slow cooker, a water filter, a fruit juicer, and a highly sculptural electric knife (1967).

Notable among Grange's other designs are a small gas cooker developed for Osaka Gas in the late 1970s, and a set of microwave containers created for Thorpac in the mid-1980s. These containers stand out for their simple, curved forms and the choice of a high-quality material—in this instance, Udel polysulfone. Grange was also called on to revitalize one of the original icons of British design, George Carwardine's *Anglepoise* lamp (1932), for a design-conscious market (plate 478).

The 2012 Olympic Games in London gave the city an opportunity to showcase the maturity of its creative industries. The design contributions of Barber Osgerby (the Olympic torch), Thomas Heatherwick (the Olympic cauldron), and Zaha Hadid (the swimming stadium) were broadcast around the world. Heatherwick was also responsible for the redesign of the *Routemaster* bus, one of the icons of postwar British design (plate 476). Meanwhile, the Victoria and Albert Museum hosted an exhibition of British design from 1948 to 2012, reinforcing the fact that the nation had established a firm place for itself in the global history of postwar design.

By the early twenty-first century, British design since World War II could clearly be seen as a series of significant shifts in emphasis, led for the most part by farsighted designers, inventors, manufacturers, and retailers, and taking place in a context of growing affluence followed by economic decline. It had alternately hung on to the past and embraced the future, like the fast-changing society whose identity it helped to define.

FRANCE

DOMINIQUE FOREST

PRECEDING PAGE
479. Pierre Paulin (1927–2009)
577 or *Tongue* chair, 1967
Made by Artifort (Netherlands)
See plate 543.

OPPOSITE
480. Eileen Gray (1878–1976)
Lounge chair for Villa E 1027 in Roquebrune-Cap-Martin, France, 1926–29
Varnished sycamore, synthetic leather, nickel-plated steel
Musée National d'Art Moderne, Centre Georges Pompidou, Paris

THE INTERWAR YEARS AND RECONSTRUCTION

In France, as in many other European countries, the interwar years were an important period of critical reflection on lifestyle and its artifacts. Furniture design remained predominantly traditional in its approach, although the modernist movement made some inroads. The heritage of fine cabinetmaking, exemplified by the great Jacques-Émile Ruhlmann, endured through the 1930s and '40s with distinguished practitioners including René Arbus, Jacques Adnet, Paul Dupré-Lafon, Jacques Quinet, and Maxime Old. At the same time, the unclassifiable Jean Royère developed the fantastical style that became his trademark. A taste for period furniture also persisted, thanks to a thriving market for antiques and, in Paris, the furniture makers of the Faubourg Saint-Antoine. However, the complacency of this bastion of tradition had been shaken during the 1910s by the independent-minded designer Francis Jourdain, son of Frantz Jourdain, the architect of the Samaritaine department store. In 1912, this committed socialist founded the Ateliers Modernes, where he devoted himself to the study of materials and functions, developing prototypes for furnishings that were simple, interchangeable, and mass-produced. Jourdain was the first to promulgate the theories of the Viennese architect Adolf Loos. In 1913, he published a French translation of Loos's *Ornament and Crime*, originally issued in 1908. Like Loos, Jourdain continued to advocate an ethic of simplicity throughout the 1920s: "Our need for comfort is increasing, and the space available to us is diminishing. We no longer have the right or the resources to waste space for purely decorative purposes. The luxury of large rooms is obsolete. How best to furnish small spaces . . . that is the distinctively modern problem."[1] Jourdain defended rationalism, as did Le Corbusier, who also called for profound intellectual change, questioning the "art" of living and furnishings, referring to the "equipment" of a house, criticizing decoration, and seeking simple, economical forms and materials.

Architecture played an essential role in this surge of fresh ideas. During the decisive decade of the 1920s, a number of newly built structures served as laboratories of modernity, combining architecture and furniture design. They included Le Corbusier's La Roche and Jeanneret villas of 1924, Robert Mallet-Stevens's Villa Noailles of 1928, Eileen Gray's Villa E 1027 of 1926–29 (plate 480), and Pierre Chareau's Maison de Verre of 1928–32. However, the concept of creating architecture, interior design, and furnishings as an ensemble remained the exception, being applied only to a very limited number of prestigious commissions. In order to emerge from the shadows and actively promote their theories, most avant-garde designers associated themselves with the Union des Artistes Modernes (UAM), established in 1929 by a group of architects, decorators, graphic designers, and artists. Whether or not they were among the founding members, almost all the advocates of the avant-garde participated in the UAM's exhibitions, including Jourdain, Le Corbusier, Mallet-Stevens, Gray, Chareau, René Herbst, Charlotte Perriand, and Jean Prouvé. The group supported the industrialization of production methods and endorsed the value of industrial expertise and materials, although their ties with business were nevertheless more tenuous than they would have liked. The iconic chaise longue designed in 1928 by Le Corbusier, Perriand, and Pierre Jeanneret has, it is true, been distributed by Thonet since 1931, but it remains the exception. Many projects remained at the prototype stage, or were produced as one-offs.

The advocates of industrialization heard little but echoes in the years between the wars, but their ideas were better received in the aftermath of World War II. Wartime devastation compelled France to embark on a massive reconstruction program, particularly for housing. The Ministry of Reconstruction and Urban Development was created in 1944, and Eugène Claudius-Petit was named its director in 1948. A close friend of Le

BELOW
481. René Gabriel (1890–1950)
Living room of a model apartment for the Le Havre reconstruction project, in *Le Décor d'aujourd'hui*, November–December 1947

OPPOSITE
482. Marcel Gascoin (1907–1986)
Child's bedroom for the Sotteville-lès-Rouen reconstruction project, presented at the Exposition Internationale de l'Urbanisme et de l'Habitat, Paris, 1947
Bibliothèque des Arts Décoratifs, Paris; Gascoin Archives

Corbusier, he was very open to modernist ideas. At the initiative of the state, the International Exhibition of Urban Development and Housing was held in the Grand Palais in Paris in 1947 to showcase a number of reconstruction projects that brought together architecture and furniture design (plate 482). Having adopted prefabrication techniques to reduce the cost of the structures themselves, the architects of these projects also enlisted designers who had expertise in simplified modes of production. Marcel Gascoin worked with the architect Marcel Lods in Sotteville-lès-Rouen, and Jacques Dumond with Pierre Sorel in Boulogne-sur-Mer. Perriand furnished the minimalist family home designed by Paul Nelson, Roger Gilbert, and Charles Sébillotte, while Le Corbusier offered models and plans of the future Unité d'Habitation in Marseille, which would be completed in 1952.

Of all the projects presented, that for the reconstruction of Le Havre, which had been largely destroyed, was a standout. The highly experienced architect Auguste Perret rose to the challenge of rebuilding an entire city. On the interiors, he collaborated with the designer René Gabriel, then aged fifty-seven, who was well versed in industrial methods of production and had created a very simple collection of furniture in 1944 for those affected by the war. In Le Havre, Gabriel imposed his vision of a highly rationalized home and established the tone of reconstruction furniture: simple, unadorned, and mass-produced (plate 481). For many, the abandonment of an artisanal approach became a vital goal, although the true democratization of modern furniture would await the arrival of new manufacturers and the gradual development of a nationwide distribution network, beginning in the mid-1950s.

Already active before the World War II, Perriand, Prouvé, and Gascoin were among the pioneers who continued, after the war, to crusade for the modern home. Prouvé was one of the few who had his own production facility: the Ateliers Jean Prouvé, which he maintained in Maxéville, a suburb of Nancy, between 1931 and 1954 (plate 486). He abandoned the metal tubing that had been favored by the avant-garde, selecting folded sheet metal as his preferred material (plates 483, 484). Perriand, also an early proponent of metal, soon turned to wood. This choice was consistent with her interest in alpine style furniture and her familiarity with Japanese traditions. When she traveled to Japan in 1940, she embraced the country's approach to housing design, just as Frank Lloyd Wright and Bruno Taut had earlier. "It wasn't a question of folkloric charm. It was a serious analysis of the timeless qualities of these houses that fascinated me with their modernity. I wanted . . .

petit Paul est méchant

ABOVE, LEFT
483. Jean Prouvé (1901–1984)
***Antony* chair**, 1954
Made by Ateliers Jean Prouvé (France)
Sheet steel, metal tubing, beech plywood
Musée des Arts Décoratifs, Paris

ABOVE, RIGHT
484. Jean Prouvé (1901–1984)
***Standard* chair**, 1951–52
Made by Ateliers Jean Prouvé (France)
Sheet steel, lacquered metal tubing, oak plywood
Musée des Arts Décoratifs, Paris

RIGHT
485. Jean Prouvé (1901–1984)
***Compas* desk**, 1958
Made by Ateliers Jean Prouvé (France)
Distributed by Steph Simon (France)
Lacquered sheet steel, Formica veneer, wood
Musée National d'Art Moderne, Centre Georges Pompidou, Paris

OPPOSITE
486. **Advertisement for furniture by Ateliers Jean Prouvé distributed by Steph Simon**, 1949
Private collection

LES MEUBLES DES ATELIERS JEAN PROUVÉ

Concessionnaire exclusif :
Steph SIMON
52, Av. des Champs-Elysées
Paris-8e / ÉLY. 45-78

to prove by example that standardized modules would not inevitably result in monotony, but rather in harmony."[2] When she returned to France, her work was marked by the materials, methods, and low profiles she had observed in Japan (plates 487, 488). Perriand had been Le Corbusier's assistant before the war, and, like Prouvé, she retained close ties with the architectural world. Both were to benefit from the country's massive reconstruction program.

Postwar France needed to make up for its inadequate school and university facilities. Huge building projects were launched in both Paris and the provinces, including the university complex in Antony, inspired by American campuses; fifteen new pavilions in the Cité Internationale in Paris; and the Aix-Marseille school of law. In their furnishings for these projects, Perriand and Prouvé focused on structure, using ordinary industrial materials (laminates, sheet metal) to achieve results at an economical price. The process of bending sheet metal continued to hold Prouvé's attention: "Steel tubes didn't satisfy me. Sheet metal was my inspiration." In his mind, furniture and architecture posed the same problems of construction. His *Compas* desk of 1949 closely reflects his thoughts on materials and structures (plate 485), and also recalls the porticoes that had appeared in his architecture since the 1930s. Widely used as a cafeteria table and office desk in university complexes, the *Compas* employed a cantilevered structure, replacing the conventional perpendicular legs with striking acute angles. This dynamic form became one of the designer's trademarks. Several other famous designs by Perriand and Prouvé originated in these university projects. For the campus in Antony, Prouvé designed conference tables with folded metal bases that evoked airplane wings. For the Maison de la Tunisie, Perriand designed the *Tunisie* bookshelf, in which metallic blocks serve both as supports for the shelves and as storage spaces. Perriand believed that "storage takes precedence," and her research with Pierre Jeanneret focused primarily on "equipping the house."

Gascoin, an early comrade of Prouvé and Perriand, had a similar point of view. Born in Le Havre, he drew inspiration from his meticulous observation of ocean liner fittings to develop a wide range of designs. These included wardrobes with doors incorporating shelves, and wall-mounted desks with bookcases. In 1951, he launched his own firm, ARHEC (Aménagement Rationnel de l'Habitation et des Collectivités), in order to industrialize the production of his "Gascoin storage systems." Gascoin proved to be a great popularizer of rationalism, and in the early postwar era, his workshop became a training ground for the next generation, including Motte, Mortier, Guariche, Richard, Philippon, Lecoq, and Paulin.

OPPOSITE
487. Charlotte Perriand (1903–1999)
***Ombre* stackable chairs, *Air France* stackable table, and built-in bookshelf (with a ceramic sculpture by Fernand Léger)**
Exhibition *Proposition d'une synthèse des arts*, Tokyo, 1955
Perriand Archives

BELOW
488. Charlotte Perriand (1903–1999)
Double-sided bookshelf for the Maison du Mexique in the Cité Internationale, Paris, 1953
Solid pine, aluminum sheet
Musée d'Art Moderne, Saint-Étienne Métropole

RIGHT
489. Atelier Le Corbusier, based on a plan by Charlotte Perriand (1903–1999)
Kitchen-bar for the Unité d'habitation in Marseille, designed 1950, built 1952
Made by Barberis and/or SITRAB
Solid oak, oak and beech plywood, molded aluminum, ceramic
Musée des Arts Décoratifs, Paris

THE POSTWAR AVANT-GARDE

The period between 1945 and 1975, known in France as *Les Trente Glorieuses*, marked a profound social transformation. From the new novel to the nouvelle vague film, neophilia was on the march, a phenomenon that was gently mocked in Jacques Tati's 1958 film *Mon Oncle*. In the domestic realm, the upheavals were driven by the process of internationalization. While Le Corbusier was revered, a school did not really form around him. Rather, in the postwar period, the younger generation of furniture designers cast a longing look overseas and were just as fascinated by Charles and Ray Eames in the U.S. as they were by Arne Jacobsen in Denmark. Florence Knoll's elegant open-plan workspaces, the Eameses' first fiberglass chairs, and Jacobsen's molded plywood chairs were powerful exemplars. Scandinavia and the U.S. were synonymous with modernity, and their design, films, and way of life exerted a universal fascination. The new magazine *La Maison française*, launched in 1946, prominently featured Scandinavian and American productions (plates 490, 491). Pierre Paulin and Olivier Mourgue traveled to Sweden, and in 1952 the Salon des Arts Ménagers organized the exhibition *Design for Use, USA*, which revealed the enviable "American way of life" to a broad audience. The opening of the first Knoll store in Paris in 1951 was both an artistic and social event, and its charismatic director Yves Vidal was a longtime proponent of American furniture.

Favorable economic conditions prevailed during this period. A newly prosperous France embarked on housing construction, and buyers benefited from an increase in their purchasing power and the development of consumer credit. To optimize the use of space in small city apartments and accommodate a new lifestyle, designers introduced new solutions, including low, armless chairs for the living room, open-space plans, various storage units, multifunctional furniture, and kitchens with appliances and storage integrated into the countertops. This type of functional kitchen, which had been introduced in Germany around 1925, gradually gained popularity in postwar France, particularly in public housing. In the late 1940s, Marcel Gascoin designed the integrated, flexible *Comera* kitchens, and in 1950, for Le Corbusier's Cité Radieuse in Marseille, Charlotte Perriand created a small but very functional kitchen that opened onto the dining room, allowing the lady of the house to converse with her guests (plate 489). During the 1960s, mass-produced versions were introduced. Philippon Lecoq's *Triennale* kitchen and Michel Buffet's *DF2000* for CEI Raymond Loewy were particularly ingenious proposals, the first featuring Formica, and the second, PVCs.

Materials were a factor in the renewal of kitchen design, as in furniture. Laminates met with unprecedented success, particularly the American brand Formica (plate 492). This remarkable material, produced by compressing paper and resin under high heat, had been invented in the early twentieth century. Its durability, color choices, and hygienic qualities strongly appealed to customers. Manufacturers of new materials enthusiastically promoted their products at the Salon des Artistes Décorateurs: Formica in laminates, Uginox in stainless steel, and Glaces de Boussois and Saint-Gobain in glass. Furnishing the raw materials, these companies organized competitions at the salon for young designers seeking to gain recognition. Roger

RIGHT
490. **Cover of *La Maison française***, No. 33, Christmas 1949
Bibliothèque des Arts Décoratifs, Paris

BELOW
491. **Living room with furniture by Alain Richard and André Monpoix**, in *La Maison française*, March 1955
Bibliothèque des Arts Décoratifs, Paris

BELOW
492. **Advertisement for Formica**,
in *Art et Décoration*,
March 1959
Private collection

OPPOSITE, TOP
493. Mathieu Matégot (1910–2001)
***Nagasaki* chair**, 1954
Made by Ateliers Mathieu Matégot (France)
Lacquered steel tubing, perforated sheet metal ("rigitulle")
Galerie Jousse, Paris

OPPOSITE, MIDDLE
494. Antoine Philippon (1930–1995) and Jacqueline Lecoq (b. 1932)
Desk, 1967
Made by E.I.B Minvielle (France)
Glass with Emauglas coating, wood, white laminate, aluminum

OPPOSITE, BOTTOM
495. Antoine Philippon (1930–1995) and Jacqueline Lecoq (b. 1932)
TV stand/record player with bar, 1958
Prototype
Formica, cherry wood, chrome-plated steel, brass
Musée des Arts Décoratifs, Paris

Fatus, Antoine Philippon, and Jacqueline Lecoq all presented pieces that were outstanding in their rigor and inventiveness, but, unfortunately, they did not find manufacturers. Meanwhile, during the war, Mathieu Matégot had discovered a form of perforated sheet metal that he called "rigitulle," and which he used to develop lightweight, imaginative furnishings and accessories (plate 493).

A number of synthetic materials were also highly successful. In 1949, the Musée des Arts Décoratifs organized a show on "plastic materials in modern life." Plastic was introduced in a variety of mass-market items for the home. Representative of this type of product were Gilac's polyethylene bowls, launched in 1953 with the slogan "Plastique Gilac, plastique miracle." More exclusive but very sensational was the "entirely plastic house" designed by Ionel Schein and René Coulon, with an interior by Alain Richard. Sponsored by the mining companies Charbonnages de France and Houillères du Nord, it was displayed at the Salon des Arts Ménagers in 1956. Despite these important inroads, synthetic materials began to be used in furniture only somewhat later, toward the end of the 1950s.

Inspired by the stimulating atmosphere of the period, the younger generation felt they had a mission to shape public taste and democratize contemporary furniture. About twenty-five years old, mostly recent graduates of the elite École Nationale Supérieure des Arts Décoratifs, these "young wolves"[3] were ready to lead the charge. Well aware that there is strength in unity, Joseph-André Motte, Pierre Guariche, and Michel Mortier established the ARP (Atelier de Recherche Plastique) to work with manufacturers receptive to new forms (plate 497): Disderot for lighting, Huchers-Minvielle for furniture, and Steiner for seating.

The founders of the ARP, along with Antoine Philippon and Jacqueline Lecoq (plates 494, 495), Roger Fatus, René-Jean Caillette, Étienne Fermigier, Alain Richard, Dirk-Jan Rol, and Janine Abraham, invented a new style of furniture that was functional, elegant, and minimal, with simple lines. Intent on creating a new way of life for as large a public as possible, they were concerned with the issue of limited space, and many also worked in the field of interior design. Their ambitions depended, however, on finding manufacturers for

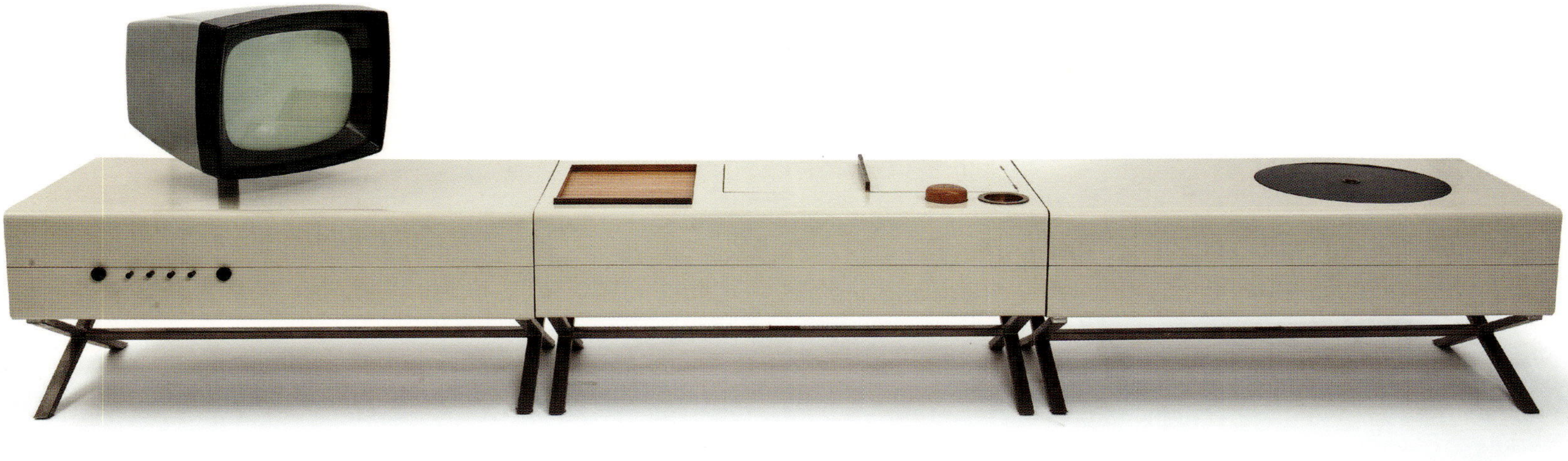

OPPOSITE, TOP
496. René-Jean Caillette (1919–2004)
Dining room at the Salon des Arts Ménagers, Paris, Grand Palais, 1961
Diamant chairs made by Steiner; table and buffet made by Charron
Les Arts Décoratifs, Paris; Documentation Center
Photograph by Jean Collas

OPPOSITE, BOTTOM LEFT
497. Joseph-André Motte (1925–2003)
Tripod chair, 1949
Made by Rougier (France)
Woven rattan, painted beech, steel
Musée des Arts Décoratifs, Paris

OPPOSITE, BOTTOM MIDDLE
498. Pierre Guariche (1926–1995)
***Tonneau* chair**, 1953
Made by Steiner (France)
Molded plywood, lacquered steel tubing, foam, imitation leather
Musée des Arts Décoratifs, Paris

OPPOSITE, BOTTOM RIGHT
499. René-Jean Caillette (1919–2004)
***Diamant* chair**, 1957
Made by Steiner (France)
Molded and painted elm plywood, chrome-plated steel tubing
Musée des Arts Décoratifs, Paris

their furnishings. The first successes came in the early 1950s: Steiner distributed Guariche's *Tonneau* chair in molded plywood (plate 498) and Caillette's *Diamant* chair (plates 496, 499); Oscar sold ingenious storage units; and, with its modular pieces, Huchers-Minvielle successfully introduced the concept of kit furniture.

While some producers remained wary of these new approaches, others underwent a spectacular renewal with the help of the younger generation of designers. Meubles TV worked with Paulin and Abraham, and particularly Alain Richard and André Monpoix. Huchers-Minvielle sought advice from the ARP, Philippon and Lecoq, Guariche, and Rol. Georges Charron founded Groupe 4 with Motte, Caillette, Richard, and Geneviève Dangles. Inheritors of a tradition of upholstered furniture, the great French chair makers, including Steiner and Roset, collaborated with younger designers and modernized their factories. Steiner was among the first to do so, in the early 1950s, enlisting the services of Guariche, Motte, and Mortier, and introducing chairs in molded plywood and polyester reinforced with fiberglass. Somewhat later, in the 1960s, the venerable firm Roset called upon Bernard Govin and Michel Ducaroy.

A number of outstanding new firms were also established at this time. Charles Bernard founded Airborne in 1951, at first using designs by the ARP. In 1959, he distributed a chair in the international style, the *Joker*, by Olivier Mourgue, still a student at the time; this marked the beginning of a remarkable collaboration that would bear its fruit in the 1960s and '70s. Like Mourgue, Pierre Paulin took up the cause of a style far removed from that which would make his reputation in later decades. He worked with Meubles TV on an original armchair, model 273, known as the *Anneau* (Ring), whose form was inspired by a folded strip of paper (plate 501). He also provided Thonet France with numerous designs for chairs and elegant little desks made from laminates and metal tubing (plate 500).

Two major new players emerged in the realm of lighting: Disderot and Biny. Motte, Guariche, Richard, Buffet, and Fatus worked with these companies, focusing on mobility and the reflective properties of light (plates 502, 504). Serge Mouille produced his own fixtures in sheet metal, with novel shapes and articulations (plate 503). In the mid-1950s, several Parisian galleries—Galerie Mai, Stef Simon, and Meubles et Fonction—supported innovative designers, as did a few independent stores in the provinces. Stef Simon backed established names, such as Jean Prouvé, Charlotte Perriand, and Serge Mouille, while Galerie Mai and Meubles et Fonction offered opportunities to younger entrants: Guariche and Paulin showed at Galerie Mai, and Fermigier and Mourgue, at Meubles et Fonction. Despite some successes for modern furnishings, especially outside Paris, the furniture trade was generally averse to change. It was not until the late 1950s that France began to develop a structured distribution network, with Claude Vassal's Meubles Pilotes, the Huchers-Minvielle stores, and later those of Roche Bobois.

BELOW
500. Pierre Paulin (1927–2009)
CM 141 desk, 1954
Made by Thonet (France)
Multi-ply oak, laminate, painted steel sheet and tubing
Fonds National d'Art Contemporain, on loan to the Musée des Arts Décoratifs, Paris

OPPOSITE
501. Pierre Paulin (1927–2009)
Chair *No. 273*, known as the *Anneau*, 1953–54
Made by Meubles TV (France)
Stainless steel, leather
Musée des Arts Décoratifs, Paris

LEFT
502. Pierre Guariche (1926–1995)
***G1/SP* floor lamp**, 1951
Made by Disderot (France)
Lacquered metal, brass, electrical components
Galerie Kreo, Paris

BELOW
503. **Advertisement for lamps by Serge Mouille sold by Steph Simon**, 1956
Galerie Downtown François Laffanour

OPPOSITE
504. Pierre Guariche (1926–1995)
***G24* desk lamp**, 1953
Made by Disderot (France)
Lacquered metal, brass, electrical components
Galerie Pascal Cuisinier, Paris

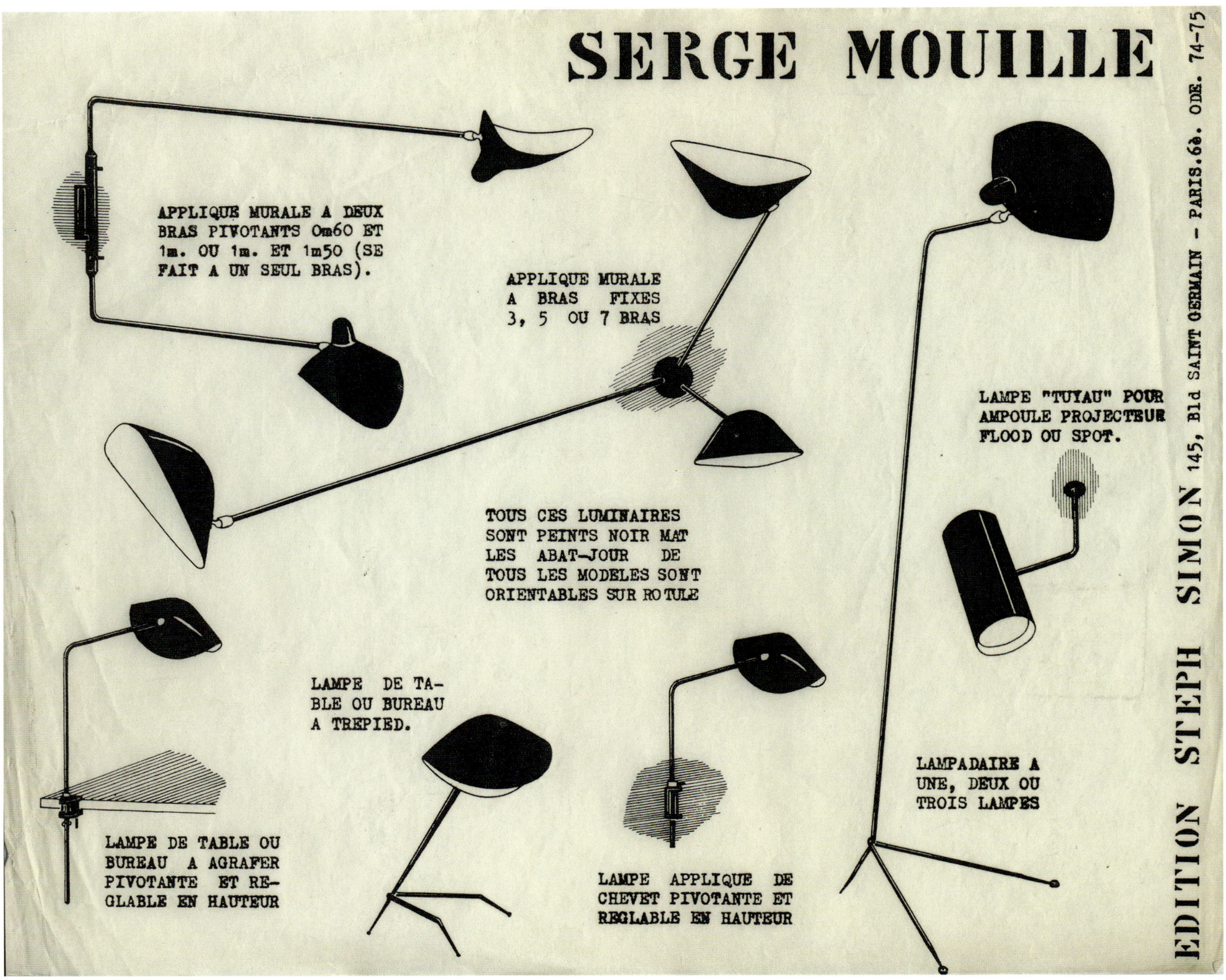

BELOW
505. **Salon des Arts Ménagers, Paris, Grand Palais**, 1956
Les Arts Décoratifs, Paris; Documentation Center
Photograph by Jean Collas

OPPOSITE
506. Jacques Nathan-Garamond (1910–2001)
Poster for the first *Formes Utiles* exhibition, at the Musée des Arts Décoratifs, Paris, 1949–50
Museum of Modern Art, New York

TOWARD MASS CONSUMPTION

In the early 1950s, everyday consumer goods, even more than costly furniture, became the symbols of comfort and well-being. In 1954, Boris Vian's *La Complainte du progrès* pointed out both the advantages and the limitations of domestic conveniences. High consumer demand was driving strong industrial growth and an economic recovery, bringing problems of design to the fore. Two distinct initiatives arose in response, one led by the "enlightened aesthetes" of Formes Utiles, a subgroup of the UAM, and the other by an individual, Jacques Viénot. Both sought to make manufacturers aware of the importance of "industrial design," a concept already recognized in the U.S. and Great Britain, but not yet in France.

Following the war, the UAM tried to recover its place by staging exhibitions, and the mood was favorable to the ideas it promoted: equipping and rationalizing the home. The architect Georges-Henri Pingusson drafted a new UAM manifesto in 1949 to replace the earlier one of 1934. Also in 1949, the UAM established the Formes Utiles division, which gradually evolved into a kind of association within the association. The architect André Hermant, the movement's theorist, wanted to foster the production of high-quality everyday objects by encouraging contacts between designers and industry, and to attract public attention to the topic by mounting themed exhibitions. The first of these, *Formes Utiles, Objets de Notre Temps*, was organized at the Musée des Arts Décoratifs in 1949 (plate 506), and later shows were held at the Salon des Arts Ménagers (plate 505).

The Formes Utiles exhibitions offered selections of different types of domestic equipment: cleaning products, clothes irons, coffee grinders, and cooking utensils, among others. In contrast to the Salon des Arts Ménagers, these presentations had no commercial aspect: they were simply intended to allow an informed public to explore trends in industrial design. Unfortunately, the pedagogical initiatives of Formes Utiles very rarely resulted in collaborations between designers and industry. Following a period of friction, Formes Utiles split from the UAM in 1957, and Hermant published the definitive work on the group in 1959. The Formes Utiles exhibitions continued beyond the postwar years, until, in 1983, a final show marked the group's activism.

While Formes Utiles engaged the debate on a theoretical level, Viénot was attempting to establish a new profession. Rejecting the English term "industrial design," in 1951 he founded the Institut d'Esthétique Industrielle, which published a review intended to promote a taste for industrial goods and impart an aesthetic value to the creations of the machine age. Viénot wanted the institute to play the same role in fostering contacts between designers and industry as the Council of Industrial Design, established in London in 1944. The very French Institut d'Esthétique Industrielle

exposition
formes utiles
union
des
artistes
modernes
nathan
décembre 1949 - février 1950
MUSÉE DES ARTS DÉCORATIFS
PAVILLON DE MARSAN, 107 RUE DE RIVOLI, PARIS Ier

BELOW
507. Jean-Louis Barrault
Hair dryer, c. 1975
Made by Moulinex (France)
ABS plastic, painted aluminum
Jean-Bernard Hebey collection, Paris

OPPOSITE
508. Éric (Castel) (1915–1997)
Poster for *Calor* clothes irons, 1966
Bibliothèque Forney, Paris

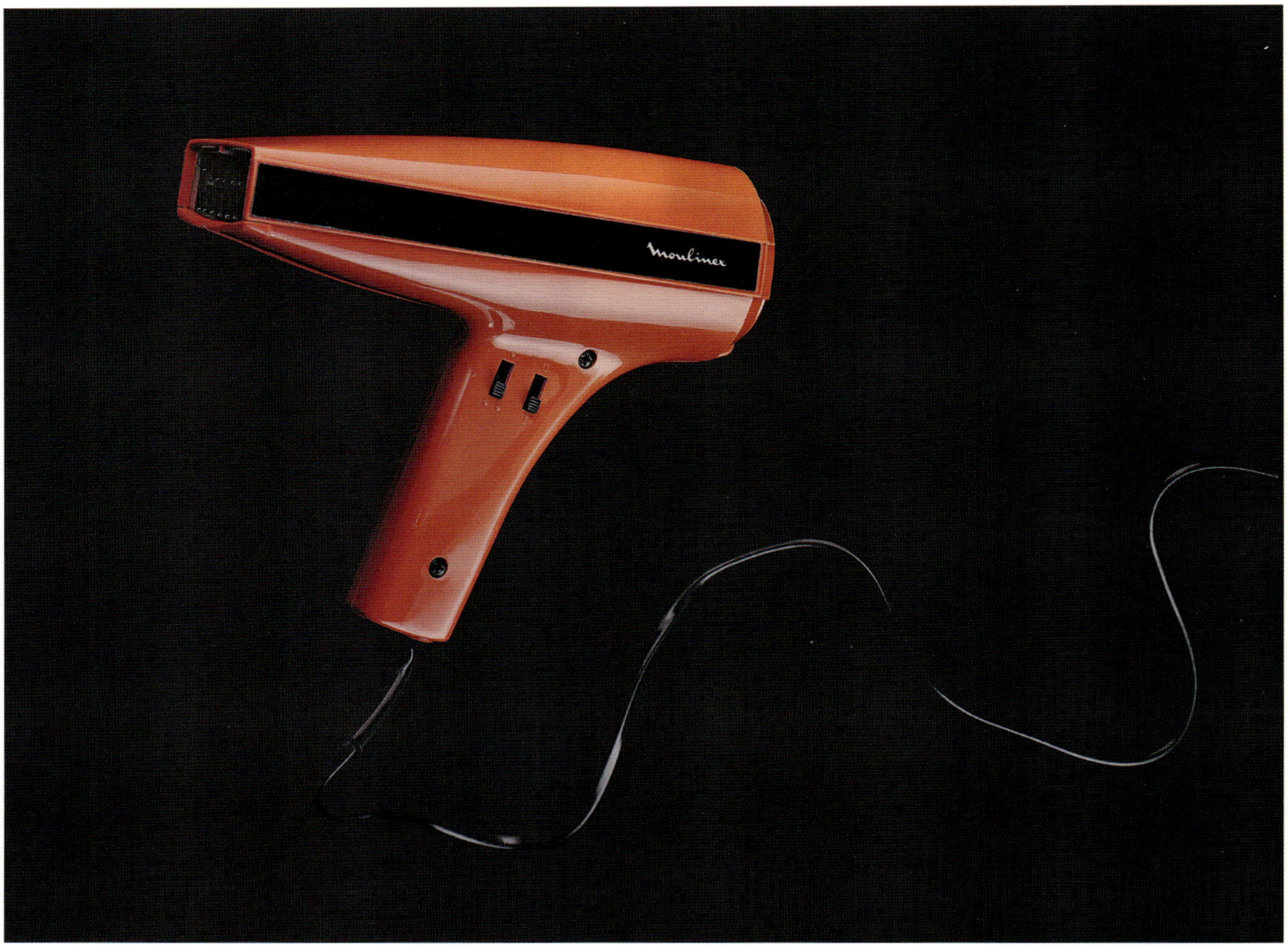

did at least successfully open the way to the establishment of international organizations in the country. It was thanks to Viénot that the ICSID (International Council of Societies of Industrial Design) gained a foothold in France in 1957.[4]

Viénot's initiative focused on the modernization of household appliances. In industrialized countries, electrification had revolutionized domestic equipment (plate 507) and created a strong demand for appliances both large (stoves, washing machines, vacuum cleaners, and refrigerators) and small (toasters, mixers, coffeemakers, and fans). This burgeoning market had an ideal showcase in Paris: the Salon des Arts Ménagers. Attended by a vast audience from all over France, this salon played a vital role in publicizing new products. First established in 1923 and suspended for the duration of the war, it reopened in the Grand Palais in 1948.[5]

In response to the modernization of contemporary consumer goods, Viénot opened the design consulting firm Technès in 1949, adopting the powerful slogan "We buy with our eyes." Viénot believed that technical progress had to correlate with aesthetic progress. At a time when industry still had little grasp of the importance of design, Technès was the only firm of its kind in France. It was staffed by two outstanding talents, Jean Parthenay and Roger Tallon, and its clients ranged from Calor, a maker of electric appliances, to Poclain, a manufacturer of heavy construction equipment. Others included Japy, a former clockmaker that had converted to producing typewriters, and the toolmaker Peugeot Frères. Switching its focus from heating fixtures and large appliances to small appliances, Calor was one of Technès's first accounts, calling on the firm regularly between 1953 and 1959. Parthenay in particular designed many items for Calor, including an electric razor, as well as the electric iron that became Calor's leading product (plate 508).

Technès was soon joined by other design consultancies specializing in different areas. At a time when the spirit of internationalism was faltering, Raymond

CALOR
FIL
COTON
LAINE
NYLON
CALOR
CALOR
FERS A REPASSER
ERic

Casserole et daubière Coquelle, Le Creuset, 1958, CEI Raymond Loewy, Paris.

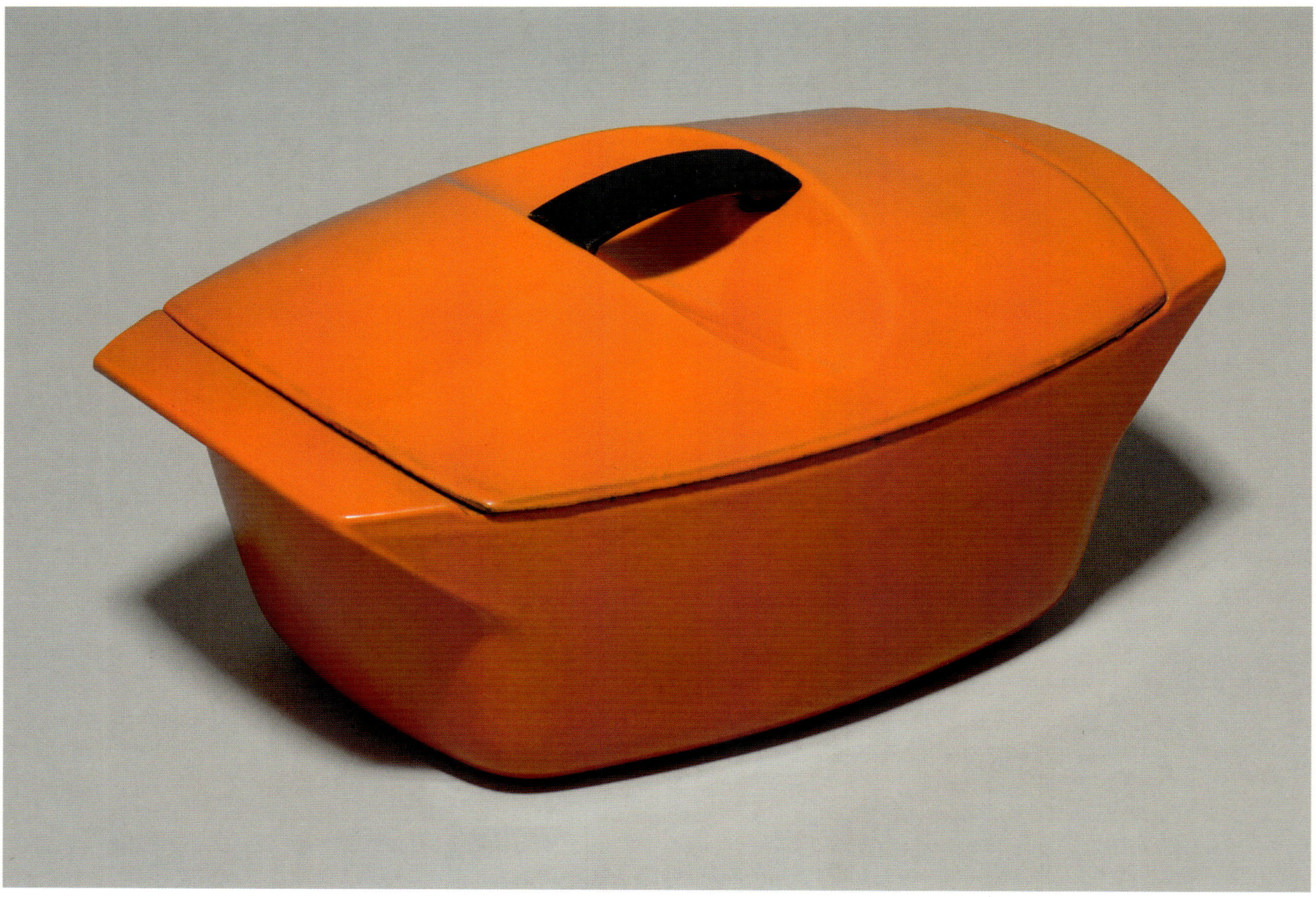

OPPOSITE, TOP
509. CEI Raymond Loewy
Kitchenware, 1959
Made by Le Creuset
Enameled cast iron
Pictured in *Esthétique + Industrie*, 1964

OPPOSITE, BOTTOM
510. CEI Raymond Loewy
***Coquelle* casserole**, 1958
Made by Le Creuset (France)
Enameled cast iron
Musée d'Art Moderne, Saint-Étienne Métropole

BELOW
511. CEI Raymond Loewy
***DF 2000* chest**, 1965
Made by Doubinski Frères (France)
Thermo-molded ABS plastic, lacquered aluminum, white laminate
Fonds National d'Art Contemporain, on loan to the Musée des Arts Décoratifs, Paris

Loewy's firm occupied a unique place between Europe and the U.S. Born in France, Loewy had a brilliant prewar career as an industrial designer in the U.S., working on trains for the Pennsylvania Railroad and cars for Studebaker, as well as on the visual identities of major brands, including Lucky Strike cigarettes and Coca-Cola. His book *Never Leave Well Enough Alone* (1951) was positively received in the U.S. but criticized in France, where it was translated under the title *La laideur se vend mal* (Ugliness Doesn't Sell). André Hermant, in particular, opined that "ugliness sells very well."

In 1952, when Loewy was at the peak of his success in the U.S., he established an office in France under the name Compagnie Américaine d'Esthétique Industrielle. Under the direction of Harold Barnett, with the assistance of Pierre Gautier-Delaye and Jacques Cooper, it concentrated at first on commercial architecture, creating interiors for the department store BHV. A wave of anti-American sentiment led Loewy to eliminate the word "American" from the venture's name, so that it became the Compagnie d'Esthétique Industrielle, also known as CEI Raymond Loewy.

The inauguration of the Common Market led Loewy to expand the operations of his French office. It had a succession of American directors, often operating jointly. One of them was Evert Endt, a graduate of the Zurich school of graphic arts, who developed the firm's practice in that field. With the arrival of the industrial designer Michel Buffet in 1957, CEI Raymond Loewy operated in three areas: commercial architecture; graphics, branding, and packaging; and product design. It focused particularly on underserved niches, such as store interiors and packaging, and won important contracts with the petroleum company BP, the baking company Lefèvre Utile (Lu), the Newman ready-to-wear label, the sewing machine maker Elna, the kitchen utensil manufacturer Le Creuset, the food conglomerate Nestlé, and the cleaning product maker Cotelle et Foucher (Mir, Javel La Croix, etc.).

The contract for Shell's visual identity, awarded at the beginning of the 1970s, marked the apogee of Loewy's reputation in France. The firm developed a comprehensive design for Shell's service stations: uniforms, ancillary products, packaging, and logo. At the

BELOW
512. **Alarm clock**, c. 1970
Made by Jaz (France)
Musée des Arts Décoratifs, Paris

same time, it was also working on the interior of the *Concorde*, the supersonic airplane introduced by the British Airways and Air France consortium in 1973. The *Concorde* commission encompassed employee uniforms and the plane's interior (everything from the seats to the glassware), as well as the airport arrival lounges. While point of sale and visual identities represented a substantial portion of the Loewy agency's business, it also had some achievements in product and furniture design: a porcelain service for Bernardaud, pans for Le Creuset (plates 509, 510), and, most notably, the *DF 2000* range of furniture and kitchens for Doubinski, made from colored ABS plastic (plate 511).

By the 1960s, manufacturers began to appreciate the value added by design consultants and enlisted their services more consistently. Moulinex, founded by Jean Mantelet in 1932, expanded its successful product line to include multifunctional food processors, including the famous *Robot Marie*. In 1962, the designer Jean-Louis Barrault began a fruitful collaboration with Moulinex, imprinting its products with his distinctive angular, interlocking forms. Barrault earned a strong reputation and went on to work for many other manufacturers of heating and cooking devices, including Rozières, De Dietrich, and Chappée.

Terraillon offers an illustration of a bold and successful business transformation. Founded as a small clockmaking operation in 1946, the firm began to produce various kinds of scales in 1956. Between 1968 and 1970, it worked with the Italian designer Marco Zanuso to develop the *T111B* bathroom scale and the *BA 2000* compact kitchen scale. The latter scale echoed the two-part form of the portable radio Zanuso had designed for Brionvega in 1964, in collaboration with Richard Sapper. Its cover, symmetrically shaped and easy to handle, can be turned over and used as a receptacle for the ingredients to be weighed (plate 514).

Like Terraillon, the Société d'Emboutissage de Bourgogne (SEB) remade itself to meet the demands of modernity. Originally a producer of household utensils in tinplate, SEB introduced the legendary *Cocotte-Minute* pressure cooker, made from stamped metal, in 1953, and some time later, in the 1960s, entered the field of small electrical appliances. In the 1970s, the company developed a number of innovative products first with Gilles Rozé, when he was employed at Technès, and then with Rozé in partnership with Yves Savinel, when both were at the Sopha agency. Playful and colorful, SEB's toaster and yogurt maker made from ABS plastic broke with the heavier, chrome-plated look of American models (plates 515, 516). Design consultancies multiplied during this period. In the 1970s and '80s, Lonsdale, Alain Carré, Jean-Pierre Vitrac, Georges Patrix, Daniel Maurandy, Jean-Louis Barrault, ENFI Design, and others modeled or remodeled the entire universe of the everyday, from the Mini-tel to the mini-oven.

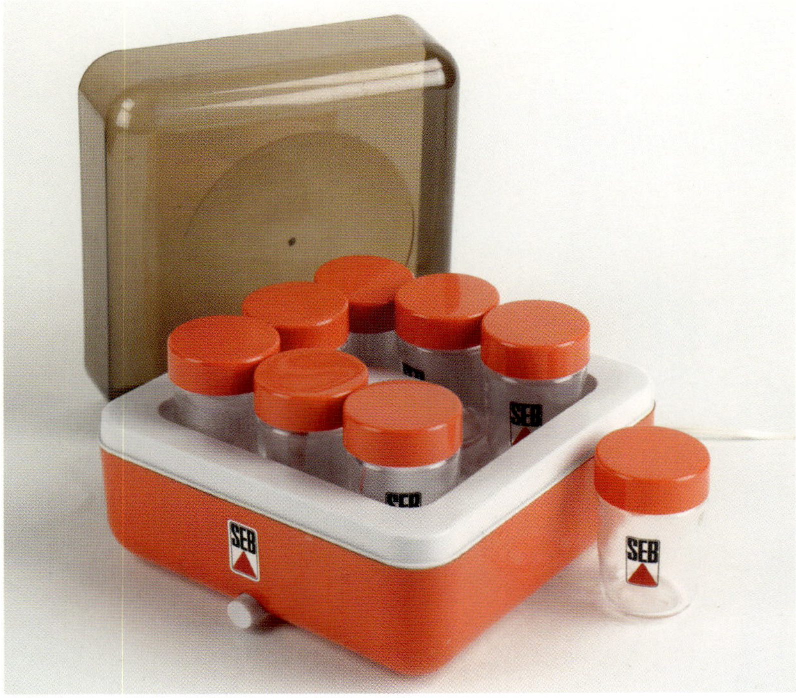

TOP, LEFT
513. **Salad spinner**, 1967
Made by Moulinex (France)
Polyethylene
Musée d'Art Moderne,
Saint-Étienne Métropole

TOP, RIGHT
514. Marco Zanuso (1916–2001)
Kitchen scale, 1970
Made by Terraillon (France)
ABS plastic, methyl
polyacrylate
Museum of Modern Art,
New York

BOTTOM, LEFT
515. Gilles Rozé (b. 1947)
for Technès
Yogurt maker, 1974
Made by SEB (France)
ABS plastic, polystyrene, glass
Musée d'Art Moderne,
Saint-Étienne Métropole

BOTTOM, RIGHT
516. Gilles Rozé (b. 1947)
for Technès
Toaster, 1974
Made by SEB (France)
Metal, ABS plastic
Musée d'Art Moderne,
Saint-Étienne Métropole

517. Guy Boucher for the Harold Barnett agency
***Saturna* table service**, 1968
Made by Duralex, Saint-Gobain Group (France)
Pressed, molded, and tempered borosilicate glass
Musée des Arts Décoratifs, Paris

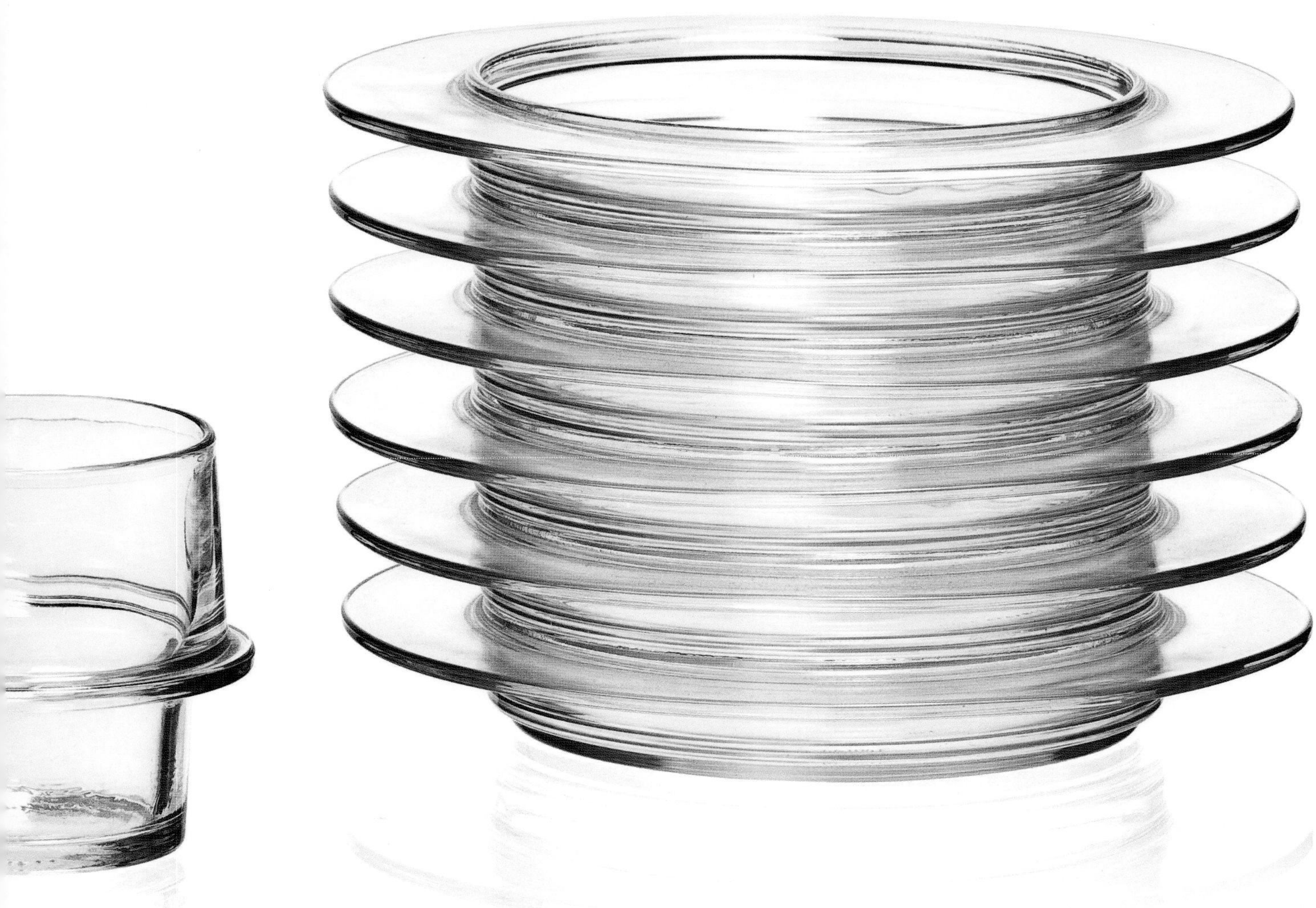

BELOW
518. Roger Tallon (1929–2011)
***Portavia 111* portable television**, 1963
Made by Téléavia (France)
Musée des Arts Décoratifs, Paris

OPPOSITE
519. Roger Tallon (1929–2011)
***Wimpy* chair**, 1957
Made by Sentou (France)
Aluminum, molded plywood
Musée des Arts Décoratifs, Paris

INDUSTRY AND DESIGN: ROGER TALLON

Roger Tallon holds a special place in French design. As the designer Marc Berthier has aptly said, "If industrial design, or design at all, exists in France, the credit goes to Tallon. If it got its second wind, that's thanks to Starck."[6] Tallon stood at the intersection of industry and design, and bridged the American and European aesthetics. This "Gallo-Ricain," to use Catherine Millet's term,[7] combined the discipline and rigor of an engineer with the creative dynamism of the U.S., a country he knew well. He worked for two American companies in the early years of his career—the heavy equipment manufacturer Caterpillar and the multinational chemical company DuPont.

But Tallon's adventure truly began in 1953, at the side of another true believer, Jacques Viénot, who hired him at Technès. There Tallon developed small household appliances for Peugeot (the multifunctional appliance *Peugimix*, coffee grinders, whisks, drills, and electric toothbrushes).[8] He also worked on dozens of other projects, including an innovative compact motorbike, *Le Taon*, for Derny, the *Sem Veronic* 8 mm camera with no visible lens, a slide projector for Kodak, a typewriter for Japy, and machine tools for Gambin. Among these undertakings, the *Gallic 16* lathe for the Belgian company La Mondiale was particularly noted for the meticulous care given to the design of an enormous machine tool. But the project that brought Tallon to the attention of a wider public was the *P111* portable television he designed for Téléavia in 1963 (plate 518). With its compact, rounded form and black and white coloring, it was made to be viewed from all angles. It broke with the aesthetic of competing models by integrating the screen, cathode ray tube, and controls into a unified shape.

Tallon worked at Technès for twenty years, all the while struggling against the concept of the "industrial aesthetic" so dear to Viénot. Tallon promoted the use of the word "design," eventually with success. A militant advocate for industrial design, Tallon was fiercely opposed to "styling," or mere "artistic window dressing," epitomized in his view by the work of Raymond Loewy. Hostile to the notion of simply camouflaging an object with a modern facade, Tallon insisted, from a very early point in his career, on an overall design concept in which the interior workings would correspond with the outside appearance. "If everything is connected, it will all hold together," was his creed. Given his philosophy, Tallon had little interest in the home. "As far as I'm concerned, design has nothing to do with the private sphere. It's something that concerns the collective interest, on the industrial level."[9]

His *Wimpy* chair, designed for the first French fast-food chain, was intended as an homage to the Eameses (plate 519). It was delivered in pairs, disassembled for easier shipping. He made very little furniture for personal use, although his rare incursions into this field were striking. They included his *TS* chairs (the initials stood for Tallon and Sentou, his distributor) and the *Module 400* series, originally made for a nightclub and later distributed by Galerie Lacloche. This series included a helicoid staircase (plate 520) and chairs padded with convoluted foam, which Tallon had discovered in the sporting goods store Au Vieux Campeur (plate 521). The project allowed him to apply his favorite principles: a comprehensive product line and the use of modules. Not only the *Module 400* series but also the *3T* table service and the *Cryptogamme* furniture line (plate 522) were built around a basic module whose size and function were modified so as to create a very large repertoire of items.

OPPOSITE
520. **Apartment with a helicoid staircase by Roger Tallon**
Photograph by Jean-Pierre Leloir

ABOVE
521. Roger Tallon (1929–2011)
***Module 400* chair**, 1966
Made by Galerie Lacloche (France)
Polished aluminum, Spazmolla foam
Musée National d'Art Moderne, Centre Georges Pompidou, Paris

RIGHT
522. Roger Tallon (1929–2011)
***Cryptogamme* stool**, 1969
Made by Galerie Lacloche (France)
Plastic, wood, polyurethane foam, jersey
Les Arts Décoratifs, Paris; Documentation Center, Roger Tallon Archives

BELOW
523. Roger Tallon (1929–2011)
***Mach 2000* chronograph**, 1976
Made by Lip (France)
Les Arts Décoratifs, Paris;
Documentation Center,
Roger Tallon Archives

OPPOSITE, TOP
524. Roger Tallon (1929–2011)
Interior of the *Corail* train for the SNCF, 1975
Les Arts Décoratifs, Paris;
Documentation Center,
Roger Tallon Archives

OPPOSITE, BOTTOM
525. Roger Tallon (1929–2011)
SNCF *TGV Duplex* on the Paris-Lyon line
Entered service in 1996

Tallon remained with Technès until 1973, when he founded his own agency, Design Programmes. That same year, he also created the *Mach 2000* line of watches and chronometers for the clockmaker Lip (plate 523), which was then embroiled in labor conflicts. With asymmetrically mounted dials and winding mechanisms and bright colors, these timepieces represented a complete break with the traditional styles that dominated watchmaking. Every element, from the dial to the wristband, was designed to maximize visual coherence. In 1984, Tallon joined ADSA, which had been established by Marc Lebailly, Maïa Wodzislawska, and Pierre Paulin.

While Tallon displayed a protean inventiveness throughout his long career, he received the most recognition internationally for his work in transport design: the Mexico City metro in 1967, the first model of the *TGV 001* in 1967, the *Corail* train in 1974–75, the *TGV Atlantique* in 1983, the *TGV Duplex* in 1991, the Montmartre funicular in 1991, the *Météor* metro line in 1991, the Eurostar in 1992, and the plans for a high-speed train in the U.S. (never realized) in 1993. Tallon was the first designer to work for the SNCF, the French national railroad company. Designed in 1971 and put into service in 1974–75, the SNCF's *Corail* train was a tremendous success (plate 524). Envisioning the train as a living space, Tallon abandoned compartments for a unified expanse, reminiscent of the interior of an airplane. Simplicity, coherence, and comfort were the essential concepts underlying the design of the *Corail*. Tallon took a comprehensive approach to planning the space and the fittings (ergonomic seats, fold-down tables in second class, air-conditioning, etc.). He also rethought the signage, devising new pictograms and new renderings of route maps.

The *Corail* was widely appreciated for its comfort and excellent performance, but perhaps most importantly, it anticipated another revolution in train travel, the high-speed *TGV*. Thanks to his charisma and technical know-how, Tallon was able to collaborate closely with the Alstom and SNCF engineers working on this groundbreaking project. While Tallon's original 1967 model for the *TGV 001* was not adopted (Jacques Cooper's design being used instead), he participated in the planning of the *TGV Atlantique* (1983) and the next-generation *TGV Duplex* (1991), for which he designed, in particular, the nose of the locomotive (plate 525). Exceeding the standard even of the *Corail*, Tallon enhanced the *TGV*'s comfort and accommodations by using lighter construction materials, adjustable armrests, and seats upholstered with comfortable and durable velour.

In 1985, Tallon was the first to receive the newly created Grand Prix Nationale de la Création Industrielle. In addition to his personal accomplishments, he established an official place for design in France by organizing courses on the subject at the prestigious École Nationale Supérieure des Arts Décoratifs in 1962.[10] Tallon, who possessed an outgoing personality, built ties

OVERLEAF
526. **Advertisement for the Renault 4CV**, 1949
Introduced in 1946
National Motor Museum, Beaulieu (U.K.)

PAGE 423
527. **New 4CVs at the Renault factory in Boulogne-Billancourt**, 1951
Photograph by René-Jacques

with his international counterparts, including Tomás Maldonado and Massimo Vignelli, and became an active member of the ICSID. He was as much interested in art as in industrial design, and was close to artists such as César and Yves Klein. The gigantic singing heads he exhibited at the 1970 Osaka World's Fair show how this highly rational designer discovered new forms and materials through his artistic contacts. With his highly original and open mind, Tallon changed the widespread perception that French design was primarily focused on furniture and decoration.

THE AUTOMOTIVE INDUSTRY

Roger Tallon was fascinated by all aspects of design and left a particularly strong mark on rail transport, but, curiously enough, this connoisseur of automobiles had little involvement with car design. Nevertheless, this field was the focal point of industrial design for the mass market. It is, of course, inseparable from the stories of the great French marques. Many prestigious names (Talbot, Delage, Delahaye) disappeared following the war, leaving France with five principal automakers: Peugeot, Renault, Citroën, Simca, and Panhard. While these firms shared a rapidly expanding market buoyed by a broad-based improvement in the French standard of living, they were confronted by significant industrial challenges. Following five years of war that had stifled innovation, virtually the entire fleet of French automobiles had to be replaced. Almost all the available models had been designed in the 1930s. The most common four- or five-seater was still the Citroën *Traction*, a model introduced in 1934, with an impeccable design by Flaminio Bertoni. Its lines were typical of the prewar era: a tall radiator grille decorated with an impressive chevron, vents along the sides of the hood, a broad front mud guard, a black body, and whitewall tires. Citroën was, in fact, still making the *Traction* in 1955, the year it introduced its new sedan, the *DS*. The *DS*, also designed by Bertoni, in collaboration with Citroën's in-house team, impressed consumers with its original styling and technical innovations. Particularly appreciated was its comfortable ride, a result of the new hydropneumatic self-leveling suspension, which caused a sensation in the automotive world. The bold design of the *DS* was soon immortalized by Roland Barthes, who wrote about it in his *Mythologies*. This vehicle represented a true industrial advance, combining a very contemporary aesthetic with impressive mechanical features (not only the suspension, but also the automatic clutch and pivoting headlights); indeed, the *DS 21* and its less powerful, more modestly

LEFT
528. **Brochure for the Citroën 2CV, designed by Delpire agency, Paris,** 1963
Musée des Arts Décoratifs, Paris

BELOW
529. **Citroën 2CV**
Introduced in 1948

OPPOSITE, TOP
530. **Renault 4 (or 4L)**
Introduced in 1961

OPPOSITE, BOTTOM
531. **Advertisement for the Renault 4 (or 4L)**, c. 1965

RENAULT 4

equipped variant, the *DS 19*, broke decisively with prewar design (plate 532). The *DS 21* was the first high-tech family car, aerodynamic and well equipped to handle any bumps in the road. Comfort and speed were now within reach in a newly prosperous France: "Don't wait any longer! Drive tomorrow's car today!" exclaimed the advertisements. The *DS* won first prize for industrial design at the 1957 Triennale di Milano, where it was mounted upright, like a rocket on a launchpad. And at the World's Fair in Brussels the following year, it was displayed outdoors beside *Sputnik 3*, creating a sensation. The *DS* was rapidly embraced by the ruling classes, as well as by a bourgeoisie that was taken with modernity and intent on keeping up appearances.

Other carmakers were also overhauling their range of family cars. Peugeot made an initial break with the past with its *403* of 1955, a compact car with "serious" lines, designed by the famed Turinese "couturier" of automobiles, Pininfarina. But despite its rounded, stylish forms, the *403* retained a slightly heavy, conventional look. It was not until five years later that Peugeot introduced a truly modernized car, the *404* (plate 533). This time, Pininfarina gave the body very structural horizontal lines that terminated in framing elements projecting from the car's front and rear. Pininfarina delivered an even more original design in 1968 with the *504*, whose style and quality kept it in production for a remarkably long time.

Renault targeted the same market for family cars with its *Frégate*, a sedan with fluid, rounded lines, launched in 1951. It was followed by the *R16* of 1965, which featured a very new design: for the first time, the conventional trunk was eliminated and replaced with a hatchback. Simca offered the *Aronde* line of family cars, which were well designed (with fenders incorporated into the body) and very popular. The *Aronde* was updated several times, losing its famous "mustache" grille in its final version, the *P60* (1958). Simca also entered the large-car market in 1954, when it purchased Ford's French plant and took over production of the *Vedette*, which it offered in various trim levels (the *Régence*, the *Versailles*, and the *Trianon*), as an estate car (the *Marly*), and in a later variant with a smaller engine (the *Ariane*). These models borrowed liberally from the conventions of American automotive styling: emphatic radiator grille, fenders, bumpers, and chrome trim; two-tone finish and upholstery; and a plastic dashboard with gold highlights.

In the meantime, Panhard, which produced relatively few models compared to its larger competitors, offered the *Dyna Z* in 1954 (plate 535) and its successor, the *PL 17*, in 1959. These models had a very novel design, with a body (initially) in aluminum set on a modular chassis made from a light alloy. The motor was air-cooled, and the windshield and rear window were made of curved glass, a new feature. The last car produced by Panhard before its acquisition by Citroën was the *24 CT* coupé (plate 534), the work of in-house designer Louis Bionier. Like Panhard's other models, it was recognized as one of the most elegant productions of the French automotive industry.

At the same time as these larger family cars were being developed, rapid economic growth created a demand for small cars among a new category of clients. At the 1946 Salon de l'Automobile, Renault, which had been nationalized, presented a compact car with a rear-mounted engine and a design that was modern, albeit with no outstanding novelties: the *4CV* (plates 526, 527). While Peugeot stayed out of the small-car market for the time being, Citroën began its TPV (Très Petite Voiture, or Very Small Car) project, targeted at a working-class clientele, as early as 1939. This car, conceived by the company's chief executive, Pierre Boulanger, and developed over an extended period by engineers under the supervision of Flaminio Bertoni, finally made its debut at the 1948 Salon de l'Automobile: the legendary

OPPOSITE
532. **Citroën *DS 19***
Introduced in 1955

RIGHT, TOP
533. **Peugeot *404***
Introduced in 1960

RIGHT, BOTTOM
534. **Panhard *24 CT***
Introduced in 1963

BELOW
535. **Three Panhard *Dyna Z* sedans**
Introduced in 1954

ELLE
SERVICE
DANS CE NUMÉRO :
La Varende,
Françoise Sagan,
Marcelle Ségal.
626-ET75
ET TOUT SUR
Votre voiture
Vos week-ends
Votre mariage

OPPOSITE
536. **The Renault *Dauphine* on the cover of *Elle* magazine**, April 1956

RIGHT
537. Philippe Petit-Roulet (b. 1953)
Advertising art for Renault's *Twingo Easy*, 1995
Private collection

BELOW
538. **Renault *R5***
Introduced in 1972

2CV was born at last! It was an astonishing car from an aesthetic point of view: large fenders in front, as in prewar designs; a small, flat windshield; a minimalist dashboard and tubular steering wheel; a canvas top instead of a roof; gray paint; and adjustable seats with a spartan appearance (plates 528, 529). It was powered by a sturdy, air-cooled "flat-twin" engine. The form of the *2CV* was determined entirely by its functional requirements, with every detail calculated to optimize simplicity, durability, and adaptability to any terrain: it was in fact an "anti-design" car. The initial *2CV* was continually enhanced over its very long production run, which extended to 1990. Like its foreign competitors, the *Fiat 500* and the Volkswagen *Beetle* (known as the "Coccinelle," or "Ladybug," in France), it was a first car for millions of households. For decades, these three vehicles symbolized their countries of origin. To replace its outdated *4CV*, and respond to the success of the *2CV*, Renault introduced its functional *R4* model, also known as the *4L*, in 1961 (plates 530, 531). It, too, was very well-received, selling 8,135,000 units over three decades.

Renault and Citroën also began to broaden their range of midsize cars. Citroën created another sensation in 1961 with its *Ami 6*, whose rear window slanted backward, in striking contrast to the norm. The boxy geometric forms introduced by Renault in the *R4* were also featured in its *R8*, as well as in the Simca *1000* and Citroën's *Méhari* (1968), an open utility vehicle with a body in ABS plastic, a novelty for the firm. Taking a very different approach, Renault targeted the urban female customer with its *Dauphine*, an elegant compact (plate 536).

The postwar era of small cars closed with the introduction, in 1972, of the Renault *R5* (plate 538). All the components of this cleanly designed subcompact (including the headlight and handles) were incorporated into its body, which featured two wide doors, a tall hatchback, and very prominent gray plastic bumpers that protected the front, back, and sides. The *R5* was a multipurpose vehicle, suitable for both city and highway driving, as was Peugeot's *205*, which followed ten years later. These models were in fact the forerunners of today's city cars, whose appearance has evolved over the years in order to maximize the interior capacity (as with a slanted windshield) and the size and accessibility of the cargo space (as with a raised back); we may speak of a "water drop" aesthetic. Meanwhile, the desire for better handling, performance, and fuel efficiency, combined with technical progress in measuring and imaging the airflow around the car's body, led manufacturers to give their family sedans long, uninterrupted lines—a style they had long cultivated in their sports cars. Citroën, which had established an early lead in this area with its *DS*, also applied aerodynamic principles to its vehicles of the 1970s and '80s, including the *SM* grand tourer, the *GS* and *BX* sedans, and the *CX* sedan of 1974, whose name is actually the French abbreviation for "drag coefficient."

The 1980s saw the introduction of an entirely new type of vehicle, the minivan. In 1983, Renault and Matra signed an agreement to jointly manufacture the *Espace*, which could hold seven passengers and had removable pivoting seats and a high ceiling. Following the success of this pioneering model, the minivan became a popular midsize vehicle, encroaching on the market for conventional sedans, and even city cars. Renault's *Twingo* (1993) was among the first very small minivans (plate 537). It was also the first vehicle to feature a playfully designed dashboard, with controls and monitors that surprised by their scale and bright colors.

Recent technical innovations in the production of electric cars have also inspired advances in design. In particular, two of Renault's electric models, the *Twizy* minicar and the *Zoé* hatchback, seem to augur a new approach to automotive design that is more free, fluid, and imaginative in its use of space.

BELOW
539. Joseph-André Motte (1925–2013)
Molded seats for the Paris Métro, 1973

OPPOSITE
540. Quasar (pseudonym of Nguyen Manh Khanh) (b. 1934)
Poster for Quasar inflatable furniture, 1968
Musée National d'Art Moderne, Centre Georges Pompidou, Paris

THE 1960S: A DECADE OF QUESTIONING

In the 1960s, many designers who had first appeared on the scene in the previous decade focused on major land use projects. The construction of the remarkable building known as the Maison de la Radio in the early '60s gave some of them an opportunity to display their talents. Pierre Paulin, Joseph-André Motte, René-Jean Caillette, Antoine Philippon, and Jacqueline Lecoq all contributed in their own distinctive styles. The colorful novelty of Pierre Paulin's *Globe* chairs responded to the functionalism and efficiency of Motte's work, or Philippon and Lecoq's.

The restructuring of the Parisian metropolitan area that began in 1964 entailed major construction projects, including the building of five new prefectures. Designers competed avidly to work on their interiors: Alain Richard won the commission for Nanterre, Pierre Guariche for Évry, and Motte for Cergy-Pontoise. Major projects were also underway in the provinces. In 1956, workers won the right to a third week of paid vacation, and this leisure time, combined with improved purchasing power, led to the rapid development of major seaside and mountain tourist complexes. Guariche worked on the new winter sports facilities at La Plagne and Isola 2000, and Charlotte Perriand, on those at Arcs. The development of seaside resorts, particularly on the Languedoc coast, had a major impact on the landscape. Keeping pace with these construction projects, the hospitality industry flourished throughout France. New hotel chains included Arcade, which called on the services of Roger Fatus, and PLM, which hired Michel Boyer.

The "culture for all" espoused by French Minister of Culture André Malraux resulted in the construction of spectacular and iconic cultural centers, including André Wogenscky's complex in Grenoble, with interiors by Richard. Le Corbusier's project in Firminy, where the interior design commission went to Guariche, is another example. Motte, one of the most gifted designers of his generation, became the dean of airport development projects in the three prosperous decades known as *Les Trente Glorieuses*. He played a key role in designing the interiors of the large airports built by a new generation of architects—Henri Vicariot (Orly Sud in 1961, Orly Ouest in 1972), Paul Andreu (Roissy 1 in 1975, Roissy 2 in 1981), and Guillaume Gillet (Satolas in 1976)—and also worked on the stations of the Paris Métro (plate 539).

Contemporary design was gaining ground. It was a sign of the times when Mobilier National, that bastion of tradition controlled by the French state, established the Atelier de Recherche et Création du Mobilier National in 1964 to encourage modern design in public spaces. Another sign of the institutionalization of the profession was the appearance, in 1969, of *CREE*, the first French journal truly devoted to the topic of design. In the same year, the CCI (Centre de Création Industrielle) was created at the Musée des Arts Décoratifs. Its

debut exhibition was aptly titled *Qu'est-ce que le design?* (What is design?). The initial goal of the CCI—to establish contacts between designers and manufacturers—was soon forgotten, however, and the institution itself was subsequently transferred to the Musée National d'Art Moderne, Centre Georges Pompidou.

With the advent of these initiatives and projects came a greater recognition of the designers who had begun their careers in the 1950s. They perfected the refined, disciplined style of their earlier creations, while also adapting it to the new spatial challenges presented by large interior commissions. At the same time, a new generation was entering the scene. Considerably younger, they had not absorbed the economizing, functionalist atmosphere of the early postwar period, and they were ready to move beyond the prevailing aesthetic of austerity. Their sources of inspiration and interests were also very different. The French designers of the 1950s, steeped in the spirit of the Bauhaus, looked toward the U.S. and Scandinavia for inspiration. Those who emerged during the 1960s, however, were fascinated by Italy: its way of life, its industries, its iconoclastic films, and its designers, nonconformists like Gaetano Pesce and visionaries like Joe Colombo. The French furniture industry was not as advanced as Italy's, but the new French designers were undaunted, and embraced the radical approach to forms that prevailed across the Alps.

The upheavals of May 1968 made themselves felt throughout society. The revolution in mores was accompanied by one in forms and materials. Furniture became "*fous, mous, doux*" (loosely, "crazy, squishy, spongy"), and the minimalism of the 1950s was forgotten.[11] Magazines featured models sprawled in pillowy chairs, even if actual sales of such furniture were weak. Retailers experimented with pieces that blurred the lines between carpeting, beds, tables, and play areas—for example, Paulin's *Déclive* for Mobilier International, Olivier Mourgue's armchair/carpet for Prisunic, Bernard Govin's *Asmara* chairs for Roset (plate 548), and Hans Hopfer's *Dromadaire* for Roche Bobois. Contemporary seating had to be modular and, above all, comfortable. This focus on comfort inspired Airborne's audacious advertisement of 1968, which pictured fifty bare bottoms above the slogan "It's all here!" Baby boomers saw nothing appealing in their parents' furniture, which seemed ponderous, and burdened with the legacy of past generations. The advent of the inflatable allowed them to purchase "furniture made to last three years," and offered an unprecedented lightness and ease of mobility. Quasar and Aérolande were the champions of inflatable furniture (plate 540). The cardboard designs of Jean-Louis Avril and Claude Courtecuisse served the same objectives.

Disposability kept products current: Bic's 1950 ballpoint pen and Jean-Pierre Vitrac's 1977 *Plack* picnic set, not to mention lighters, razors, and paper tissues, put an end to the notion of the permanent object. This was no minor revolution. Another revolution, this time on paper, occurred when Prisunic mailed out its first catalog in April 1968. "Buy the beautiful for the same price as the ugly" was the slogan of this new venture that sought to democratize nonconformist furniture (plates 541, 542). Inspired by the experience of Terence Conran and his Habitat line in England four years earlier, Prisunic was led by the dynamic duo of Denise Fayolle and Maïmé Arnodin. Francis Bruguière, Yves Cambier, and Michel Cultrutra oversaw the selection of merchandise, and Andrée Putman, succeeded by Janine Roszé, supervised the settings (rugs, fabrics, etc.). The audacious spirit of Prisunic's founders was evident in its offerings, which included Mourgue's armchair-carpet

OPPOSITE
541. Danielle Quarante (b. 1938)
***Balthazar* chairs**, 1970
Made by Prisunic (France)
Pictured in the October 1971 Prisunic catalog
ABS plastic
Bibliothèque des Arts Décoratifs, Paris

BELOW
542. **Cover of the Prisunic catalog**, 1969
Bibliothèque des Arts Décoratifs, Paris

la vida misma en forma de
DJINN:la vie même en forme de sièges. DJINN:life itself in the form of seats. DJINN:Ausdruck modernen Lebens

OPPOSITE, TOP
543. Pierre Paulin (1927–2009)
577 or *Tongue* chair, 1967
Made by Artifort (Netherlands)
Steel, polyester
Musée National d'Art Moderne, Centre Georges Pompidou, Paris

OPPOSITE, BOTTOM
544. **Airborne catalog illustrating *Djinn* chairs by Olivier Mourgue**, c. 1965
Daniel Bernard collection

BELOW
545. **Advertisement for the *Tongue* chair by Pierre Paulin**, 1967

and Marc Held's low bed molded from a single piece of plastic. Nearly all the avant-garde furniture designers were involved in this popularizing experiment, including Mourgue, Held, Conran, Courtecuisse, Gae Aulenti, and Marc Berthier. Regrettably, it was all too short-lived, and the last catalog was published in 1976.

Two figures stood out as representatives of a new approach to furniture that emphasized comfort and visual appeal. Independent and forceful personalities, Pierre Paulin and Olivier Mourgue are widely recognized as the most inventive French designers of their generation. Both received major accolades in 1971: Mourgue from his peers, when he was invited to succeed Verner Panton and Joe Colombo in creating a model home for the Visiona display at the Cologne furniture fair, and Paulin from the state, when French president Georges Pompidou asked him to redecorate his private apartments in the Élysée Palace. Moreover, both served as the equivalent of creative director to a major retailer: Mourgue worked with Airborne, the formidable company that Charles Bernard had established in Tournus, while Paulin, unable to find a suitable French backer for his work, joined forces with the Dutch company Artifort. For this firm, Paulin developed a series of colorful chairs—including the *Mushroom*, *Tongue*, *Oyster*, and *ABCD* (plates 543, 545)—whose flowing lines and absence of visible structural elements established his reputation. Contrary to popular belief, Mourgue's work for Airborne went far beyond the famous series of *Djinn* chairs selected by Stanley Kubrick to furnish the space station in his 1968 film *2001: A Space Odyssey* (plates 544, 546). Mourgue also designed collections for a broader market, such as *Whist*, and limited-edition chairs, such as *Bouloum* and *Tric-Trac*.

Innovations in the use of foam facilitated many novel furniture designs. Airborne perfected the technique of covering injected foam with a synthetic knit jersey fabric. Roset also found a variety of applications for foam, in particular using cut foam to make lavishly rounded chairs, such as Bernard Govin's *Asmara* and Michel Ducaroy's *Adria*, *Kali*, and *Togo* (plate 549). Roche Bobois, too often thought of as merely a distributor, also produced furniture. The firm's artistic director, Hans Hopfer, played with foam structural elements and modular blocks in his ensembles *Lounge*, *Entr'acte*, and *Dromadaire*, the last being designed as a ring. In its new factory in Noisy-le-Grand, Steiner employed the talented young designers Claude Courtecuisse, Roger Landault, Jacques Famery, and Kwok Hoï Chan to shake up its rather traditional image. Chan, originally from

BELOW
546. Olivier Mourgue (b. 1939)
***Djinn* chair**, 1964–65
Made by Airborne (France)
Bent steel tubing, polyurethane foam, jersey
Musée National d'Art Moderne, Centre Georges Pompidou, Paris

OPPOSITE
547. Marc Held (b. 1932)
***Culbuto* chair**, 1966–67
Made by Knoll International (U.S.)
Polyester reinforced with fiberglass, leather
Musée National d'Art Moderne, Centre Georges Pompidou, Paris

BELOW
548. **Advertisement for Bernard Govin's *Asmara* seating, made by Roset**, 1966
Roset Archives

OPPOSITE
549. Michel Ducaroy (1925–2009)
***Togo* chair and ottoman**, 1973
Made by Roset (France)
Polyester foam, polyester wadding, fabric
Musée des Arts Décoratifs, Paris

Hong Kong, became Steiner's premier designer, establishing a signature style of rounded, generous forms, in contrast to the aesthetic of the firm's earlier reformers, who had belonged to the ARP in the early 1950s.

Also increasingly common was plastic and fiberglass furniture, whose method of production encouraged simplified forms without any strong relief. These sleek pieces were designed by the dozens and included Held's *Culbuto* armchair (plate 547), Courtecuisse's *Soleade* chair, Landault's *Unibloc IV*, Christian Daninos's *Bulle* armchair, Christian Germanaz's *Half and Half* chairs, Maurice Calka's *Boomerang* desk, Annie Tribel's cafeteria furniture, and Berthier's *Ozoo* line. Asked by *Le Monde* to characterize the '70s look, the artist and humorist Roland Topor alluded to the "smooth style" and even a "suppository" aesthetic: "From the Place de la Concorde to the Tour Montparnasse, we're caught up in an aerodynamic mental universe. Our clothes irons look like they're made to break the sound barrier, our telephones are low-riders like Porsches, our calculators look like UFOs, and our pens resemble torpedoes. Anything that stands out—that's the enemy. Anything that creates friction—that's the danger. . . . Outside, featureless autoroutes take us faster and faster between one place and another. Inside, we have seats down on the floor, low coffee tables, molded plastic beds, and that blasted wall-to-wall carpeting everywhere."[12]

In furniture, however, the smooth aesthetic was dealt a sudden blow by the oil crisis of 1973. The price of petroleum soared, and so did that of the plastics made from it. Plastic no longer seemed to be the almost magical substance that Raymond Queneau had described in Alain Resnais's short film of 1957, *Le Chant du Styrene*, and it lost its privileged place in design.

BELOW
550. Philippe Starck (b. 1949)
***Max le Chinois* colander**, 1987
Made by Alessi (Italy)
Stainless steel
Museum of Modern Art, New York

OPPOSITE, TOP LEFT
551. Philippe Starck (b. 1949)
***Dr Skud* flyswatters**, 1998
Made by Alessi (Italy)
Thermoplastic resin
Museum of Modern Art, New York

OPPOSITE, TOP RIGHT
552. Philippe Starck (b. 1949)
Toothbrushes, 1989
Made by Fluocaril (France)
Thermoplastic resin
Musée National d'Art Moderne, Centre Georges Pompidou, Paris

OPPOSITE, BOTTOM
553. Philippe Starck (b. 1949)
***Juicy Salif* citrus squeezer**, 1988
Made by Alessi (Italy), 1990
Cast aluminum, thermoplastic rubber
Museum of Modern Art, New York

STARCK: A BREATH OF FRESH AIR

The designers of the 1980s—with Philippe Starck foremost among them—achieved an unprecedented degree of recognition, popularizing a discipline that was still little known among the general public. Starck's preeminence may be attributed not only to his extraordinary energy, his rhetorical skills, and his ability to approach any project, whether for a toothbrush or a motorbike, without preconceived notions, but also to his entrepreneurial passion and technical know-how. His father, an aeronautical engineer, impressed on him the importance of research in all fields of endeavor.[13] The passion for creativity and technology that he inherited from his father would be the foundation for his entire career, which has been marked by very fruitful collaborations with business leaders, primarily Italian: Piero Gandini of Flos, Giulio Castelli and Claudio Luti of Kartell, and Alberto Alessi of Alessi. In France, he developed close contacts with two companies where he later became artistic director: the furniture maker XO and the Thomson electric group. Although he worked for fewer French companies than he might have wished, he consoled himself by thinking on a European scale. In 1983, Starck was aware "of being at the beginning of a style of design. . . . There's a sense of urgency, of the need to rekindle people's interest in their lives, in their homes."[14]

Starck studied at the École Camondo, which specializes in interior design. He first attracted notice with his interiors for two Paris nightclubs: La Main Bleue in 1976 and Les Bains Douches in 1978. However, the two projects that truly made his reputation were his renovation of the private apartments of the president of the French Republic in 1983, and the café near Les Halles that he designed for the Costes brothers in 1984 (plate 556). The Café Costes created a sensation, with its remarkable staircase, narrow at the base, that broadened as it ascended; an enormous clock like one in a railway station; enveloping tripod chairs; and spectacular restrooms. Style and panache were the hallmarks of Starck's interior design.

His furnishings, in contrast, began in a more modest fashion. The earliest, some of which were produced by Starck Product, the firm he had founded in the late 1970s, were reduced to their essentials. The *Easy Light* portable lamp, designed in 1979 for Electrorama, consisted of a neon tube and two end pieces. His *Dr Bloodmoney* and *Miss Wirth* chairs, designed in the 1980s, are comparably restrained, black with very simple lines. This tendency toward minimalism has persisted in much of Starck's furniture, from the *Richard III* armchair of 1981 (plate 554), which recalls classic club chairs but without the weight of a heavy frame, to the light and transparent *La Marie*, designed for Kartell in 1998. In the 1990s, Starck was still exploring archetypes: the *Miss Sissi* lamp for Flos is a single unit, in a single material; the *Starck 1* bathroom fixtures for Duravit recall barrels and buckets; and the *Bohême* stools for Kartell echo the silhouettes of ceramic vases.

Starck's very distinctive designs are easily recognized by the public. A provocateur, he often gives his works imaginative titles, sometimes employing simple wordplay, sometimes drawing inspiration from historical figures or the novels of Philip K. Dick. Starck enjoys mockery, and does not hesitate to transgress conventional standards of good taste. He called the stool-tables he designed for Kartell *Attila*, *Saint-Esprit*, and *Napoléon*.

Some of his most iconic objects are virtual UFOs in the universe of design: the *Ara* lamp for Flos, and the *Max le Chinois* strainer (plate 550) and the *Juicy Salif*

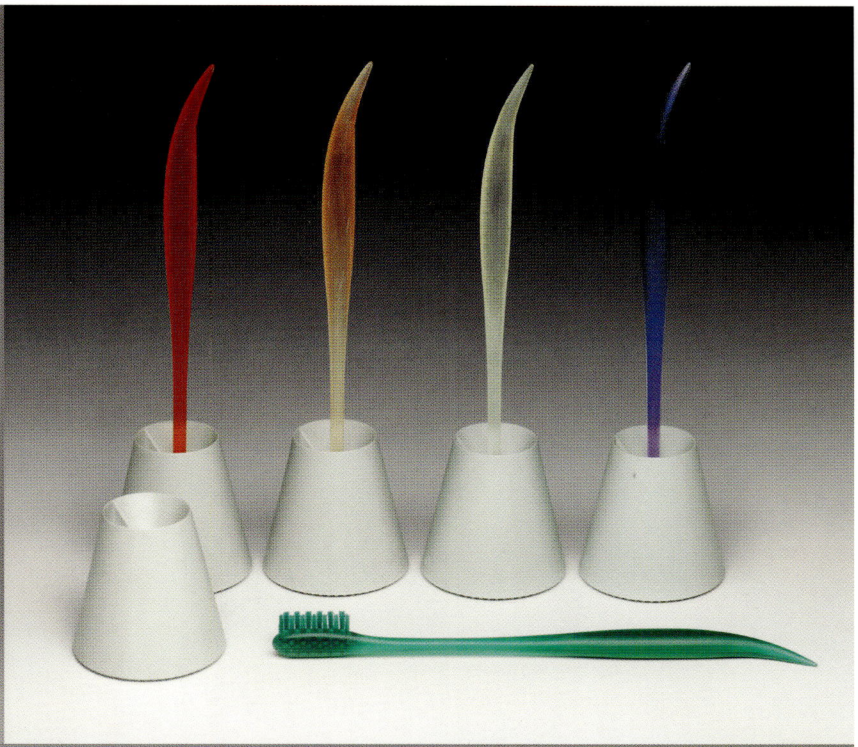

BELOW
554. Philippe Starck (b. 1949)
***Richard III* tripod chair**, 1981
Made by Baleri (Italy), 1984
Polyurethane, Dacron, leather
Musée des Arts Décoratifs, Paris

OPPOSITE
555. Philippe Starck (b. 1949)
***Louis Ghost* chair**, 2001
Made by Kartell (Italy), 2002
Polycarbonate
Musée des Arts Décoratifs, Paris

BELOW
556. Philippe Starck (b. 1949)
Interior of the Café Costes, Place des Innocents, Paris, 1984 (closed 1994)

OPPOSITE
557. Philippe Starck (b. 1949)
Interior of the Félix restaurant, Peninsula Hotel, Hong Kong, 1994

citrus squeezer (plate 553) for Alessi. The image of the *Juicy Salif* inevitably appears in every publication about Starck. Indeed, this piece exemplifies the way design has gained an audience by offering a solution that seems remote from the intended purpose. Starck is also fond of oversize versions of daily objects, as can be seen in both *Max le Chinois* and the *Archimoon K* lamp for Flos.

Starck went to Thomson "to try to find a way to get the image and the sound out of that boring gray box" and escape from "the old mandates of the 1960s and '70s that came out of America and Japan." In the mid-1990s, Starck worked on the identity of Thomson's eponymous brand, as well as those of its subsidiaries Telefunken and Saba, totally reconsidering the forms, colors, and controls of their devices. The *Moa Moa* radio assumed the shape of a Brancusi head, the *LaLaLa* radio resembled a little microphone, and the *Jim Nature* portable television was daringly covered with resin-impregnated sawdust as if from a construction site. The influence of the Italian architect and designer Achille Castiglioni, who had introduced Starck to Flos, is very apparent: there is the same humor, the same regard for the everyday. Starck has taken the lesson from Castiglioni's *Snoopy* and *Bip Bip* lamps—and gone even farther. He has constantly advocated for design that is humanized, and objects that are "right before being beautiful." Even if Starck the polemicist is not always entirely convincing, his projects and technological research are intensely relevant.

Working with Kartell, he was particularly enthusiastic about new advances in plastics. Molded in a single piece, his chair *La Marie* (1998) emancipated itself from Kartell's customary rounded forms. This is a case where he obtains "the maximum using the minimum"—with less material, less visual presence due to transparency, and less cost. The *Louis Ghost* chair that followed was also a major success in every market (plate 555). In the *Bubble* series, he borrowed the technique of molded polypropylene from kayak-making, using it to produce colorful, light, sturdy, and reasonably priced furniture. The *Louis XX* chair for Vitra was not only a stylistic reinterpretation but also—and more importantly—the first piece of furniture to be made by the process of blow molding, more often used in the manufacture of water bottles.

Starck takes an interest in all aspects of a product. "All that concerns me at the moment are costs: the price, the value of things, the appropriateness of the price for the object," he asserted in 1994. He also raised the issue of distribution: "We cannot say that a piece of furniture is modern if it is distributed in an archaic manner at an inaccessible price." Following this logic, he worked with the catalog retailer 3 Suisses in the early 1980s. His culminating experiment in this area, again with 3 Suisses, came in 1994, when he sold a prefabricated "House in a Box" by mail order. (It was excoriated by the critics.) The *Good Goods* catalog of 1998, created for La Redoute, featured products designed or recommended by Starck, ranging from champagne to a four-headed teddy bear, while not omitting table settings. In yet another bid for the mass market, he created a line of fifty-one highly utilitarian objects for the American chain store Target in 2002.

Starck is at the forefront of a number of fields: furniture and accessories, industrial design, interior design, and finally architecture, which he began to practice later in his career. In the 1980s, fashionable hoteliers began to commission interiors from well-known designers and architects; for example, Andrée Putman worked on the Morgans Hotel in New York, and Jean Nouvel on the Hôtel Saint James in Bouliac, near Bordeaux. Following a series of discussions, Starck, too, was drawn into projects of this type, and he created a sensation from Miami to Madrid with his stunning hotel interiors. He designed several hotels for Ian Schrager in the U.S. and, in 1988, embarked on a worldwide series of hotel and restaurant commissions: the Saint Martins Lane Hotel in London, the Teatriz restaurant in Madrid, and the restaurant of the Peninsula Hotel in Hong Kong (plate 557). In France as well, projects for boutiques and restaurants followed in succession: the Hugo Boss store, the Jean-Paul Gaultier store, the Baccarat museum and headquarters, the Bon and Kong restaurants, and, more recently, the restaurants in the Meurice and Royal Monceau hotels. In these designs, decorum meets spectacle.

Starck has always advocated for an affective architecture that stimulates, and perhaps even unsettles, the visitor. He accomplished exactly that in Japan, with the Asahi Beer Hall (1990) and the Nani Nani and Le Baron Vert office buildings (1989 and 1992); in the Netherlands, with the Groningen Museum (1993); and in France, with the Laguiole factory (1987). More recently, in Bilbao, he turned away from this formalism, adopting a simple and even slightly severe style for his conversion of the Alhóndiga, a former wine warehouse, into a cultural center (2010).

Resisting the trend toward limited production runs, Starck remains faithful to the idea of accessible design, manufactured in large quantities and affordable for all (plates 551, 552). Prolific in words and deeds, he has succeeded in identifying his name with contemporary design. However, there really is no "Starck style" as such: again and again, his undeniable stature has allowed him to impose his latest vision on manufacturers. Relishing provocation and challenges, he has embarked on more personal ventures in recent years. In 2007, he was named the artistic director of Virgin Galactic, the first private aerospace company, founded by the British entrepreneur Richard Branson. Starck is once again taking up the cause of democratization—this time in space! In 2012, he unveiled the *Venus*, a yacht designed with and for Steve Jobs, in a minimalist style eminently suited to Apple's image.

BELOW
558. Élisabeth Garouste (b. 1949) and Mattia Bonetti (b. 1953)
***Enfer* cabinet**, 1998
Made by Galerie Néotù (France)
Painted cast iron, enameled ceramic
Centre National des Arts Plastiques/Fonds National d'Art Contemporain, on loan to the Musée des Arts Décoratifs, Paris

OPPOSITE
559. Élisabeth Garouste (b. 1949) and Mattia Bonetti (b. 1953)
***Barbare* chair**, 1981
Made by Galerie Néotù (France)
Patinated iron, calfskin, leather thongs
Centre National des Arts Plastiques/Fonds National d'Art Contemporain, on loan to the Musée des Arts Décoratifs, Paris

THE LIBERATION OF DESIGN

Just as functionalism was on the verge of being left behind, it experienced a temporary resurgence, thanks to the decorator Andrée Putman's reissues of classic designs. Reacting against the styles of the moment, she founded Écart International in 1978 to distribute reproductions of furnishings and light fixtures from the 1930s, by Pierre Chareau, Félix Aublet, Robert Mallet-Stevens, and Eileen Gray. In this way, the work of the founders of French functionalism was belatedly revealed to the public at large.

Today, however, functionalism truly is in retreat, to the point that the legibility of design is sometimes obscured by a jumble of references and an insistence on individuality. We may trace the roots of this development to 1966, when the American architect Robert Venturi published *Complexity and Contradiction in Architecture*. To Mies van der Rohe's assertion that "Less is more," Venturi retorted, "More is not less, less is a bore." This postmodernist slogan rings very true. Outside France, the advent of the 1980s marked a turning point. The Italian Gaetano Pesce opined that "the form/function duo" had long since become inadequate, and he added experimentation and differentiation as essential components of design. The establishment of Memphis in Milan in 1981 introduced an entirely new language to the debate. The stature of its founder Ettore Sottsass gave Memphis an aura that obliterated conventional boundaries. A number of French designers, including Martine Bedin and Nathalie du Pasquier, affiliated themselves with the Milanese movement, whose designs were shown in Paris in 1982 by Nestor Perkal. Meanwhile, in Lyons, the designers working with the Totem studio also adopted the Memphis aesthetic.

The new expressive and cultural relations espoused by Memphis unseated the received dogmas of design not only in Italy, but also in France (where, however, the break between form and function may be placed somewhat later in the 1980s). The gradual shift from the "industrial" to the "iconic" product was accompanied by a stronger sense of individualism among designers. The parameters of design were becoming less rigorous, and traditional materials, decorative elements, and smaller production runs were making an appearance in a realm that had formerly abandoned them. Ornament was no longer a crime. A unique piece with an improbable function now had as much claim to the title of design as any icon of mass production. Shaken up, but recognized by a growing public, contemporary design was welcomed into François Mitterrand's refurbished

apartments in the Élysée Palace and Jack Lang's Ministry of Culture, as well as the hotels designed by Andrée Putman, Jean Nouvel, and Philippe Starck.

While computer-aided design and manufacturing were regularly used in the development of prototypes, artisanal techniques and traditional materials returned to favor. In 1981, at the Jansen Gallery, Élisabeth Garouste and Mattia Bonetti exhibited their *Barbare* chair (plate 559), which gave its name to an entire style. This reaction against industry was organized by Frédéric de Luca, the representative and spokesman of a small group of sculptors and decorators he called "En Attendant les Barbares" (Waiting for the Barbarians). As two key participants, Garouste & Bonetti, forthrightly acknowledged in 1984, "We are against design."

Garouste & Bonetti expressed their stance particularly through the use of materials that were unusual in furniture design, including bronze, papier-mâché, stone, and animal skins (plate 558). They were, moreover, very cultivated, basing their style on unexpected artistic and literary sources. Journeying through time, they revisited the Oriental dream with the *Topkapi* sofa, and French history with the *Prince imperial* chair. For the headquarters of haute couturier Christian Lacroix, they created one of their most exquisite designs, half Marie Antoinette boudoir, half *Alice in Wonderland*

RIGHT
560. François Bauchet (b. 1948)
***H6 L3* bookcase**, 2001
Made by Winner-Pro (France)
Sold by Galerie Kreo (France)
Polyester resin
Centre National des Arts Plastiques/Fonds National d'Art Contemporain, on loan to the Musée des Arts Décoratifs, Paris

OPPOSITE, TOP
561. Martin Szekely (b. 1956)
***Heroic Carbon Desk Loop* desk**, 2010
Made by Galerie Kreo (France)
Carbon fiber, resin, honeycomb aluminum

OPPOSITE, BOTTOM
562. Sylvain Dubuisson (b. 1946)
***T2/A3* clock**, 1986
Stainless steel, electroluminescent diodes, epoxy composite circuit
Musée National d'Art Moderne, Centre Georges Pompidou, Paris

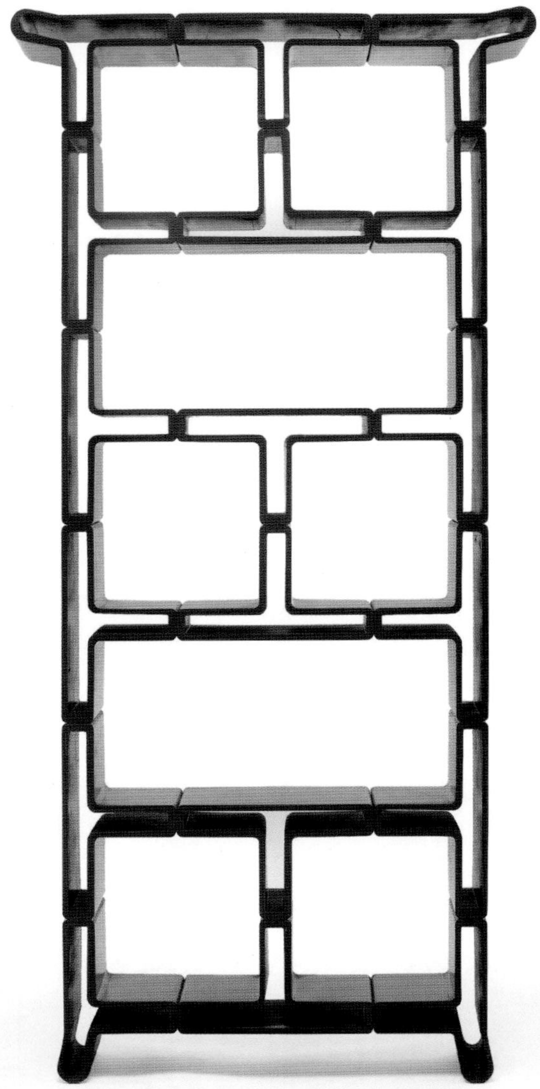

antechamber. They indulged in a fantastical vision of a magical natural setting that included tree-trunk stools and cabana-like fitting rooms. Several other designers who appeared on the scene in the 1980s also displayed this kind of imaginative flair and handmade look: Olivier Gagnère, Éric Schmitt, Kristian Gavoille, Pucci de Rossi, and André Dubreuil.

At this time, the art dealer Pierre Staudenmeyer played a key role in helping many young designers launch their careers. An enthusiastic proponent of psychoanalysis who humorously observed that the era was "neo-everything," he opened the Néotù Gallery in 1985 with Gérard Dalmon. They gave an entire new generation the opportunity to express itself without prerequisites, encouraging designers as diverse as Garouste & Bonetti, Gagnère, Schmitt, Gavoille, de Rossi, Martin Szekely, and François Bauchet, as well as foreigners like Jasper Morrison.

The leading participants in French design chose very divergent paths. One atypical and cultivated designer, Sylvain Dubuisson, bases his work on skillful geometric calculations. Resisting the banalization of everyday objects, Dubuisson creates sophisticated, symbolic designs such as the *T2/A3* clock (plate 562) and the *1989* desk that the Minister of Culture Jack Lang used in his office. Martin Szekely searches for universality through the sparest of forms (plate 561), while employing the latest techniques and materials: carbon fiber, Ductal concrete, silicon carbide, and crystal that has been frozen for several months. A perfectionist who claims "not to draw anymore," he strives to eliminate any trace of the extraneous from his pieces. On the other hand, drawing is the guiding principle for two designers trained at the École Nationale des Beaux-Arts: Bauchet, with his stripped-down clarity (plate 560), and Pierre Charpin, with his unifying use of color (plate 567).

Large production runs are no longer the only way for designers to gain visibility and recognition. Artisanal techniques are recovering vitality not only in private companies (the glassmakers Daum and Baccarat, and the porcelain maker Bernardaud, for example) but also through government initiatives (CRAFT, Centre de Recherche sur les Arts du Feu et de la Terre in Limoges; CIRVA, Centre International de Recherche sur le Verre et les Arts, in Marseille; and the national manufactory in Sèvres) and partnerships with towns, such as Vallauris.

At the level of industry, the alliance of design and manufacturing remains a constant. Like the carmakers, many companies have their own team of designers; the sporting goods firm Decathlon, for example, makes designers an essential part of the business. Design also helps differentiate the products of manufacturing giants like Ricard and the SEB Group. The latter enterprise, which became the world's largest manufacturer of household appliances around 2000, encompasses a wide range of brands, including some former German competitors: Moulinex, Tefal, Calor, Rowenta, Krups, and All-Clad, among others. At the initiative of Jacques Alexandre and Frédéric Beuvry, the SEB Group has given a number of French and international designers their first chance to work on industrial products. As SEB's director of design, Beuvry himself defined a strong image for each brand before calling on outside designers.[15]

Many high-tech firms have also worked with independent designers in the last decade. Parrot, a maker of multimedia devices, worked with Starck, and Schneider Electric, an energy management firm, with Normal Studio and Elium Studio. La Cie, an American data storage company with a production site in France, called on Starck, 5.5 Designers, and Constance Guisset, among others. Of course, French designers help define the image of firms in a whole variety of industries, including furniture (Cinna, Roset), mass-market distribution/retail (Carrefour, Decathlon), and fine tableware (Baccarat, Bernardaud).

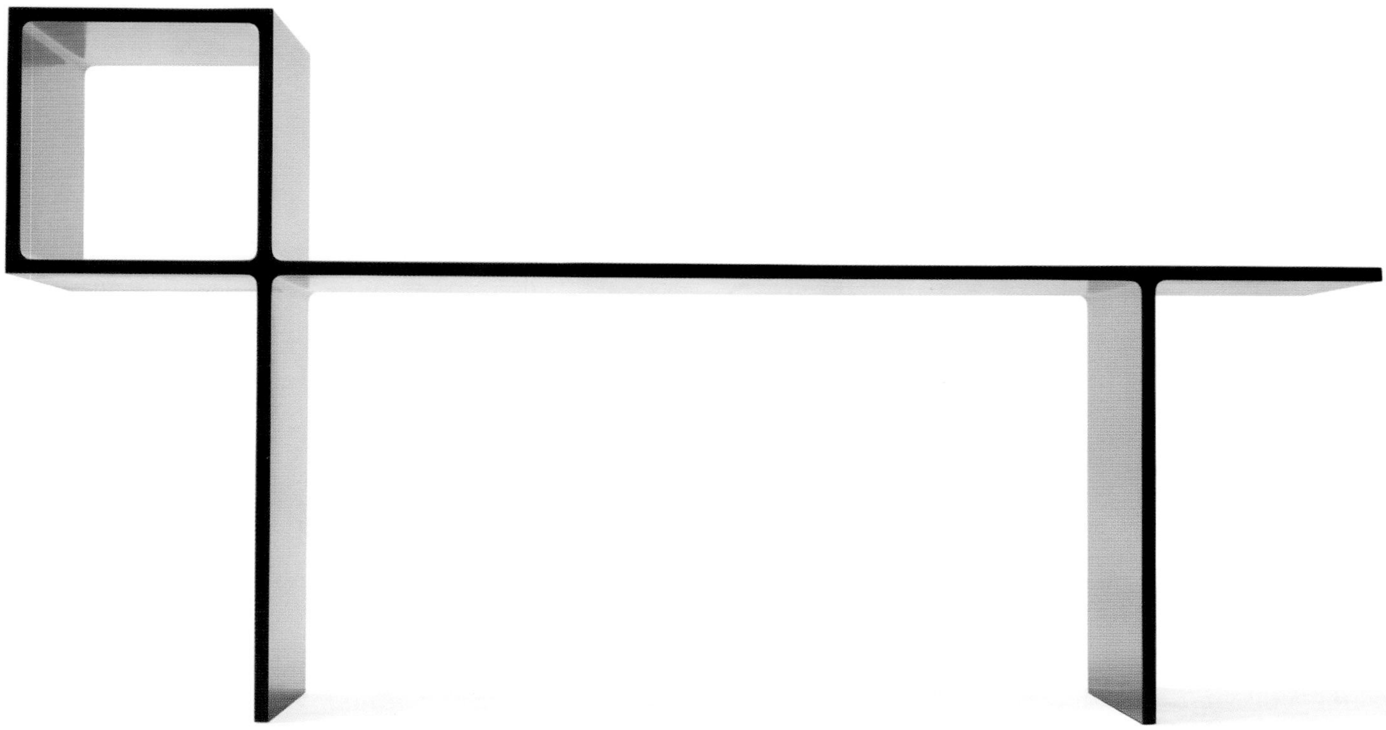

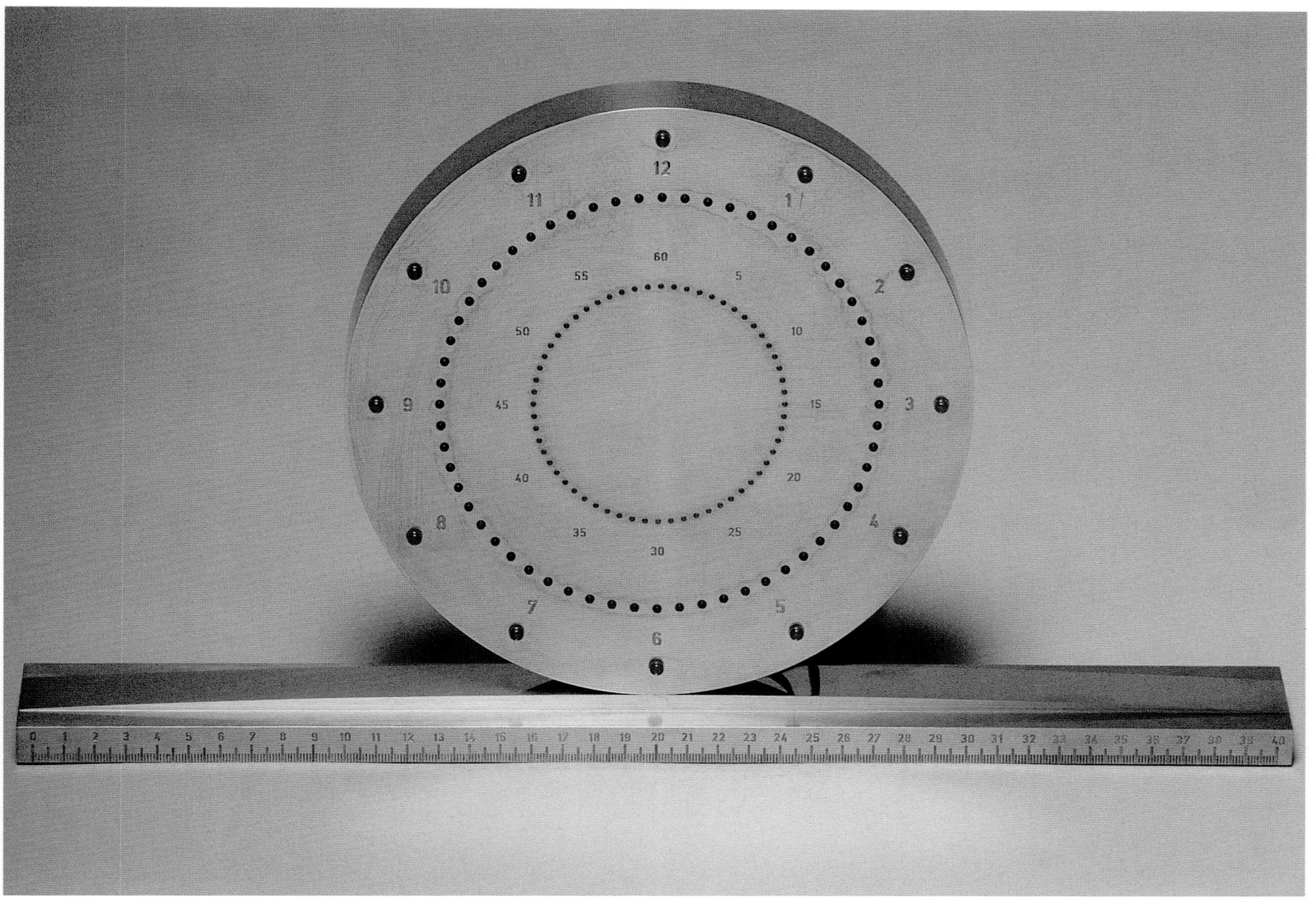
12 1 2 3 4 5 6 7 8 9 10 11
60 5 10 15 20 25 30 35 40 45 50 55
0 1 2 3 4 5 6 7 8 9 10 11 12 13 14 15 16 17 18 19 20 21 22 23 24 25 26 27 28 29 30 31 32 33 34 35 36 37 38 39 40

BELOW, LEFT
563. Matali Crasset (b. 1965)
***Quand Jim monte à Paris* guest bed**, 1998
Made by Domeau & Pérès (France)
Batting and bi-density high resilience foam, cotton inner and outer mattress cover
Centre National des Arts Plastiques / Fonds National d'Art Contemporain, Paris

BELOW, RIGHT
564. Elium Studio
Smart Baby Monitor, 2011
Made by Withings (France)

OPPOSITE
565. Patrick Jouin (b. 1967)
***Solid C2* chair**, 2004
Made by Materialise MGX (Belgium)
Epoxy resin polymerized by stereolithography
Musée National d'Art Moderne, Centre Georges Pompidou, Paris

Around 2000, several galleries (Kreo, Tools, BSL) and numerous smaller distributors (Domeau & Pérès, ENO, Ymer&Malta, Moustache, Petite Friture, and Artuce) appeared on the scene, offering greater visibility to the new generation. A host of young designers are committed to developing an efficient mode of design for the mass market, including Matali Crasset (plate 563), Patrick Jouin (plate 565), Christophe Pillet, Patrick Norguet, Christian Ghion, Christian Biecher, Éric Jourdan, Frédéric Ruyant, Normal Studio, Arik Levy, Jean-Marie Massaud, and Inga Sempé. They have ties to various brands (Pillet with Lacoste, Normal Studio with Tolix, Jourdan with Roset, and Didier Gomez with Cinna), but these are not exclusive relationships. Rather, this cohort of designers engages in various fields of activity and adopts varied approaches: Radi Designers exercises a surrealist bent at the confluence of art and design; Crasset emphasizes graphics; François Azambourg focuses on technological advances; and Guisset and Sempé express a poetic sensibility.

Like all these designers, the brothers Ronan and Erwan Bouroullec are deeply interested in "concrete" design. However, in the early 2000s, they distinguished themselves by rethinking space (plate 570). In contrast to their eminent predecessors Charlotte Perriand, Jean Prouvé, and Roger Tallon, their work refers less to the culture of engineering and more to experimental research and their own generation's way of life. Their experience at Cappellini and later at Vitra marked their introduction to the industrial world and their debut on the international stage, and now companies in both France (Roset, Tectona) and Italy (Kartell, Flos, Magis, Alessi) frequently approach them for advice. Their carefully considered use of partitions and modularity are the most striking aspects of their work, from the *Nuages* and *Algues* modules for Vitra (plate 569), to the *Tuiles* for Kvadrat, and the *Cabane* and *Lit clos* produced in a limited edition by Galerie Kreo. In a more traditional spirit, they focus on comfort in their work for Roset, which includes the graphic *Facett* sofa, with origami-inspired folds, and the voluptuous *Ploum* sofa, upholstered with ultrasoft foam.

If the Bouroullec brothers continue in a certain tradition, the functional transformations in the field of

BELOW
566. Cédric Ragot (b. 1973)
***Hyper Fast* vases**, 2003
Made by Ymer&Malta (France)
Solid surface
Galerie Ymer&Malta, Paris

BOTTOM
567. Pierre Charpin (b. 1962)
***Écran* vases**, 2000
Made by CIRVA (France)
Distributed by Galerie Kreo (France), 2005
Blown and molded glass

OPPOSITE
568. Benjamin Graindorge (b. 1980)
***IkebanaMedulla* vase**, 2010
Made by Ymer&Malta (France)
ABS plastic, lacquered polyamide, fiberglass
Galerie Ymer&Malta, Paris

OPPOSITE
569. Ronan Bouroullec (b. 1971) and Erwan Bouroullec (b. 1976)
Algues, 2004
Made by Vitra (Switzerland)
Injected polyamide
Museum of Modern Art, New York

ABOVE
570. **View of the exhibition *Ronan et Erwan Bouroullec. Momentané***, 2013
Musée des Arts Décoratifs, Paris

571–72. Jean-Marie Massaud (b. 1966)
Renderings of the *Manned Cloud* dirigible hotel, 2005
Studio Massaud

design are increasingly apparent. Three exhibitions—*D Day* at Beaubourg in 2005, *Safe* at MoMA that same year, and *Made 4 You/Design für den Wandel/Design for Change* in 2012 at MAK in Vienna—have revealed how the focus of design is shifting from considerations of comfort and visual appeal to a greater concern for ecology and even for survival. In France as elsewhere, the early years of this century have witnessed a desire to "go beyond design." Massaud's project for an airborne hotel in the form of a dirigible to avoid the pollution caused by the hospitality industry is among the boldest proposals (plates 571, 572). Jean-Louis Fréchin specializes in new products and the digital world; Mathieu Lehanneur is interested in medicine and well-being; Cédric Ragot, in the representation of natural forces (plate 566); Olivier Peyricot, in the concept of preservation; and Benjamin Graindorge, in our relationship with nature (plate 568). These designers may be alarmist, utopian, or visionary, but their voices resonate together on the issues confronting our society. Creative and spatial boundaries are now more porous than ever: Paris is attracting world-renowned designers, including the Australian Marc Newson and the Briton Jasper Morrison, who has opened a studio here.

Now more than ever, design is an international phenomenon. And, now more than ever, the concept of design has extended its scope. We may well recall the prescient words of Jean Baudrillard, in his 1968 book *Le Système des objets*: "Should we classify the vast forest of objects like flora or fauna with tropical and arctic species, abrupt mutations, and species in the process of extinction? Our urban civilization will see new generations of products, devices, and gadgets appear in increasingly rapid succession. Compared to these ephemeral objects, humans seem to represent an exceptionally stable species." Forty years later, the proliferation of products described by Baudrillard has only accelerated, and his question remains, now and always, revelant: "Can we hope to classify a world of objects that is changing before our very eyes?"

JAPAN

PENNY SPARKE

PRECEDING PAGE
573. Sori Yanagi (b. 1915)
***Butterfly* stool**, 1956
Made by Tendo Mokko (Japan)
See plate 611.

BELOW, LEFT
574. Junzo Sakakura (1901–1969)
Sketch of a bamboo chair for the Low Cost Furniture Design competition sponsored by the Museum of Modern Art, 1950
Museum of Modern Art, New York

BELOW, RIGHT
575. **The Japanese pavilion at the 1937 Exposition Internationale des Arts et Techniques in Paris**
Pictured in *L'Illustration*, May 29, 1937
Private collection

OPPOSITE
576. **Charlotte Perriand with a folding wooden chaise longue, Japan, 1940**
Archives Charlotte Perriand, Paris

Japon. — (Sakakura, archit. ; collaborateur français : Danis, archit.)

Japan has nurtured a distinctive and extraordinarily successful modern design culture since it opened itself to the West at the end of the nineteenth century. Consciously sustaining links with its past, Japan has also used design to bridge the worlds of tradition and technology.

An account of Japan's modern design movement must begin at the moment when the country embraced the challenge of not simply catching up with but actually leapfrogging the technological, social, and economic achievements it was encountering for the first time. Through the 1870s and '80s, Western dress and chairs began to replace the kimono and the practice of sitting on tatami mats. Many of the Japanese manufacturing companies that would drive the post-1950 economic miracle were also created in the late nineteenth century.

One of the strongest reactions to industrialization came in the ideas and achievements of the Japanese mingei, or folk craft, movement, which gathered momentum in the 1920s. The Japanese philosopher and art critic Soetsu Yanagi, together with the potters Shoji Hamada and Kawai Kanjiro, sought a return to traditional Japanese craftwork. The promoters of mingei valued the functionality of craft objects above all else, considering them an integral element of everyday life.

Japanese and Western architects and designers certainly did not work in ignorance of each other. As soon as Japanese art and design became accessible to them, Westerners began to borrow heavily from it. In turn-of-the-century Europe, the French and Belgian art nouveau movement was particularly influenced by Japonisme. Also, along with the numerous Japanese students who went to be educated in the West, young Japanese architects traveled to Europe and the U.S. to work alongside the masters of modernism. Junzo Sakakura and Kunio Maekawa, for example, both spent time in the studio of Le Corbusier (plate 574). Sakakura went on to create a pavilion for the Paris International Exposition of 1937 that combined elements of Eastern and Western architecture (plate 575). In 1933, the German modernist architect Bruno Taut was invited to Japan by the Japan Architects Association and ended up spending three years there. Le Corbusier's collaborator Charlotte Perriand also lived in the country between 1940 and 1943 (plate 576).

AFTER 1950

By 1950, Japan was poised to engage fully with the modern world, both technologically and culturally. For a short period in the early 1950s, it seemed as if tradition might be eclipsed by the large Japanese corporations' desire to mass-produce goods, sell them at the lowest

TOP
577. ***TR-55* portable transistor radio**, 1955
Made by Sony (Japan)

BOTTOM
578. **Tape recorder**, 1950
Made by Totsuko (Sony) (Japan)

OPPOSITE
579. ***TX8-301* portable television set**, 1959
Made by Sony (Japan)
Museum of Modern Art, New York

SONY
AC
DC
OFF

BELOW
580. **Advertisement for the Seiko *Digital Quartz LC Chronograph*,** 1975

OPPOSITE
581. ***Astron* analog quartz watch,** c. 1969
Made by Seiko (Japan)

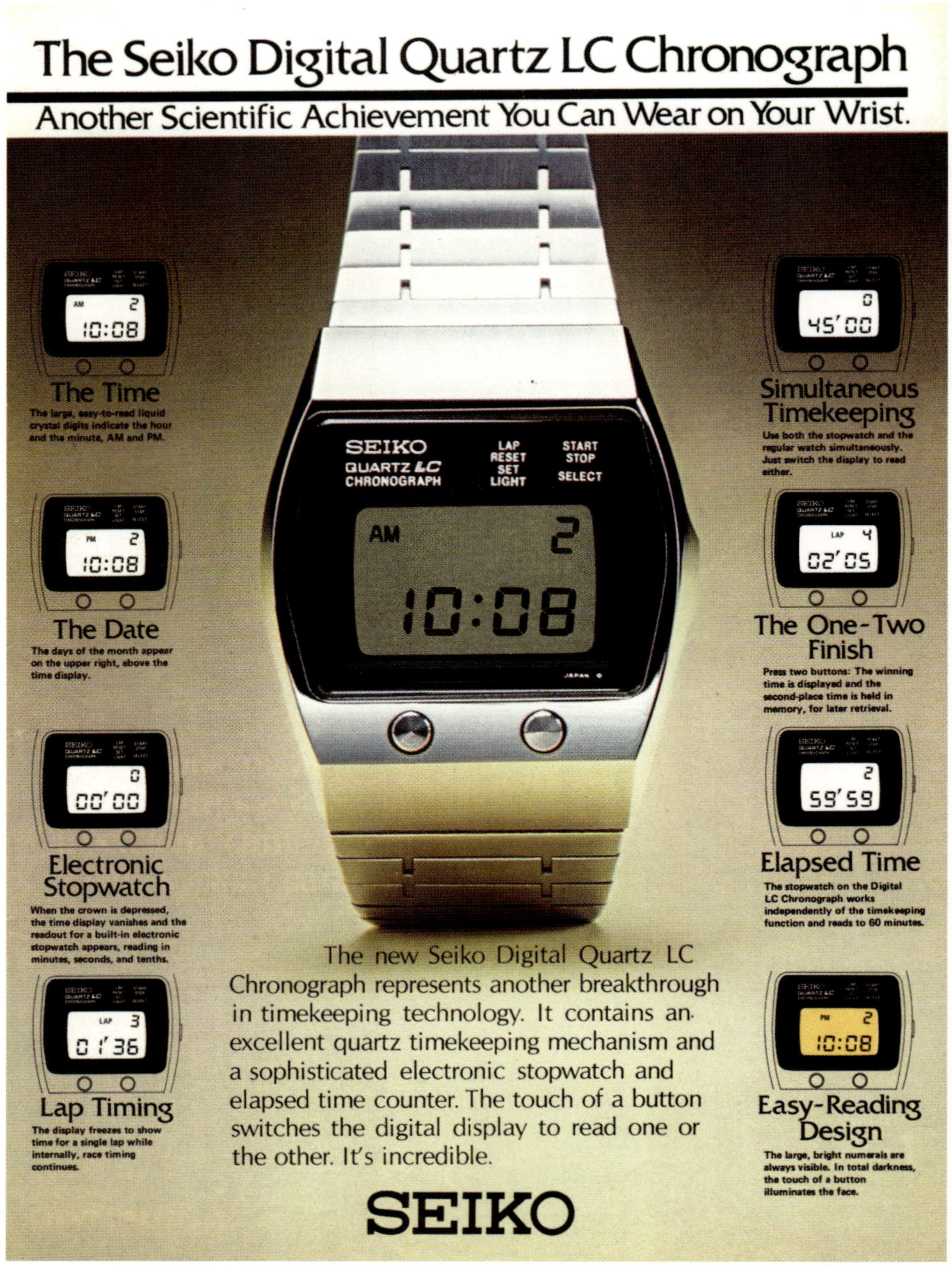

possible price, and use them as tools for rapid westernization. Within a decade, however, it became clear that this was not to be the way forward: design began to play an important role in bringing technology and culture together and ensuring that, in the rush to modernize, the country's traditions were not completely forgotten.

Several manufacturers that had already been in existence for some decades came to the fore at this time, turning their attention to the production of high-tech consumer goods. The roots of the Seiko watch company (plates 580, 581), for example, went back to the late nineteenth century, as did those of many of the postwar automakers—Toyota, Mazda, and Datsun among them. Founded in 1912 as the Hayakawa Electric Company, the Sharp Corporation evolved into a leading manufacturer of electrical and electronic goods in the postwar years, although its early consumer products were more technically than aesthetically appealing. The same could be said of a number of other goods produced at the same time, like Sony's bulky *G* tape recorder of 1950 (plate 578), which owed much to wartime developments but had not yet been transformed into a desirable consumer product. (Unlike some of the companies mentioned above, the Sony Corporation was established after the war.)

In the 1950s, Japanese manufacturers had to create a home market for their goods before they could address the challenge of overseas sales. To that end, in 1955 Toshiba created an electric rice cooker that rapidly became an essential item on all newlyweds' gift lists. As the number of servants in Japanese homes continued to decline, the demand for electrical and electronic gadgets grew apace. In 1960, Sharp offered Japanese consumers a new color television set, a bulky wooden object with Detroit styling applied to its metal control panel. Two years later, the same company launched another crudely designed product: a domestic microwave oven. The use of advanced technology was enough to engage new consumers with this novel object, and it was not necessary to add any other levels of consumer desire. As Henry Ford had understood when he created the *Model T*, the mere possession of such a technologically innovative object was sufficient to bestow a high status on its owners.

The presence of the U.S. army on Japanese soil stimulated a strong program of Americanization, and many of the first products to be manufactured after the war—transistor radios, automobiles, motorbikes, and tape recorders among them—were indebted to the American idiom of streamlining, which was used as a way of engaging consumers with new, unfamiliar goods. The control panel of Sony's little *TR-55* transistor radio, for example, had a distinctly Detroit look to it (plate 577),

SEIKO
Quartz
13

SONY
UHF
21 30 40 50 60 69
CCIR
819
F
625
F & B
VHF
UHF
TRANSISTOR TV

LEFT
583. ***BP 156* portable transistor radio**, c. 1970
Made by Sharp (Japan)
Victoria and Albert Museum, London

BELOW
584. ***R-72* radio wristband**, 1972
Made by Panasonic (Japan)
ABS plastic
Museum of Modern Art, New York

OPPOSITE
582. ***9-306UM* portable television set**, 1975
Made by Sony (Japan)
Musée National d'Art Moderne, Centre Georges Pompidou, Paris

Pentel
manufacturers of the original Sign Pen
Pentel Sign Pen
MADE IN JAPAN

CASIO
FILM CARD
SOLAR CELL
SL-800
7 8 9 ÷ %
√ 4 5 6 × MR
C 1 2 3 − M−
AC 0 · = + M+

SONY

OPPOSITE, TOP
585. ***Sign Pen S 520* felt-tipped pen with acrylic fiber tip**, 1963
Made by Pentel (Japan)

OPPOSITE, MIDDLE
586. ***SL-800* solar-powered calculator**, 1982
Made by Casio (Japan)
Museum of Modern Art, New York

OPPOSITE, BOTTOM
587. **Microcassette recorder**, 1978
Made by Sony (Japan)
Musée National d'Art Moderne, Centre Georges Pompidou, Paris

BELOW
588. ***HDW-700* HD camcorder**, 1997
Made by Sony (Japan)

as did the Toyota *RS Crown* automobile of the same period (plate 589). Tokyo's famous bullet train, introduced in 1964, also had a dynamic streamlined profile (plates 590, 591).

However, Japanese product designers quickly moved beyond the idea of surface styling as a means of making goods appealing, and began to use technology itself as a stimulant for innovative design and a powerful selling tool. Learning from the past—in particular, the Japanese tradition of living efficiently in a small space, using shoji screens to transform the open area as needed and storing futons and other items when not in use—a number of Japanese high-tech manufacturers began to develop sophisticated products characterized by their portability, flexibility, and miniature scale (plates 582–84, 587, 588).

In 1959, Sony developed the world's first solid-state television receiver; it had an eight-inch screen and weighed only thirteen pounds (plate 579). Items like these, which combined technical virtuosity with clever design, began to give Japanese electronics a strong reputation in the international marketplace. Miniaturization became a key strategy of Japanese industry, culminating, by the 1970s, in wafer-thin pocket calculators. Numerous designs, such as that of Casio's little film-card solar calculator of 1982 (model *SL-800*; plate 586), demonstrated Japanese manufacturers' ability to produce seemingly weightless electronic gadgets. Japanese designers also excelled in the creation of other miniature items, such as sewing kits and Mitsubishi's *Electric Micro-Shaver* of 1980. Such objects showcased the country's advanced capabilities in molding plastics, and also built on its tradition of careful packaging and effective storage in the domestic sphere.

The concept of miniaturization went far back in Japan's history. It was rooted in a number of phenomena, including the country's limited size and lack of resources. Small things were not only cheaper to make and export but also reflected a unique aesthetic of flexibility, portability, and multifunctionality. In their home life, and particularly in their use of textiles, the Japanese had long been accustomed to activities such as folding, pleating, and stacking, and now they sought to create small products by packing miniature components into tiny shells. The small digital camera of the early twenty-first century was a Japanese innovation, for example, pioneered by Canon, Nikon, and Pentax. In 2000, Canon introduced the neat little *Digital IXUS*, the first in a line of so-called ultracompact cameras (plate 593). Its boxlike shell contained a vast number of components in a very limited space, and by 2004 the company had moved from using CF to SD memory

OPPOSITE
589. ***Toyopet Crown* car**, 1959
Made by Toyota (Japan)

BELOW, TOP
590. **Locomotives of *Shinkansen* trains, Tokyo Station**, 2010

BELOW, BOTTOM
591. ***Shinkansen 700 series* high-speed train**, manufactured 1997–2004, Japan

cards to create even thinner cameras. More recently, the Canon *Digital IXUS 675* camera reveals a commitment not just to compactness but also to elegant body design, manifested in its curved form, its chic metallic finish, and the distinctive black ring around its lens.

In the early 1980s, Sony's *Walkman* showed how miniaturization could be used not only as a marker of technological virtuosity but also to link products to the body and even transform daily life (plate 592). This was an indication that the close historical relationship between material culture and lifestyle in Japan continued to inform the design of high-tech products. As conceived by Japanese designers, such devices were meant to support life, not the other way around. The *Walkman* was to utterly transform daily activities like commuting on the train and going for a jog. It is significant that the *Walkman* project was generated by Sony's marketing department, rather than its engineers. Working with designers, the marketers succeeded in conceptualizing not just a new product but a new set of human behaviors. It was this shift in thinking that would eventually lead Apple to develop its famous range of high-tech lifestyle products, from the *iPod* to the *iPad*.

Miniaturization did not manifest itself only Japan's high-tech products, however. It was also apparent in the careful attention given to the design of small items, such as pens and pencils. Pentel, founded in 1946, excelled in this field, creating a number of iconic, yet very simple and inexpensive, writing instruments that became ubiquitous worldwide. These included the world's first felt-tip pen, the *Sign Pen*, introduced in 1963 and known for being used on a U.S. space mission in 1966 (plate 585). Ten years later, Pentel created its classic green *R50*, the world's first water-based roller pen. Its fine-tipped counterpart was the orange *S570*, manufactured in France. The simple *P200* mechanical pencil, used in design offices worldwide, has also made the Pentel name familiar to millions of people.

As well as embracing American commercial styling, Japanese designers also began to respond to modern European design in the early postwar era. Through the work of critic Masaru Katsumi and others, the ideas developed at the Bauhaus in the interwar years were transplanted to Japanese soil, and a number of new design schools embraced its pedagogy, including the Creative Arts Education Institute (1951), the Kuwasawa

OPPOSITE
592. ***Walkman***, 1979
Made by Sony (Japan)
Musée d'Art Moderne,
Saint-Étienne Métropole

ABOVE
593. ***Powershot S110* digital camera**, 2001
Made by Canon (Japan)
Museum of Modern Art,
New York

Design School (1954), and the Visual Art Education Center (1955). Conscious that it needed to train its own designers to meet the demands of its rapidly developing manufacturing sector, Japan sought to emulate European educational practices and align its design culture with the modernist ideals underpinning them. The Industrial Arts Research Institute also moved into design education in those years, with the furniture and product designer Isamu Kenmochi heading its industrial design department. Although Japan was not ready to participate in the tenth Triennale di Milano in 1954, three years later it felt confident enough to play a part in the eleventh.

The 1950s also witnessed the emergence of such organizations as the Japan Advertising Artists Club (1951); the Japanese Industrial Designers Association, founded by Isamu Kenmochi, Riki Watanabe, and Sori Yanagi (1952); the Japan Design Committee (1953); the Japan Designer-Craftsman Association (1956); the Industrial Design Promotion Council (1957), which introduced the G-Mark prize, modeled on Italy's Compasso d'Oro; and the Japan Interior Designers' Association (1958). Collectively these organizations, which all mirrored Western equivalents, provided a supportive framework for the further development of a modern Japanese design culture.

TRANSPORT DESIGN

One area in which Japan advanced rapidly in the postwar years was transport design. Many of the key companies in this field, including Honda, Toyota, and Nissan, already had long histories. The war with Korea saw developments in the design of military vehicles, and with the end of hostilities these were transferred into a peacetime context. An early success story was Honda's little *Super Cub* motorcycle of 1958. Just as the Italian *Vespa* was designed for the back streets of Milan, the 50cc *Super Cub* was created with Japan's urban alleys and narrow tracks through rice paddies in mind.

Like its high-tech products, Japan's passenger cars of the late 1950s and '60s moved from an American aesthetic to a more European one. This trend was illustrated by the Nissan *Bluebird*, introduced in 1959, and the Toyota *Crown*, introduced in 1955, which owed much to the cars produced by Hillman in Britain. Gradually large sedans were joined by smaller models, and by the late 1960s and '70s Japanese cars were competing in the international marketplace. The Honda *Civic*, launched in 1972, was one of the first Japanese compacts to make a significant impact in the West (plate 595); it was designed with greater attention to visual detail than many other cars of the same period. A decade later followed the Honda *City*, whose dramatically sloped hood and generous headroom represented a breakthrough in Japanese automotive design. One important factor in the rise of Japanese compact cars was the oil crisis of the 1970s.

The success of the Japanese motorcycle industry is also part of this story. In addition to Honda (plate 597), Suzuki, Kawasaki, and Yamaha played important roles (plate 598). Yamaha, which originally produced only musical instruments, diversified into motorcycles after the war, collaborating with Kenji Ekuan's GK Industrial Design Associates on a number of striking models (plates 603, 604).

OPPOSITE
594. **Datsun *2000* roadster**, 1968

BELOW
595. **Honda *Civic***, in production 1972–79 (first generation)

BOTTOM
596. **Toyota *Prius* (gasoline-electric hybrid)**, 2010

ABOVE
597. **C50 motor scooter**, 1965
Made by Honda (Japan)

OPPOSITE
598. **Z1 four-cylinder 900cc motorcycle**, in production 1973–76
Made by Kawasaki (Japan)

KAWASAKI
DOHC

BELOW, LEFT
599. Kenji Ekuan (b. 1929)
Soy sauce dispenser, 1961
Made by Kikkoman Shoyu Corporation (Japan)
Glass, plastic
Museum of Modern Art, New York

BELOW, RIGHT
600. Makio Hasuike (b. 1938)
Toilet brush holder, 1976
Distributed by Gedy S.p.A. (Italy)
ABS plastic, rubber
Museum of Modern Art, New York

CORPORATE DESIGN

While much support was provided by the government, it was the large corporations that were mainly responsible for Japan's progress in modern design. However, unlike American firms, which made a virtue of naming the famous consultant designers they patronized—men like Raymond Loewy, Walter Dorwin Teague, Norman Bel Geddes, and Henry Dreyfuss—Japanese companies did not lionize individuals but depended on anonymous in-house design teams that collaborated with the engineering and marketing departments. In line with Japanese tradition, designers were seen as faithful employees working in a collective; it was the brand name that was important. Still, the corporations did act as training grounds for many of the young designers who, in later years, were to become much more visible internationally and be credited for their individual contributions. For example, Makio Hasuike, who went on to set up his own design agency in Italy in 1968, began his career working, at various times, for Seiko, Mitsubishi, Mazda, and Honda (plate 600), while Naoto Fukasawa, who came to prominence in the 1990s, also spent some time with Seiko, in the 1980s (plates 601, 602). Although Japanese designers remained "humble servants" in the 1960s, they would soon come to be recognized, like their counterparts in the West, as magicians who could transform the nature and meaning of objects with—it appeared to those outside the profession—the mere wave of a wand.

BELOW, LEFT
601. Naoto Fukasawa (b. 1956)
Mobile phone, 2003
Made by KDDI Corporation (Japan)
Museum of Modern Art, New York

BELOW, RIGHT
602. Naoto Fukasawa (b. 1956)
8-inch LCD portable television, 2003
Made by ±0 and IO (Japan)

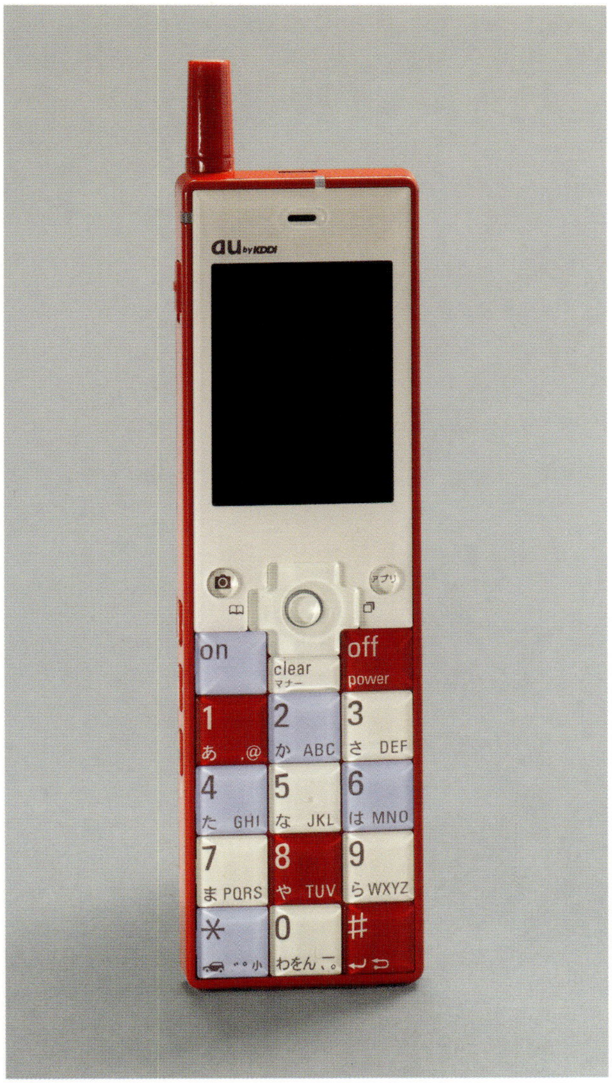

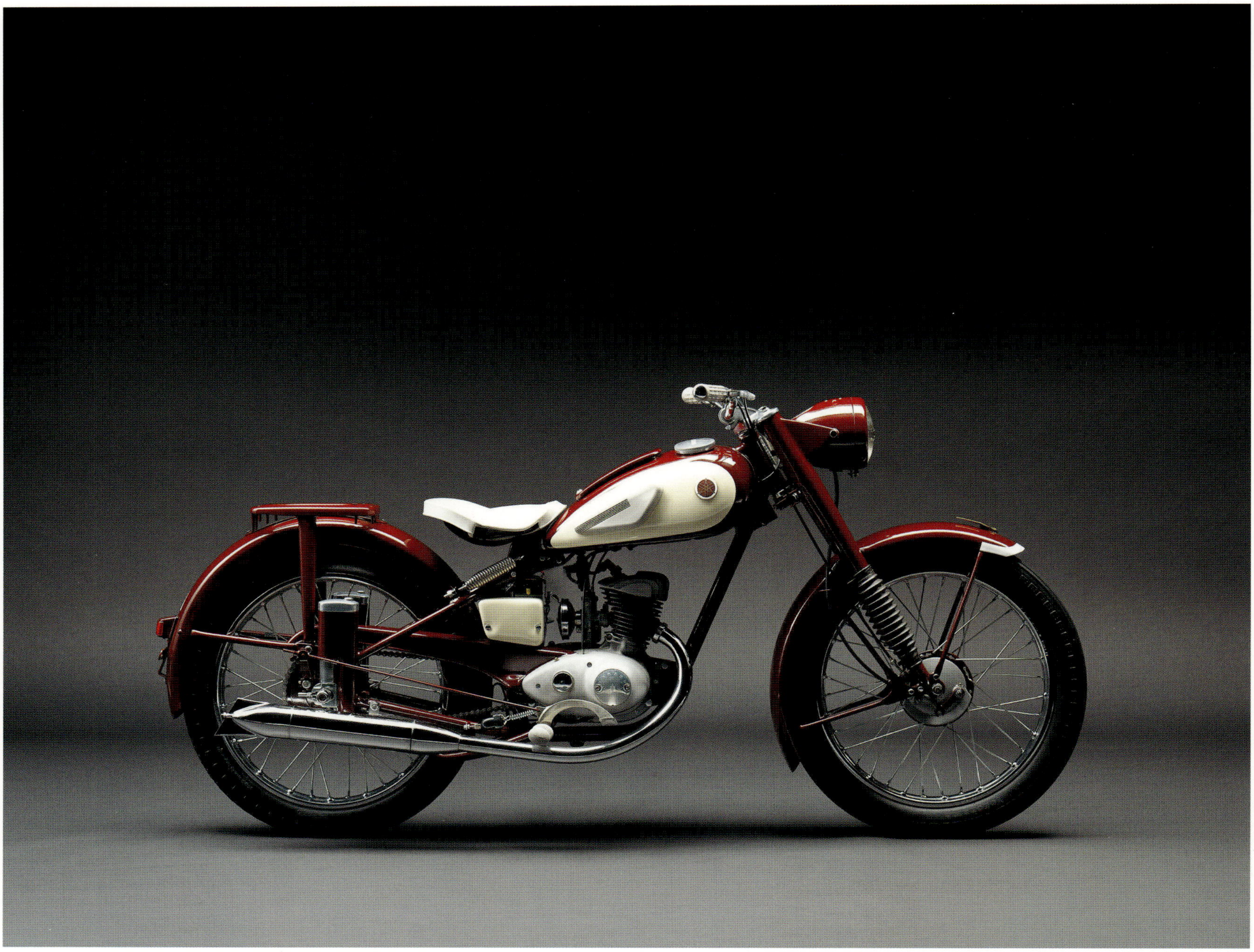

ABOVE
603. Kenji Ekuan (b. 1929)
YA-1 motorcycle, 1955
Made by Yamaha Motor Co.
(Japan)

OPPOSITE
604. Kenji Ekuan (b. 1929)
***Y125-MOEGI* motorcycle**, 2011
Made by Yamaha Motor Co.
(Japan)

THE PIONEERING GENERATION

Kenji Ekuan was one of those magicians. A founding member of the GK Design agency, he was one of the first Japanese designers to work independently, outside the corporations. Prior to his education as a designer at Tokyo's National University of Fine Arts and Music—where he met the cofounders of GK Design—Ekuan trained as a Buddhist priest. Throughout his long career as a consultant designer, during which he has been responsible for many successful products, including Yamaha motorcycles and the Akita bullet train, he has sought to maintain links between traditional Japanese aesthetics and contemporary design. According to his book *The Aesthetics of the Japanese Lunch Box* (1998), he believes that all objects contain a "spirit of form" and represent an "aesthetic ideal." His simple 1961 design for the Kikkoman soy sauce dispenser—one of Japan's best-known but most commonplace artifacts—demonstrates how that philosophy can be applied to an everyday object (plate 599).

Although many of the designers employed by Japanese corporations in the 1960s and '70s remained anonymous, a handful did succeed in making their individual contributions known. They were all members of a pioneering generation that consolidated the concept of modern Japanese design and opened the door for younger designers to follow after them. Isamu Kenmochi has already been mentioned in connection with the Industrial Arts Research Institute. Having begun his career as a furniture designer under the German architect Bruno Taut, who came to Japan in the 1930s, Kenmochi formed his Design Laboratory in 1955 (plate 605). He is perhaps best known for his collaboration with the Japanese-American sculptor and designer Isamu Noguchi, whom he met in the office of the architect Kenzo Tange in 1950 (plates 608, 609). The pair collaborated on the *Bamboo Basket Chair*, which was to become an international icon of modern Japanese design (plate 606). Although this chair famously employed traditional Japanese bamboo basket weaving, its form reflected the organic aesthetic embraced by Western furniture designers in those years. Because it thus combined Japanese and Western elements, and because the chair itself was still a relatively new object in Japan, Noguchi and Kenmochi's design made a powerful statement about the role and meaning of modern design in that country. The original chair no longer exists, but replicas of it have been produced. A later Kenmochi design, his wooden *Kashiwado* chair of 1961, has also achieved iconic status (plate 607). Most importantly, Kenmochi's extensive travels, during which

BELOW
605. Isamu Kenmochi (1912–1971)
Armchair, 1958
Made by Yamakawa Rattan Company (Japan)
Rattan
Musashino Art University Museum & Library, Tokyo

OPPOSITE, TOP
606. Isamu Noguchi (1904–1988) and Isamu Kenmochi (1912–1971)
Bamboo Basket Chair, 1950
Replica by Kenmochi Design Associates (Japan) in association with Shosaku Kondo and YMK Co, 2007
Bamboo, iron
The Noguchi Museum, Long Island City

OPPOSITE, BOTTOM
607. Isamu Kenmochi (1912–1971)
***Kashiwado* armchair**, 1961
Made by Tendo Mokko (Japan)
Lacquered Japanese cedar
Philadelphia Museum of Art

PAGE 484
608. Isamu Noguchi (1904–1988)
***Akari 1AG* lantern**, 1960
Mulberry paper, lacquered steel wire
Musée des Arts Décoratifs, Paris

PRECEDING PAGE
609. Isamu Noguchi (1904–1988)
***Akari 25N* lantern**, 1960
Mulberry paper, lacquered steel wire
Musée des Arts Décoratifs, Paris

BELOW
610. Sori Yanagi (b. 1915)
***Elephant Stool* stackable stools**, 1954
Made by Kotobuki Company (Japan)
Polyester, fiberglass
Musée des Arts Décoratifs, Paris

OPPOSITE
611. Sori Yanagi (b. 1915)
***Butterfly* stool**, 1956
Made by Tendo Mokko (Japan)
Molded plywood, metal
Museum of Modern Art, New York

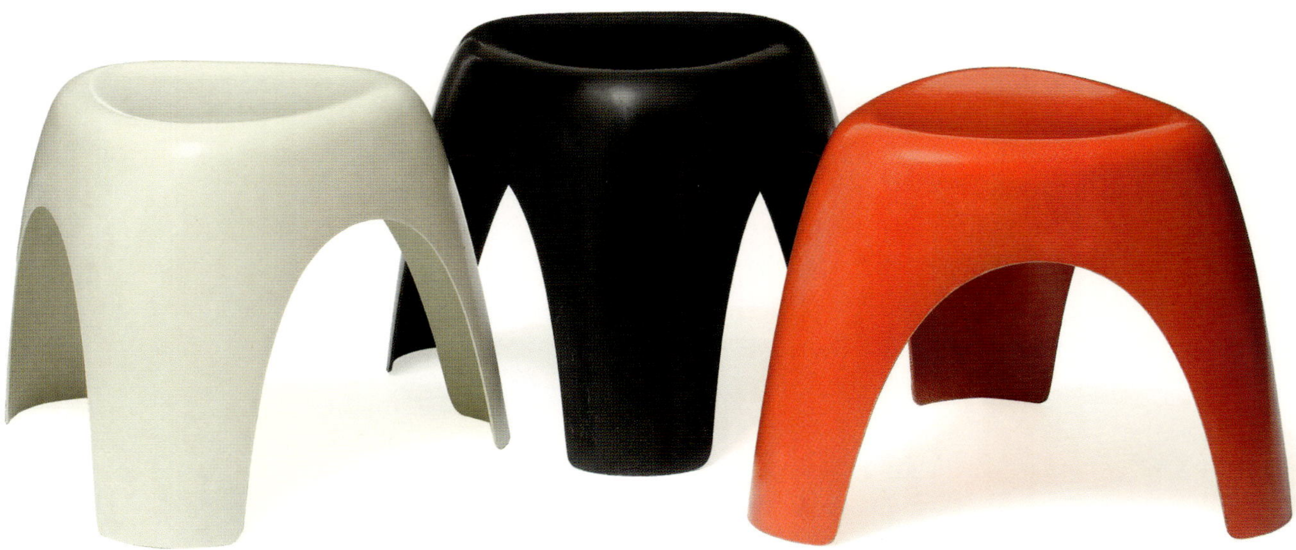

he collected many objects and images, helped expose other Japanese designers to modern Western design.

Along with Kenmochi, Sori Yanagi—son of Soetsu Yanagi, the leading promoter of the mingei movement—cofounded the Japanese Industrial Designers Association. Sori Yanagi also played a key role in the early years of the formation of a modern Japanese design movement. Between 1940 and 1942, he had worked in Charlotte Perriand's studio, learning about European modernism at first hand. Originally trained as an artist, he began to study industrial design in 1947, and he opened his own design studio in Tokyo in 1952. Like Kenmochi, Yanagi brought Japanese tradition and modern Western design principles into contact with each other. Nowhere is this more apparent than in his elegant little *Butterfly* stool of 1954, which recalls a Japanese shrine (plate 611). This object combines Japanese aesthetics with molded plywood, an industrial woodworking technique developed by the American designer Charles Eames.

Yanagi worked in both furniture and product design, and in the early 1950s Sony hired him to design its *H* tape recorder; this was among the first instances in which an electronics company collaborated with one of the new Japanese designers who were seeking to realign tradition with innovation. Another iconic Yanagi design was a stackable plastic stool that came to be known as the *Elephant Stool* because of its resemblance to that animal's chunky feet (plate 610). His stools and kitchen pots remain among his most successful pieces.

Yanagi went on to create a wide range of products in the "Japanese modern" style, from ceramics to lamp shades to children's toys to subway stations to cars and motorbikes. In 1964, he designed the torch for Tokyo's Olympic Games, an event that brought the burgeoning Japanese design movement to the eyes of the world. Thirteen years later, he became the director of the Japan Folk Crafts Museum in Tokyo, which his father had helped found four decades earlier. Yanagi's contribution to postwar Japanese design lay in his ability to combine Western ideas (he greatly admired Le Corbusier) with a traditional Japanese approach to artifacts, entailing simplicity, the use of organic forms, and a strong respect for materials.

Kazuhide Takahama and Reiko Tanabe are both members of the generation of Japanese architects and furniture designers who carried on the work initiated by Kenmochi and Yanagi in the 1950s and '60s.

612. Reiko Tanabe (b. 1934)
Stool, 1961
Made by Tendo Mokko (Japan)
Molded plywood
Museum of Modern Art,
New York

Takahama had studied architecture in Tokyo between 1949 and 1953, before working in the studio of Kazuo Fujoka. Like many of his contemporaries, he traveled to Milan in 1957 to see the eleventh Triennale, and there he was recruited by the Italian furniture manufacturer Dino Gavina. Takahama thus became one of the first Japanese designers to stay and work in Italy, producing numerous items for a variety of firms, including a lamp for Sirrah that won the Compasso d'Oro in 1979. His use of stretched fabric, which gave this object the appearance of a rice-paper lantern, spoke to his Japanese origins.

Reiko Tanabe was also trained in architecture and worked in the areas of interior and furniture design. Like Yanagi before her, she employed molded plywood, and her *Murai* stool stands out as perhaps her greatest achievement in that material (plate 612). Also like Yanagi, Tanabe's work straddled the worlds of Japanese tradition and Western modernism, exemplifying the new synthesis that defined Japanese design in the 1960s and '70s.

The best-known Japanese furniture designer in that pioneering generation, however, was Shiro Kuramata. Indeed, it has been said that Kuramata was the first Japanese designer working in the modern style to be recognized and admired outside Japan. His connection with the group of designers described above, who sought to reconcile Japanese tradition with Western modernism, was less obvious, however. While he also built upon traditional Japanese forms, combining them with new materials like metals and plastic, his contribution was more individual. Operating in the gap between design and fine art, Kuramata moved beyond the "tastefulness" of modernist design to embrace a new postmodern approach. He developed a multifarious grouping of forms, materials, and images with which to express his inner thoughts and preoccupations.

Kuramata's background was in woodworking, which he studied in Tokyo, and furniture making, followed by training in interior design that exposed him to many Western principles. He first applied them in his work for the interiors department of the Matsuya department store. After opening his own office in 1965, he created an enormous body of work, eventually including more than 180 pieces of furniture. He also designed a number of store interiors throughout his career, notably boutiques for the fashion designer Issey Miyake in Paris, Tokyo, and New York. Characterized by the use of terrazzo and metal mesh (the latter also employed in the Matsuya department store), these boutiques brought together two hugely talented designers. Regrettably, they were only temporary spaces, although Miyake claimed actually to enjoy the impermanent nature of retail shops.

The preoccupations that Kuramata displayed in his interior work were mirrored in his furniture designs (plate 616). From his wavy chests of drawers of 1970 (plate 613), he went on to create the famous *How High the Moon* armchair, made entirely of steel mesh, in 1986 (plate 614) and the *Miss Blanche* chair in 1988 (plate 615). Inspired by the corsage worn by the main character of Tennessee Williams's play *A Streetcar Named Desire*, the latter chair epitomized Kuramata's skill in combining transparency, spatial articulation, and storytelling in a single piece of furniture. His versatility as a designer allowed him to embrace a wide range of objects, including flower vases and an umbrella stand.

Besides raising the profile of Japanese design generally, Kuramata showed that the country could excel not only in high tech but also in architecture, interior design, and fashion—fields that offered abundant opportunities to combine traditional Japanese aesthetics with the challenges of global modernity. While the design of many Japanese stereos, televisions, and cars was

OPPOSITE
613. Shiro Kuramata (1934–1991)
***Side 2* chests of drawers**, 1970
Made by Fujiko (Japan) and after 1986 by Cappellini (Italy)
Lacquered bentwood ash, brushed steel
Musée National d'Art Moderne, Centre Georges Pompidou, Paris

BELOW
614. Shiro Kuramata (1934–1991)
***How High the Moon* armchair**, 1986
Made by Vitra (Switzerland)
Nickel-plated steel mesh
Victoria and Albert Museum, London

OVERLEAF
615. Shiro Kuramata (1934–1991)
***Miss Blanche* armchair**, 1988
Made by Ishimaru Company (Japan)
Paper flowers, acrylic resin, aluminum
Museum of Modern Art, New York

PAGE 493
616. Shiro Kuramata (1934–1991)
Acrylic Stool with Feathers, 1990
Made by Ishimaru Company (Japan)
Acrylic, aluminum, feathers
Musée des Arts Décoratifs, Paris

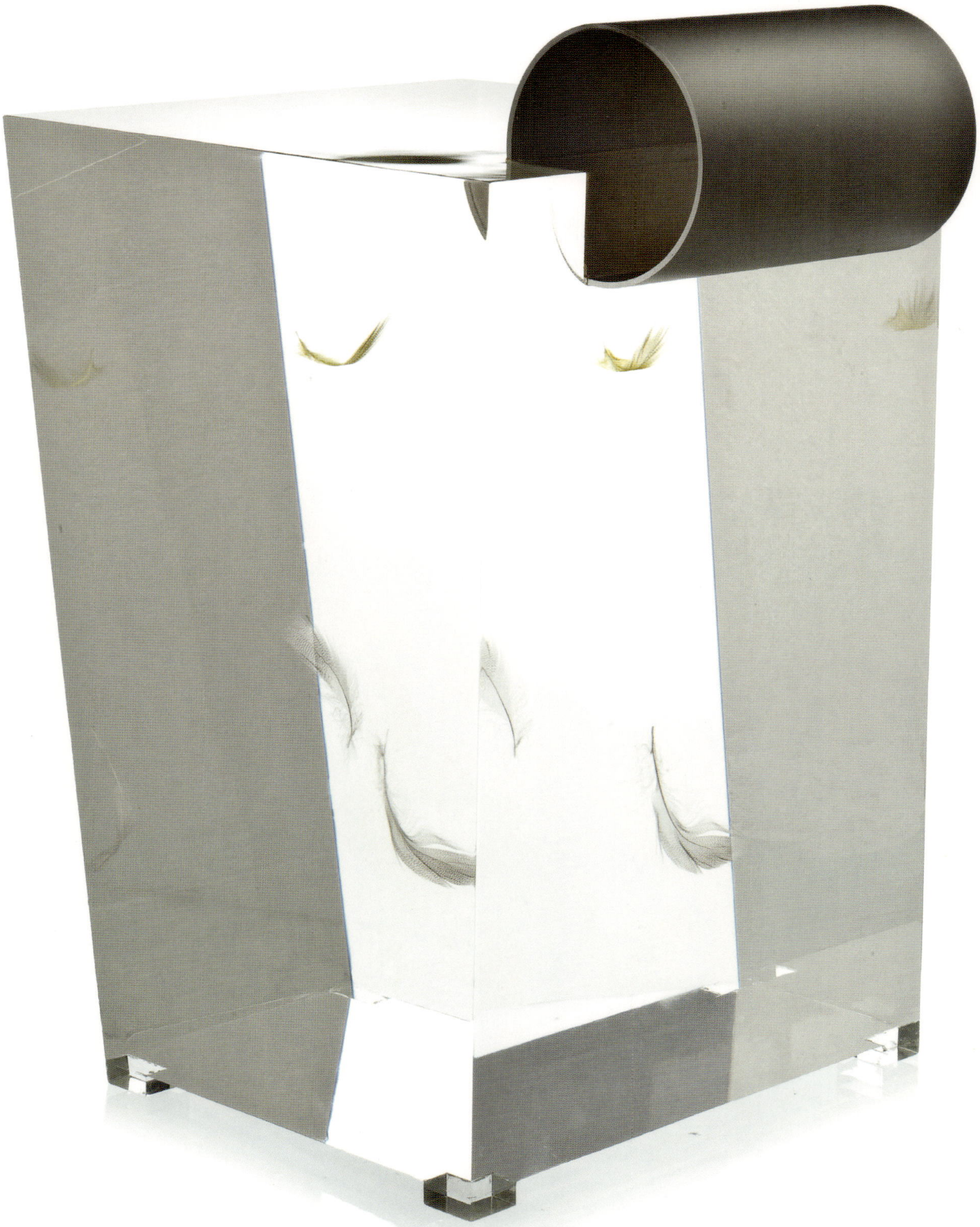

LEFT
617. Masayuki Kurokawa (b. 1937)
***GOM* Pen**, 1992
Made by Fuso Gomu Industry Company (Japan)
Synthetic rubber, stainless steel
Museum of Modern Art, New York

BELOW
618. Masayuki Kurokawa (b. 1937)
***GOM* ashtray**, 1973
Made by Fuso Gomu Industry Company (Japan)
Synthetic rubber, stainless steel
Museum of Modern Art, New York

OPPOSITE
619. Arata Isozaki (b. 1931)
***Marilyn* chair**, 1973
Made by Tendo Mokko (Japan)
Varnished wood coated with polyurethane

linked to traditional concepts of portability, flexibility, and the importance of everyday life, the craft-based fields of architecture, interior design, fashion, textiles, and ceramics could be linked directly to a range of other Japanese traditions. This is seen especially in the Zen Buddhist–inspired minimalism that, from the early 1980s, became one of the world's favored styles of interior design. Existing alongside the more decorative and complex forms of postmodernism, this neomodern style has provided a level of stability in a world increasingly characterized by change. Within Japan, both architects and furniture designers embraced the possibilities it offered. One of its foremost practitioners is Takashi Sugimoto, who established Super Potato Design in 1973. Since then, his team has been responsible for many remarkable minimal spaces that emphasize natural materials such as wood and granite; a recent example is the interior of Seoul's Park Hyatt Hotel (2005). Studio 80, founded by Shigeru Uchida in 1981, was also responsible for many stunning minimal interiors, including boutiques for Yohji Yamamoto and Issey Miyake. Drawing on traditional ideas of space, Uchida promoted new ways of tea-drinking and created a number of rooms for that purpose.

The story of modern Japanese architecture paralleled that of design in the years after 1945. It was driven by Kenzo Tange and his students and colleagues—including Kisho Kurokawa, Arata Isozaki, Fumihiko Maki, and Yoshio Taniguchi—as well as by Kazuo Shinohara, Mayumi Miyawaki, Hiroshi Hara, Toyo Ito, and Tadao Ando. In Japan as elsewhere, architecture in fact had a direct impact on design, as many architects also engaged in design projects.

Arata Isozaki, a member of the same heroic generation as Ekuan, Miyake, and Kuramata, is among the best-known of the Japanese architect-designers. In 1973, he created his famous *Marilyn* chair (plate 619), which owes much to one of his heroes, the Scottish architect and designer Charles Rennie Mackintosh. This piece quickly became an icon of modern Japanese furniture, earning Isozaki a strong reputation as a designer as well as an architect.

Masayuki Kurokawa, another architect-designer born in the prewar years, graduated from the Department of Architecture at the Nagoya Institute of Technology and set up his architectural practice in 1967. He has produced some of Japan's most enduring products of

BELOW
620. Mitsuo Maki (b. 1948)
Mechanical pencil and ball-point pen, 1980
Made by Platinum Pen Company (Japan)
Stainless steel
Museum of Modern Art, New York

OPPOSITE
621. Makoto Komatsu (b. 1943)
***Crinkle Super Bag* vase**, 1974
Made by Ceramic Japan Company (Japan)
Porcelain
Museum of Modern Art, New York

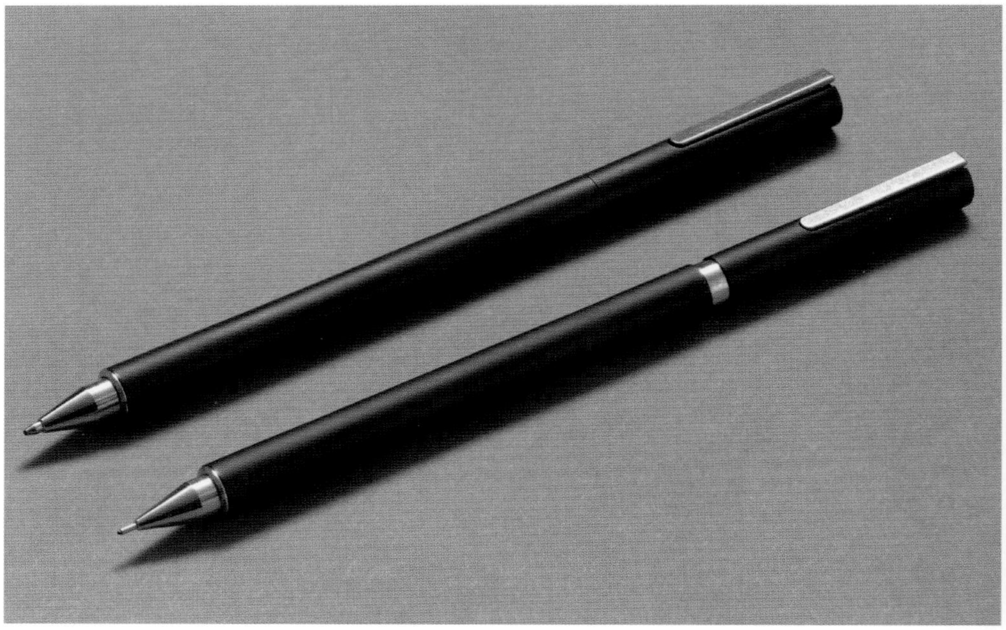

the last four decades, ranging from furniture to clocks to watches to lighting fixtures to drinking glasses to cutlery to jewelry. All his work is deeply indebted to the culture of Kanazawa, a city with one of the oldest continuous craft traditions in Japan. In particular, the set of desk objects in stainless steel and rubber that he created in the mid-1970s—including ashtrays, containers, and a pen—epitomized the minimal approach of traditional Japanese craftspeople (plates 617, 618). Perhaps because of the importance of calligraphy in their culture, the pen is in fact a favorite object of Japanese architects and designers; another noteworthy example was designed by Mitsuo Maki in 1980, for the Platinum Pen company (plate 620).

Through the 1970s and '80s, the boundaries between architecture and interior, product, furniture, fashion, and textile design progressively eroded, extending the idea of the unified designed environment, or *Gesamtkunstwerk*, that had underpinned traditional Japanese craft and design. According to the old Japanese idea that a space is defined by the smallest objects it contains, a pen requires as much attention as, say, a folding screen. It was in this spirit that Kuramata and Uchida worked with Miyake to create many of the stunning retail spaces in which he placed his garments. Collaborations such as these grew more frequent in the 1980s, with architects, interior designers, and industrial designers working together on innovative projects that respected traditions even as they sought to transcend them by engaging with new technologies.

But even as certain types of design began to blend together in Japan, others—such as ceramics—remained more firmly entrenched in their traditional territories. There was still a considerable amount of innovation in those areas, however. The crumpled "paper bags" that Makoto Komatsu began making from porcelain and glass in 1975 reflected a meeting of craft traditions with a Pop sensibility (plate 621). Japan has always embraced the popular and the everyday while also celebrating craft skills, and Komatsu's artifacts have succeeded in extending that alliance into the present. Komatsu, who spent some years with the Swedish porcelain company Gustavsberg, has said that his goal as a ceramist is to make home and social life more comfortable.

BELOW
622. Toshiyuki Kita (b. 1942)
***Wink* chaise longues**, 1980
Made by Cassina (Italy)
Expanded polyurethane foam, Dacron, steel
Philadelphia Museum of Art

OPPOSITE
623. Toshiyuki Kita (b. 1942)
***Wink* chaise longue**, 1980
Made by Cassina (Italy)
Expanded polyurethane foam, Dacron, steel
Museum of Modern Art, New York

THE 1980S: GLOBALISM AND POSTMODERNISM

By the 1980s, modern Japanese design was fully formed and wielded a growing international influence, as the country's architecture, interiors, high-tech products, crafts, textiles, fashion, and graphics were featured in the design press and in exhibitions. Western designers traveled to Japan to learn about its deeply rooted aesthetic traditions and work in its exciting contemporary design culture, which had warmly embraced the opportunities offered by new technologies. However, the expansion of globalism meant that nothing could be contained within a country's borders any longer, and a fluid movement across boundaries began to encourage the development of a hybrid culture. Where design was concerned, that hybridization provided a backdrop for enhanced creativity and innovation, and as a result, increasing numbers of Japanese designers sought to work outside their own country for a period of time. Many returned at a later date, bringing their knowledge of global culture back with them.

The phenomenon of Japanese designers working abroad was not new, of course. As we have already seen, many of them traveled to France and Germany in the interwar years to learn at the feet of the modernist masters. After 1945, some went to the U.S. and Scandinavia as well, but, increasingly, Italy became the favored destination of Japanese designers. The Triennales in Milan were, as we have seen, the first attraction. The furniture designer Kazuhide Takahama did not return from his visit to the Triennale of 1957. In 1963, he was joined in Italy by Makio Hasuike, who had previously worked for a number of Japanese companies, including Seiko, where he had designed clocks for the 1964 Tokyo Olympics. On moving to Italy, Hasuike spent five years working for a number of other designers, including Rodolfo

Bonetto, before establishing his own consultancy. In the early 1980s, he founded the bag and luggage company MH Way, for which he created, among other products, a popular line of plastic briefcases and the telescoping PVC *Zoom* tube for carrying drawings. The question inevitably arises as to whether or not Hasuike's designs can be described as "Japanese," but they do clearly display aesthetic characteristics—purity, clarity, simplicity, transparency, portability, a love of materials—that owe much to Japanese traditions. Equally his family of desk accessories—pens, pencil cases, and diaries among them—have much in common with the simple everyday objects created by the "no-name" brand Muji, which became a global Japanese success story in the 1990s.

Another highly successful Japanese designer, Toshiyuki Kita, also made his reputation on Italian soil, arriving there in 1975. He had studied at Naniwa Design College and opened his own office in Tokyo in 1967. From 1969 onward, Kita's work took him to Italy as he began to collaborate with a number of manufacturers there. His first "Italian" design of note, the famous *Wink* chair, was launched by Cassina in 1980 (plates 622, 623). Its bright colors and Mickey Mouse ears betray its Pop allegiances, but its flexibility and adjustability can be seen as Japanese in inspiration. After the success of *Wink*, Kita went on to design a series of Japanese-style lamps and, from his base in Italy, initiated a campaign to protect traditional Japanese crafts—including papermaking, lacquerwork, bamboo work, and basket making—from extinction.

The first Memphis show, held in 1981 in Milan, represented a key moment in the story of Japanese

OPPOSITE, LEFT
624. Arata Isozaki (b. 1931)
Sketches for the *Fuji* cabinet, 1981
Victoria and Albert Museum, London

OPPOSITE, RIGHT
625. Arata Isozaki (b. 1931)
***Fuji* cabinet**, 1981
Made by Memphis (Italy)
Wood

BELOW
626. Tomoyuki Sugiyama (b. 1954)
***Bubble Boy* speakers**, 1986
Made by Inax Corporation (Japan)
Ceramic
Museum of Modern Art, New York

designers working in Italy. Led by the veteran Italian designer Ettore Sottsass, the Memphis group sought to renew the language of contemporary design. Sottsass's close friend Arata Isozaki participated in this experiment, as did Kuramata and Masanori Umeda. The Japanese designers' work was among the most innovative on display, demonstrating that their country was no longer following the European avant-garde but running alongside it and arguably even leading it (plates 624, 625).

Japanese design did not have to travel to Milan to become postmodern, however, as the rapid social, cultural, and technological changes that were occurring on its home soil encouraged a new aesthetic. In addition to the more exclusive market for craft-based goods, a new consumer society was emerging that favored products aligned with pop culture. Increasingly, household gadgets became linked with the concept of lifestyle, and pastel shades were introduced to a variety of appliances, from washing machines to television sets, in emulation of American goods of the 1950s. Sharp's *QT50* portable radio-cassette player of 1986 was offered in a variety of soft colors, including pink, which became a marker of goods aimed at young girls and housewives. Even cameras, like Canon's *Snappy* range, sported bright colors, and products that retained more sober hues often took on at least a "cute" shape, like Tomoyuki Sugiyama's *Bubble Boy* speakers (plate 626). In a continuation of the two parallel levels of material culture that had long been in place in Japan, "cheap and cheerful" goods and images designed to fulfill popular requirements (manga, Tamagotchis, Ninja Turtles, and so on) coexisted with other objects and spaces that sustained deeply rooted traditions.

This "postmodern" phase of Japanese design, which lasted through the 1980s brought new design values into the marketplace and created a new consumer culture: products were given evocative names like "Holiday," suggesting that they contributed to a new lifestyle characterized by leisure rather than work. Where postmodern design was concerned, Japan led the way and was emulated by countries in the West. A high level of product differentiation, which allowed designers to cater to diverse and changing tastes, was made possible by the introduction of flexible manufacturing using high-tech machinery.

POWER
GAME BOY pocket
Nintendo
SELECT START
B
A

OPPOSITE
627. ***Game Boy Pocket* transparent gaming console**, 1996
Made by Nintendo (Japan)
Centre National des Arts Plastiques/Fonds National d'Art Contemporain, Paris

BELOW, RIGHT
628. ***AIBO ERS 110* robot dog**, 1999
Made by Sony (Japan)
Height 10½ in. (26.6 cm)

BELOW, LEFT
629. ***SDR-3X* toy android robot**, 2000
Made by Sony (Japan)
Height 19⅝ in. (50 cm)

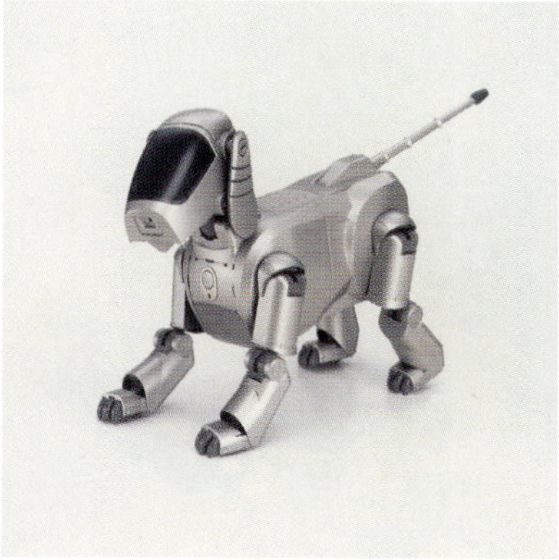

THE 1990S AND BEYOND

The economic recession of the 1990s was to change the landscape of Japanese manufacturing and design irrevocably, however. With the slowing down of the manufacturing-driven economic miracle that had underpinned the emergence of Japan's postwar design culture, a new sensibility came to the fore. The pendulum swung away from the ephemerality of gadgets and pop culture and back to a more universal, craft-based mode of design. Architecture and interior design took on a new importance and reemphasized their roots in Japanese tradition. The anonymity of the corporate designer was replaced by a more individualized design culture in which, following the Western example, designers came to be revered almost as fine artists, capable of acts of creativity and poetry. A new generation of designers emerged in the 1990s, many of whom had studied and worked with the leading figures of the 1960s and '70s, including Kuramata and Miyake. The collaborations that had taken place a decade or so previously—such as those between Kuramata, Isozaki, and the textile designer Junichi Arai—were inspirational to the new generation. The premature death of Kuramata in 1991 was an occasion for Japan and the rest of the world to reflect on the past and future of Japanese design. Several exhibitions held around this time—both of Kuramata's work and of postwar Japanese design in general—offered ideal opportunities for such an assessment.

The processes of internationalization and globalization that had already been set in motion accelerated after 1990, and designers became increasingly peripatetic. More and more Japanese designers worked outside their homeland—although many later returned—while increasing numbers of Western architects and designers came to Japan, including the Britons Nigel Coates, Jasper Morrison, George Sowden, John Pawson, Norman Foster, and Ron Arad; the Italians Ettore Sottsass, Michele De Lucchi, and Aldo Rossi; the Australian Marc Newson; and the Frenchman Philippe Starck.

One of the strongest fields of Japanese design—product design—took on a new identity in the postcorporate era. Individual designers became prominent, and they increasingly embraced a wide spectrum of goods, from high tech to lighting and furniture. (Some companies, it should be said, continued to work successfully with in-house design teams, including Mazda, Canon, Nintendo, and Sony; plates 627–29.) Following

BELOW
630. Naoto Fukasawa (b. 1956)
CD-1 CD-ROM reader, 1999
Made by Muji (Japan)
Centre National des Arts Plastiques/Fonds National d'Art Contemporain, Paris

OPPOSITE
631–32. **Pages from the Muji catalog**, 2008

in the footsteps of craft-based designers like Michio Hanyu, who had created simple metal flatware in the 1960s, many of the new product designers aligned themselves more closely to Japanese craft traditions than to high tech. Among those born in the 1940s were Shun Takaoka, who designed an influential alarm clock in the early 1980s; Hiroyuki Tazawa, whose creations in recycled paper include a rescue board (1998) and a snow grabber (2001); and Kazuo Kawasaki, who has made a significant impact with his socially conscious designs. Kawasaki, originally employed at Toshiba, was injured in a car accident in 1977 and has used a wheelchair ever since. That life-changing event inspired him to focus on design for people with specific needs; he realized, for example, that by designing a wheelchair to meet his own physical requirements, he could create a functional product for others with a similar disability. He founded Kazuo Kawasaki Product Designs in 1979.

Among the product designers born in the 1950s were Hiroaki Kozu, who created an innovative pair of speakers for Yamaha in 1989, and Kosuke Tsumura, who crossed the boundary between fashion and product design. Naoto Fukasawa has undoubtedly been the most successful and influential designer of that generation, however. In 1989, he decided to leave his job at Seiko and move to San Francisco, where he joined the firm that was soon to become IDEO. In 1996, he returned home to set up the Japanese branch of that prestigious agency, with which he remained for the next six years. In 2003, he formed his own company and became one of Japan's leading product designers, taking on a number of roles, among them creative director of Muji. The following year he launched the home product brand ±0 (Plus Minus Zero). The wall-mounted CD player he designed for Muji (plate 630), along with his many other creations—including mobile phones, a humidifier, household appliances, and furniture—have earned him numerous prizes. Fukasawa's philosophy is rooted in the traditional Japanese belief that design should be a part of everyday life, making it both more beautiful and more efficient. He has explained that, in Japan, the relationship between the object and its environment is more important than the object itself, and that his work has become less about making interesting shapes and more about such relationships.

The Muji retail company, whose philosophy is that its goods should be unbranded and unadvertised, was founded in 1980 with a collection of just forty products (plates 631, 632). It has been a huge global success story and has recently diversified its activities into cafés, home furnishings, and housing construction while maintaining its original involvement with stationery, storage items, and clothing. In addition to eschewing the concept of branding, Muji also produces high-quality, low-cost goods, is very environmentally aware, and embraces a simple yet distinctive aesthetic with a minimal color range. The Muji concept was developed by the company's first art director, the graphic designer Ikko Tanaka, together with the interior designer Takashi Sugimoto. Kenya Hara took over as art director in 2001 and has developed the original concept to suit the twenty-first century.

Although Muji presents its goods anonymously, it has collaborated with a number of significant international designers. As noted above, Fukasawa has been closely linked with the company, as has the British designer Jasper Morrison, whose work is in strong sympathy with the simple objects sold by Muji. James Irvine,

オーク材ラウンジアームチェア

7118438　税込40,950円

幅68×奥行61.5×高さ70（座面高41）cm
背面・座面：積層合板、オーク材突板
脚部：オーク積層材・ポリウレタン樹脂塗装

※積み重ねて収納できます。

積み重ね時

オーク材ダイニングアームチェア

7112610　税込30,450円

幅56×奥行50.5×高さ75.5（座面高44.5）cm
背面・座面：積層合板、オーク材突板
脚部：オーク積層材・ポリウレタン樹脂塗装

※積み重ねて収納できます。

積み重ね時

積み重ね時

オーク材ラウンジチェア

6652939　税込27,300円

幅60×奥行62×高さ68.5（座面高41）cm
背面・座面：積層合板、オーク材突板
脚部：オーク積層材・ポリウレタン樹脂塗装

座面が低く、くつろいで座れる一人がけのチェアです。リビングだけでなく、書斎や寝室にもお使いいただけます。

※積み重ねて収納できます。

積み重ね時

オーク材ダイニングチェア

6606475　税込19,950円

幅49×奥行47.5×高さ75.5（座面高44.5）cm
背面・座面：積層合板、オーク材突板
脚部：オーク積層材・ポリウレタン樹脂塗装

脚部に曲げ木を使用し、座面も座ったときに収まりの良い曲線に仕上げています。

※積み重ねて収納できます。

3

4

オーク材ラウンジチェア用シートクッション

3　6777182　ベージュ　税込6,825円

4　6777199　ブラウン　税込6,825円

幅52.5×奥行52.5×厚さ2cm
【洗濯不可】

オーク材ラウンジチェア専用のクッションです。

6

5

オーク材ダイニングチェア用シートクッション

5　6777045　ベージュ　税込4,725円

6　6777052　ブラウン　税込4,725円

幅42.5×奥行42.5×厚さ2cm
【洗濯不可】

オーク材ダイニングチェア専用のクッションです。

an English designer based in Milan, and the German designer Konstantin Grcic have also designed products for Muji. Morrison, whose designs for Muji include a 2008 wall clock, has collaborated with a number of other Japanese firms as well, making the *Lightwood* chair for Maruni (2011) and a set of cast-iron cookware for Oigen (2012), in association with japan creative, an organization set up to protect the country's indigenous craft traditions. This cookware is strongly indebted to traditional Japanese aesthetics, especially in its simplicity, dependence on a strong sense of craftsmanship, and attention to detail.

Product designers born in the 1960s and '70s include Kazuhiro Yamanaka, who created a range of lamps in the early 2000s, and Tokujin Yoshioka, one of Japan's most innovative young designers. A graduate of the Kuwasawa Design School, Yoshioka worked with both Kuramata and Miyake, for whom he created a number of store interiors before going freelance in 1992; he made his Milan debut a decade later. Yoshioka is an interdisciplinary practitioner who combines retail, exhibition, and furniture design. He employs technology—especially fiber optics—in a highly creative way, combining it with a sophisticated use of light. He states, "I like something which is surrounded by an aura, something that appeals to one's heart, something that enlightens one's soul, and something that brings an emotional experience." He compares his design philosophy to Japanese cuisine, whose simple appearance conceals the hard work that has gone into preparing it. His work includes interior designs for a number of carmakers, including Toyota, Peugeot, and BMW; furniture for the Italian companies Cassina, Driade, and Kartell; and the *Tofu* lamp for Yamagiwa. His interest in space and light has led him to think of furniture in a new way, less as fixed objects and more as fluid presences (plate 634). A group of optical glass projects included a "chair that disappears in the rain," while the iconic *Honey Pop* chair is made entirely of honeycomb paper (plate 633). His approach also lends itself to installations, of which he has created several, including displays for Lexus, Swarovski, and Moroso at the Milan furniture fair. For an exhibition at 21_21 Design Sight in Tokyo, he showcased an object called *Venus*, which he described as a "natural crystal chair," based on his ideas about applying the laws of nature to design.

The last few years have seen a much greater interest in the role of materials and the concept of lightness and air in Japanese design. In addition to Yoshioka's explorations of these ideas, the architect Junya Ishigami has created installations on the theme of Architecture as Air. On one level, this is a new manifestation of minimalism, trying to link the visual and material world back to the elements from which they came. A significant contribution in this area has also been made by the Nendo group, led by Oki Sato and formed in Tokyo in 2012, in response to a visit to the Milan furniture fair. Its name means "modeling clay," and Sato has said that he wants the group's designs to have that same sense of fluidity. Nendo's output ranges from one-offs to mass-produced items, and it has recently launched a project called 1%, consisting of objects produced in one hundred examples each. One Nendo design, the *Cabbage Chair*, is made from a roll of pleated paper left over from the production of clothes by Issey Miyake (plate 635). That kind of lateral thinking, combined with a

OPPOSITE
633. Tokujin Yoshioka (b. 1967)
***Honey-Pop* armchair**, 2000
Paper
Museum of Modern Art,
New York

BELOW
634. Tokujin Yoshioka (b. 1967)
Pane Chair, 2003
Polyester fiber shaped in
a steam mold
Museum of Modern Art,
New York

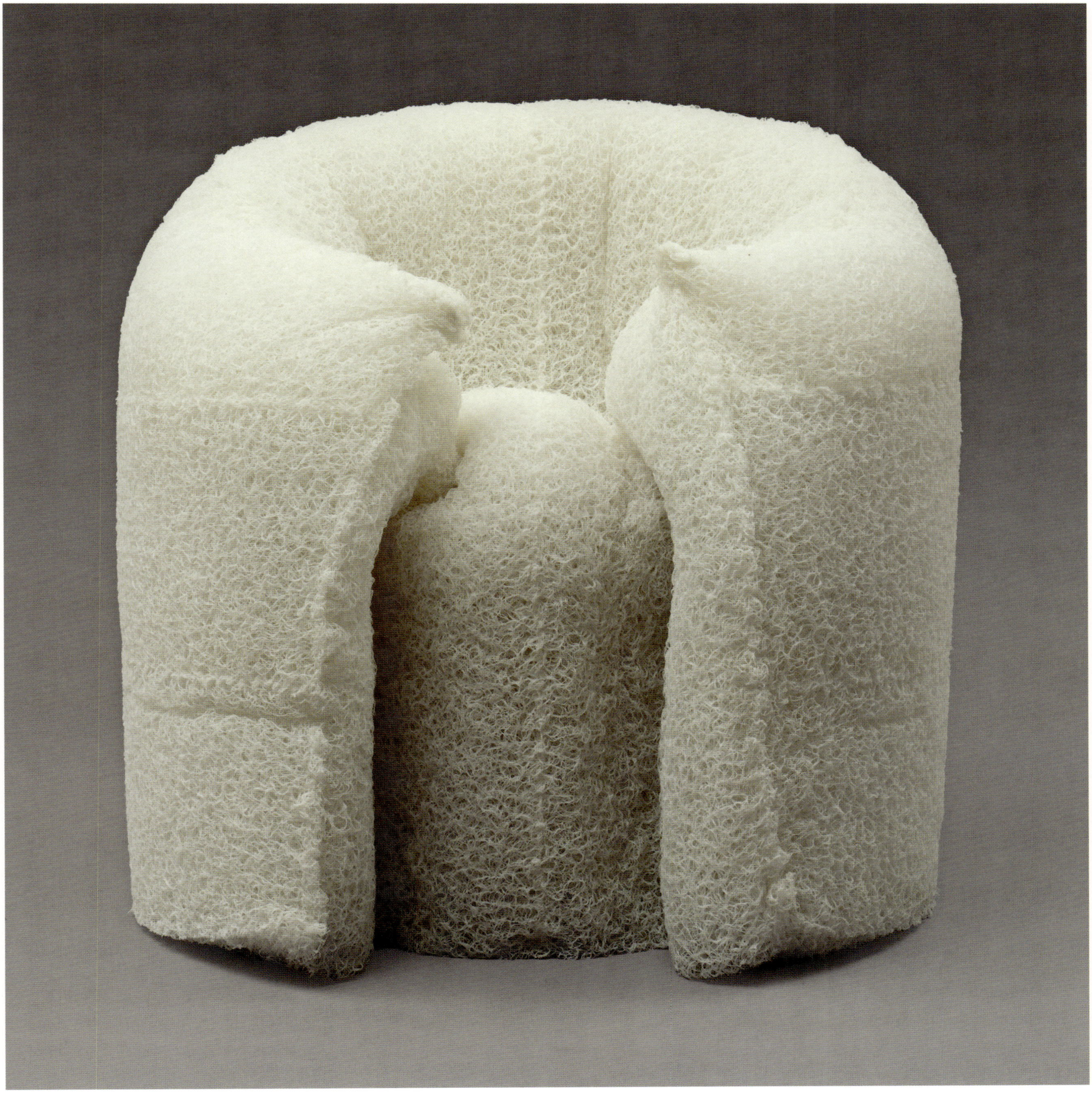

635. Oki Sato (b. 1977) for Nendo
Cabbage Chair, 2008
Paper, fabric, resin
Musée des Arts Décoratifs, Paris

commitment to the environment, typifies the group's work. Another example is its *Hanabi* lamp, made from a shape-memory alloy, which opens up like a flower when it is turned on.

Along with product, furniture, and environmental design, Japanese crafts have also gone from strength to strength over the last two decades. In the field of contemporary ceramics, Shigeyoshi Morioka, among others, has shown that the work of the early twentieth-century potters has not been forgotten, and that traditional forms can still make a lasting impact. Indeed, as the twenty-first century advances, it becomes increasingly apparent that Japan's real contribution to modern design lay not in its enthusiastic embrace of new technologies but in the way it brought its traditions to bear on the present.

CONCLUSION

In the early twenty-first century, Japan's economy may continue to lag, but the country is still a serious contender for world leadership in design innovation, having not only absorbed lessons from the West but also taken several extra steps forward on its own. Tokyo now boasts its own design museum—21_21 Design Sight, led by Miyake and Fukasawa—and hosts its own annual design shows on the European model. Indeed, the urban fabric of Tokyo itself, epitomized by the upscale shopping street Omotesando Dori, pays homage to international architecture and design. Young designers from around the world travel to Japan to see the innovative projects on display, while young Japanese designers—including the Tonerico trio, two of whose members emerged from Shigeru Uchida's office, and Rieko Miyata, who makes lamps out of ribbons—exhibit internationally. Other Japanese designers live abroad; Shin and Tomoko Azumi, for example, are based in London. The globalism that began to emerge several decades ago is now a reality, and Japanese design can be found anywhere in the world.

A few Japanese corporations continue to produce world-class designs: the strikingly minimal LCD projector designed for Sony by Takuya Niitsu and the Toyota *Prius* stand out in this context. The latter, one of the first gasoline-electric hybrids, combines environmental awareness with a stylish look (plate 596). For the most part, however, Japanese design is the work of talented individuals who challenge conventions and cross boundaries, while calling upon tradition for a sense of continuity. There can be no doubt that the giants of modern Japanese design—Kuramata, Miyake, and Isozaki in particular—were instrumental in bringing the past to bear on the present. Thanks to them, in the early twenty-first century Japanese design stands on its own two feet, providing a beacon for the future. And while Japan may be a relative newcomer to the international design scene, the fact that it developed its own modern idiom on the basis of its traditional culture has made it, in many ways, a leader in this field—indeed, it may be that the West owes more to Japan than the other way around.

BELGIUM AND THE NETHERLANDS

MIENKE SIMON THOMAS

PRECEDING PAGE
636. Maarten Van Severen (1956–2005)
MVS chaise longue, 2000
Made by Vitra (Switzerland)
See plate 678.

RIGHT
637. Henry van de Velde (1863–1957)
***Bloemenwerf* chair**, c. 1898
Made by Van de Velde & Co (Belgium)
Elm, leather, nails
Vitra Design Museum, Weil am Rhein, Germany

OPPOSITE
638. Hendrik Petrus Berlage (1856–1934)
Dressing table, 1896
Oak, white metal
Gemeentemuseum, The Hague

As yet there is no book that traces the development of Belgian and Dutch design together. Moreover, existing histories of design almost always devote more space to the Netherlands than to Belgium. This is remarkable. These are two small European countries that share a long border and even, in part, a common language. So why should there be such a disparity?

In fact, industry developed quite differently in each country. After Great Britain, Belgium was the first European country to industrialize—starting in the late eighteenth century—and it soon became a truly industrial state. In the Netherlands, the process began more than half a century later and progressed more gradually. Around 1900, however, industrial development in the two countries started to move along slightly more parallel lines. A brief description of the first half of the twentieth century in both places is essential to understanding the design sector that was established after World War II.

ART NOUVEAU AND NIEUWE KUNST

At the beginning of the twentieth century, the most famous designers in Belgium and the Netherlands, Henry van de Velde and Hendrik Petrus Berlage respectively, were also fierce rivals. This is interesting, because both were driven by the same social commitment, namely the creation of a new concept of design to suit modern times, resulting in products that would be affordable for all and contribute to a better society. In addition, both claimed to pursue a logical, constructivist style, and their common source of inspiration was the British Arts and Crafts movement.

Originally a painter, Van de Velde soon evolved into an all-around architect and designer with many international contacts. In the milieu of Belgian art nouveau, where Van de Velde was a central figure, the removal of traditional boundaries was a central tenet. The turn-of-the-century Belgian style was rich in organic and asymmetrical elements, as can also be seen, for example, in Victor Horta's interior designs and Philippe Wolffers's silver.

Berlage disliked the international art nouveau movement. When the Arts and Crafts gallery, founded in The Hague in 1898, began to sell Van de Velde's graceful yet simple chairs (plate 637), along with the work of other Belgian designers, Berlage made no attempt to hide his displeasure. He had studied architecture and would always continue to exercise that profession, but his talents extended well beyond, to furniture, textiles, glassware, and ceramics. The dressing table he designed for the Fentener van Vlissingen family in 1896 exemplifies his ideals (plate 638). It is a sturdy piece of furniture in locally harvested oak, with restrained decoration and a construction based on traditional Dutch furniture, including metal fittings that explicitly proclaim their process and function.

It became fashionable to refer to this specifically Dutch version of art nouveau as Nieuwe Kunst (New Art). In response to the Arts and Crafts gallery in The Hague, Berlage founded the 't Binnenhuis showroom in Amsterdam. Here on display—in addition to Berlage's own work—were pieces by Jac. van den Bosch, Willem Penaat, Jan Eisenloeffel, Chris van der Hoef, and Chris Lebeau. Their designs focused on simplicity, affordability, and reliability and were decorated—almost reluctantly—with highly abstract or geometric ornaments.

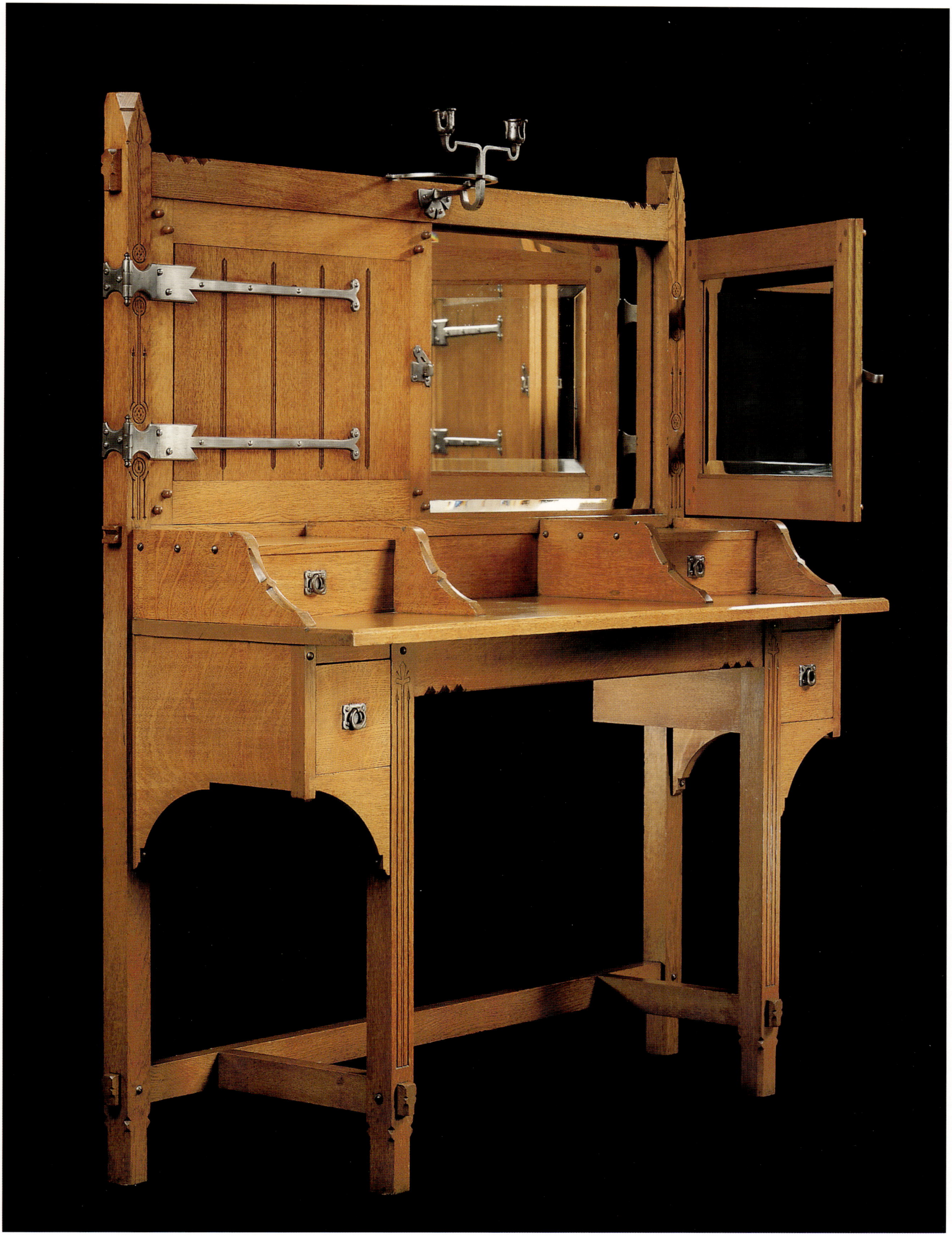

OPPOSITE
639. Gerrit Rietveld
(1888–1964)
***Beugel* chair**, 1927
Made by Metz & Co
(Netherlands), 1930
Plywood, steel
Victoria and Albert Museum,
London

THE AMSTERDAM SCHOOL AND DE STIJL

Before World War II, strong ties existed in the Netherlands between the world of design and that of art and architecture. The Vereeniging voor Ambachts- en Nijverheidskunst (VANK; Association for Craft and Industrial Art), founded in 1904, was the first professional association of Dutch designers. It is significant that, along with "art," this organization included the words "craft" and "industrial" in its name. In principle, it was open to cooperation with industry, but in fact "industrial design" was barely known before 1945.

At the same time, the Amsterdam architects' association, Architectura et Amicitiae (A et A), was active in the field of design. Around 1915, a group of young A et A members voiced their opposition to the extremely sober and—in their eyes—unimaginative style of Berlage and his followers. Their number included Michel de Klerk, Piet Kramer, and Jo van der Meij, who were working at that time on the construction and decoration of the Scheepvaarthuis in Amsterdam. This large office building is considered the first typical example of the style of the Amsterdam school, with its expressive forms and innovative sculptural ornament.

De Klerk continued to exercise his remarkable talent not only in architecture and interior decoration for wealthy private clients but also for the new workers' homes that were being built under the Socialist-led Amsterdam City Council. The varied and fanciful architecture of the "workers' palaces" in Amsterdam-Zuid still bears witness to this period.

The work of Jaap Gidding, a Rotterdam-based designer of ceramics, glass, mosaics, and tapestries, also reflected this new expressive style. In 1921, Gidding orchestrated the exuberant and colorful interior of the Tuschinski Theater, which, virtually unchanged, is still a popular movie house in central Amsterdam.

Another noteworthy initiative of A et A was the magazine *Wendingen* (Turnings), whose motivating force was the architect H. T. Wijdeveld. This widely informed magazine, with its exceptionally beautiful layout and Wijdeveld's distinctive typography, was published monthly from 1919 to 1932, each cover being designed by a different artist.

The Netherlands were represented in the Amsterdam school style at the 1925 Exposition Internationale des Arts Décoratifs et Industriels Modernes in Paris. Only a single item from the Dutch display belonged to the now much better known De Stijl movement, which takes its name from a magazine published between 1917 and 1931. The founder and driving force behind this publication was the artist Theo van Doesburg, while Piet Mondrian was its theorist, an advocate of the unification of painting and architecture in the "Nieuwe Beelding," or neoplasticism. (It should be noted that De Stijl was not an actual group or association; many of the artists and architects who wrote for the magazine did not even know each other.) Gerrit Rietveld (plate 639), J. J. P. (Bob) Oud, Vilmos Huszár, and Bart van der Leck were the other key contributors, covering either design or architecture. Rietveld designed his red-blue chair (plate 640)—initially without color—in 1919 and published it that same year in *De Stijl*. This was a materialization of Mondrian's theories, a chair in which you really could sit, yet at the same time a spatial, visual, and conceptual object. In 1924, there appeared in Utrecht the asymmetrical white faces and red-, blue-, and yellow-painted woodwork of the Rietveld Schröder House, one of the best-known examples of the De Stijl in the Netherlands (plate 641).

OPPOSITE
640. Gerrit Rietveld
(1888–1964)
Red and blue armchair, 1918
model
Painted wood
Museum of Modern Art,
New York

BELOW
641. Gerrit Rietveld
(1888–1964)
Interior of the Rietveld Schröder House in Utrecht,
1924

BELOW
642. Willem Gispen (1890–1981)
Desk lamp, 1928–38
Made by W.H. Gispen & Co (Netherlands)
Nickel-plated brass
Los Angeles County Museum of Art

OPPOSITE
643. Mart Stam (1899–1986)
***B33* chair**, c. 1926
Made by Thonet (Germany)
Chrome-plated tubular steel, leather
Musée des Arts Décoratifs, Paris

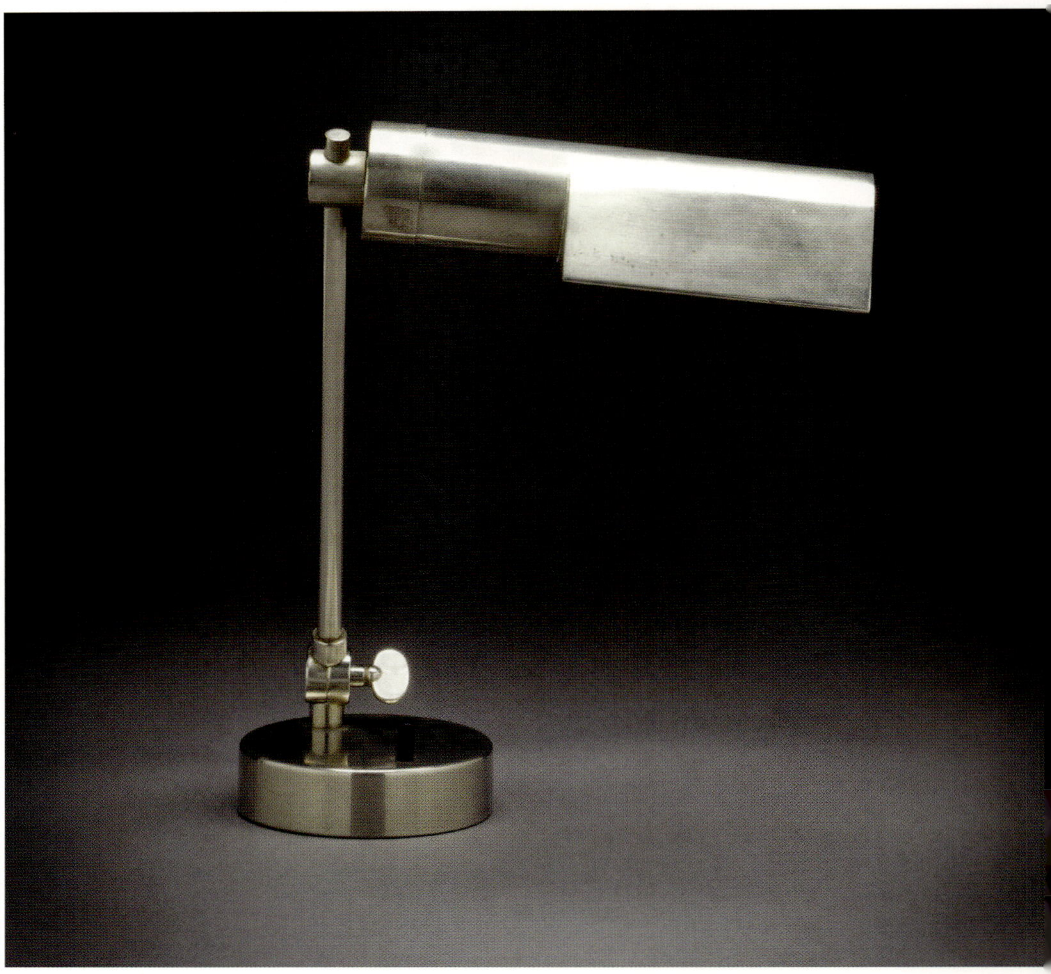

MODERNISM AND ART DECO

Oud and Rietveld distanced themselves from the De Stijl after just a few years and became involved in the more socially conscious movement known as Nieuwe Zakelijkheid (New Objectivity) or Nieuwe Bouwen (New Architecture). In 1927, together with the architect Mart Stam and the designer-manufacturer Willem Gispen, they contributed to the exhibition of model homes at the Weissenhof Estate in Stuttgart, an initiative of the Deutscher Werkbund. This was the first time that Stam exhibited his revolutionary cantilevered chair in tubular steel (plate 643), and Gispen also presented his innovative lamps (plate 642).

Also in 1927, the group known as De 8 was founded by progressive architects in Amsterdam who also resisted the artistic and elitist forms of the Amsterdam school and wanted to promote a modern, technocratic mode of building. Their bright, open, functional interiors were intended as the first step toward an ideal they called the Nieuwe Wonen (New Living).

Two years later, the Opbouw association was set up in Rotterdam, leading to some interesting collaborations. With support from Stam, Opbouw members Michiel Brinkman and Leen van der Vlugt designed the Van Nelle coffee and tea factory in Rotterdam, the iconic embodiment of the Nieuwe Bouwen style in the Netherlands. The factory was decorated with furniture by Gispen, while Jacob (Jac) Jongert developed a modern style for the company's packaging and advertisements (plate 644).

Developments in Belgium were, of course, quite different during this period. Unlike the Netherlands, Belgium did not have the option of remaining neutral in the Great War, as it was occupied by Germany, with the lines of the Western Front falling squarely within its territory. It is almost surprising that after the violence and devastation, and all the casualties, so many interesting Belgian designers emerged so soon after the war. In addition to Van de Velde, who had returned to Belgium in 1924, after an absence of more than twenty years, the silversmith Philippe Wolffers is a good example of a designer whose style evolved over a quarter-century from the most characteristic art nouveau to the finest art deco. His *Gioconda* coffee and tea set displayed at the art deco exhibition in 1925 was one of the high points of this period.

Two other interesting figures from these years are Albert Van Huffel and Elisabeth de Saedeleer. At the beginning of his career as an architect, Van Huffel was influenced not only by his compatriot Van de Velde but also by the latter's Dutch rival, Berlage. Reconstruction after World War I earned Van Huffel many commissions. His life's work, the Koekelberg Basilica in Brussels, was dedicated to the memory of the hundreds of thousands of war victims.

In the 1920s, De Saedeleer's weaving workshop near Brussels was an important venue where artists, designers, and architects worked together. De Saedeleer had spent her youth—during the war—in Wales

RIGHT
644. Jacob (Jac) Jongert (1883–1942)
Coffee and tea box, 1930
Made by Van Nelle (Germany)
Printed metal
Museum of Modern Art, New York

OPPOSITE
645. Huib Hoste (1881–1957)
Dining room furniture, 1927
Made by Vynckier (Belgium)
Painted wood
Design Museum, Ghent

and learned weaving from William Morris's daughter. Her workshop produced knotted and woven carpets by Albert Van Huffel starting in 1925, as well as designs by Paul Haesaerts, Edgard Tytgat, Kees van Dongen, Sonia Delaunay, Paul Klee, Marc Chagall, and many others. De Saedeleer also collaborated with the Dutchman Jaap Gidding around 1927–28.

During World War I, the Belgian designer Huib Hoste fled to the Netherlands, where he came into contact with Dutch art and architecture. Originally inspired by Berlage, Hoste soon became interested in the modern movement and contributed to *De Stijl* in 1918, its second year of publication. When he returned to Belgium, he emerged as the main advocate of modernism there. The influence of De Stijl is evident in his designs of the 1920s (plate 645). His *bureau-fumoir*, displayed next to Wolffers's silver, was one of the most startling of the Belgian entries in the 1925 Paris Exposition. While the Belgian pavilion was otherwise unimpressive, it did inspire many new initiatives, such as the creation of the design school at Ter Kameren (La Cambre), also known as the Belgian Bauhaus, with Henry van de Velde as its first director. Hoste shifted course toward functionalism in the late 1920s, developing furniture in tubular steel. During the 1930s, the architect Gaston Eijsselinck became the most radical, and successful, designer of modern furniture and interiors in Belgium. Victor Bourgeois was another prominent modernist architect, designing one of the model homes in the Weissenhof Estate exhibition in Stuttgart and maintaining close contacts with his Dutch colleagues.

The production of tubular steel furniture in Belgium and the Netherlands shows that some collaborations between industry and architects or designers did take place even before World War II. In this, the Deutscher Werkbund was a source of inspiration. In 1924, the Bond voor Kunst in Industrie (BKI; Federation for Art in Industry) was established as a select group of companies, never more than about thirty, that shared an idealistic commitment to working with artists—although this was, of course, also good publicity for their products.

Members of the BKI included the Van Dissel linen factory in Eindhoven, where damask designs by Chris Lebeau had been produced since 1905. In the 1930s, Kitty van der Mijll Dekker, a Dutchwoman educated at the German Bauhaus, also worked for Van Dissel. Glasfabriek Leerdam was prominent in the BKI as well. In 1915, its director, P. M. Cochius, invited the architect Karel de Bazel to design nine sets of drinking glasses in an austere style based on his theosophical beliefs. Other designers followed him, including Chris Lanooy, Gidding, and Berlage. From the late 1920s, Leerdam's chief designer was Andries Copier, who created hundreds of vases, bowls, and drinking glasses individually and in sets, some mass-produced and others unique pieces blown using traditional techniques. His 1929 *Gildeglas* is a functionalist icon, designed according to guidelines provided by the wine tasters guild. The socialist publishing house Brusse also belonged to the BKI, employing the typographers Sjoerd de Roos and Jan van Krimpen, the graphic designers Piet Zwart and Paul Schuitema, and many others. Gispen's metal-furniture and lamp factory was of course invited to join, as was the luxury furniture store Metz & Co, which played an exceptional role in the prewar years, promoting the work of progressive Dutch designers such as Rietveld, Oud, Stam, and Bart van der Leck. It was also the first store in the Netherlands to offer furnishings by Marcel Breuer, Le Corbusier, and Alvar Aalto, while the proprietor's French artist friend Sonia Delaunay designed original fabrics for sale.

RIGHT
646. Friso Kramer (b. 1922)
***Revolt* chair**, 1953
Made by De Cirkel (Netherlands)
Ciranol, steel
Galerie Catherine Houard, Paris

OPPOSITE
647. Friso Kramer (b. 1922)
***Revolt* chair with armrests**, 1953
Made by De Cirkel (Netherlands)
Plywood, steel
Galerie Catherine Houard, Paris

A NEW POSTWAR ÉLAN

World War II inflicted greater damage on the Netherlands than on Belgium, although both countries faced significant problems of reconstruction, as well as a lack of raw materials that created shortages of all kinds of goods. The aid provided by the Marshall Plan between 1948 and 1952 was therefore most welcome, and stimulated vigorous activity despite the difficult conditions.

Even during the conflict, some designers who had been forced to remain in the Netherlands were considering their postwar future. Together with the economist Jan Bouman, Willem Gispen and Piet Zwart formulated a vision of officially sanctioned positions for designers. They foresaw little room in a modern industrial society for the individualistic, craft-oriented, artistic approach to design that had characterized the prewar era. In 1950, their report led—though not without heavy criticism—to the creation of the Instituut voor Industriële Vormgeving (IIV; Institute for Industrial Design), with which the still-extant BKI then merged.

The IIV brought designers into contact with many new industries and drew leading American and Italian designers to the Netherlands by arranging exhibitions and establishing a permanent showroom. It issued booklets that promoted industrial design as a means to achieve "increased purchasing power" or as a "weapon" to be used in the international market and touted the usefulness, durability, and attractiveness of Dutch products.

The IIV also published a monthly bulletin and coordinated the Dutch presence at the Triennale di Milano. The 1954 Triennale was especially successful for the Netherlands. Here the international public discovered Friso Kramer's *Revolt* chair, produced by De Cirkel (plates 646, 647). This chair did not use steel tubing but rather steel sheet folded into a U shape, which gave the designer far more options. An industrially produced—and thus more affordable—version of Copier's *Gildeglas* was also on display. The Gero metalware factory contributed beautiful cookware and flatware by Dick Simonis; the Mosa ceramics factory, new dinnerware by Edmond Bellefroid; and Artifort, Theo Ruth's *Congo* chair.

In Belgium, an Institute for Industrial Design was founded in 1956, and Henry van de Velde was appointed its honorary president in the year before his death. The institute, based in Liège and Brussels, had a mainly documentary and archival function, although it also published regular bulletins. The more influential Belgian Design Center was established in 1964.

The U.S. was the strongest influence in this period. In 1953, the Dutch Ministry of Economic Affairs appointed a select group to travel there and investigate the latest design practices. They were Karel Sanders, director of the IIV; the designers Wim Gilles, René Smeets, Jaap Penraat, and Karel Suyling; and the journalist Rein

BELOW
648. Friso Kramer (b. 1922)
Sofa, 1960
Made by Wilkhahn
(Netherlands)
Steel, plastic, synthetic leather
Galerie Catherine Houard,
Paris

OPPOSITE
649. Friso Kramer (b. 1922)
***Repose* armchair**, 1960
Made by De Cirkel
(Netherlands)
Steel, fabric, synthetic leather
Galerie Catherine Houard,
Paris

Blijstra. Gilles, who had attended a technical school and had no specific training in art, was then a designer at the DRU metalworks. He was the first Dutch designer to draft a systematic methodology of design based on an in-depth analysis of the market and the product. He then created a "rational, mathematical flowchart" that made the actual design process as objective, and free of artistic considerations, as possible. Shortly after returning from the U.S., Gilles designed a teakettle that is a good illustration of this process; its shape was conceived in part to minimize boiling time (plate 650).

In 1950, Smeets had become the director of the first Dutch school of industrial design, in Eindhoven, which developed into today's internationally renowned Design Academy. Two other members of the delegation, Penraat and Suyling, were freelance designers. Penraat was involved in many products of a technical nature; one of his designs, in collaboration with Friso Kramer, was for a new tram system for Amsterdam. Suyling was a graphic designer who created a long-running advertising campaign in the Netherlands for the French automaker Citroën.

In the U.S., the six Dutchmen saw that industrial design had already been fully incorporated into the manufacturing process. They took note of the high regard in which designers were held and the high fees they commanded, and explored the new concept of ergonomics, which interested them greatly. However, they also discovered the extent to which commerce and a superficial, modish kind of "styling" were gaining importance in the U.S. The Dutch, who were still rather idealistic, found this distasteful and remained convinced that they could persuade the consumer to appreciate quality design beneficial to society as a whole.

OPPOSITE
650. Wim Gilles (1923–2002)
Kettle, 1954
Made by Diepenbrock & Reigers (Netherlands)
Enameled iron, aluminum, Bakelite
Museum Boijmans Van Beuningen, Rotterdam

RIGHT
651. **Cover of the magazine *Goed Wonen***, no. 6/7, July–August 1948

BELOW
652. Andries Dirk Copier (1901–1991)
***Gildeglas* glasses**, 1929
Made by Glasfabriek Leerdam (Netherlands)
Museum Boijmans Van Beuningen, Rotterdam

BETTER LIVING THROUGH DESIGN

The same idealism underlay an organization that influenced Dutch design in the 1950s and '60s: the Stichting Goed Wonen (Good Living Foundation). This initiative was conceived during World War II, although its roots can be traced back even further, to the philosophy of the interior designers of the Nieuwe Bouwen. Formally established in 1946, Goed Wonen brought together designers, manufacturers, distributors, and the government in the cause of a better "domestic culture" and for "style against unemployment, material scarcity, and housing shortage." Its ultimate goal was a harmonious society in which every individual could develop his or her own potential. The home, and especially the interior, was at the heart of this effort.

It soon became apparent that the peaceful cooperation of its four constituent groups was itself a utopian dream, so in 1954, Goed Wonen became a consumer association, though one in which the designers held a leading position. In the early years, Mart Stam and Johan Niegeman were the major policymakers. Stam was director of the Instituut voor Kunstnijverheidsonderwijs (IvKNO; Institute for Arts and Crafts Education, renamed the Rietveld Academy in 1968) in Amsterdam and hired Niegeman to teach there. At the IvKNO, this progressive duo attempted to set up a design training program similar to that of the Bauhaus, where Niegeman had briefly taught in the 1930s, working alongside Walter Gropius and Hannes Meijer. Afterward he spent a few years in Siberia, where he contributed to the construction of the communist worker city of Magnitogorsk. Among the first graduates of the IvKNO were Friso Kramer, Kho Liang Ie, Hein Stolle, Cora Nicolaï-Chaillet, Wim de Vries, and Dick Simonis (plate 654), all of whom would subsequently play important roles in the Goed Wonen movement.

Goed Wonen tried to spread its ideals through its eponymous magazine (plate 651), a showroom, model homes, lectures, and training courses. It also awarded a quality label to products that met its high standards. All manner of products were eligible, including Cees Braakman's storage cabinets in blond wood for Pastoe; chairs by Friso Kramer for Ahrend and by Gerrit Rietveld's son Wim for Gispen; and table services by Bellefroid for Mosa and by Wim de Vries for Fries. A simple door handle designed by Wim Gilles for Nedap, a dish drain by the Tomado company, and even a stainless steel sink bowl by Gero were also illustrated in *Goed Wonen* and praised for their aesthetic and functional forms.

Goed Wonen was published for exactly twenty years, from 1948 to 1968. Browsing through the volumes, one observes a gradual shift toward a more casual and democratic tone, while the didactic aspect fades into the background. Articles were devoted to the interiors of

ordinary individuals, home handyman projects, comfortable armchairs and other luxuries, and eventually even oriental rugs and plants growing in macramé hangers.

A very similar association known as Nouvelles Formes/Nieuwe Vormen was active in Belgium during this period. This organization worked with designers, manufacturers, and distributors to encourage the production of quality inexpensive furnishings "for the people," and to promote the style of interior in which this modern furniture was best presented. Unlike Goed Wonen, Nieuwe Vormen did not publish a magazine, but instead concentrated on holding exhibitions and setting up model homes. It would be interesting to investigate which group was more effective, the Dutch or the Belgian. In any case, because the Dutch magazine provides an informative and accessible source for design historians, Goed Wonen is better known today. The study of Nieuwe Vormen has begun only more recently.

Among the leading designers in this idealistic Belgian movement were Willy Van Der Meeren, Marcel Baugniet, Jules Wabbes, Jos de Mey, Emiel Veranneman, and Alfred Hendrickx. Van Der Meeren, trained as an architect by Victor Bourgeois, had his roots in the prewar Nieuwe Bouwen style. As a furniture designer, he collaborated with the Tubax metalworks in Vilvoorde. Metal furniture was already popular in Belgian homes for its lightness, simplicity, and affordability, and by experimenting with its mass production, Van Der Meeren made it even more attractive from an economic standpoint. Also popular in 1950s Belgium—more so than in the Netherlands—was wood furniture in the organic Scandinavian style. Hendrickx designed very sophisticated pieces in this mode for the Belform factory in Mechelen. De Mey, meanwhile, was the central figure in the Ghent design world. His work was produced by the firm of Van den Berghe-Pauvers and later by Luxus, based in Kortrijk.

In the 1950s and '60s, museums also played an important role in the dissemination and appreciation of contemporary design, in both Belgium and the Netherlands. Institutions such as the Stedelijk Museum in Amsterdam, under the directorship of Willem Sandberg, and the Museum voor Sierkunst en Industriële Vormgeving in Ghent (Museum of Decorative Art and Industrial Design, now the Design Museum), under A. L. J. Van de Walle, offered space to modern furniture and textiles and mounted exhibitions on the contemporary home. For example, from 1955 through 1958, Van de Walle organized the Nationale Salon voor Modern Sociaal Meubel (National Fair for Modern Social Furniture) at his museum.

OPPOSITE
653. Guillaume Marie Edmond Bellefroid (1893–1971)
***Wilma* coffee service**, 1950
Made by Mosa (Netherlands)
Porcelain
Museum Boijmans Van Beuningen, Rotterdam

ABOVE
654. Dick Simonis
Plate and flatware, 1972
Made by Gero (Netherlands)
Stainless steel, ash
Museum Boijmans Van Beuningen, Rotterdam

RIGHT
655. Gerrit Rietveld (1888–1964) and Wim Rietveld (1924–1985)
***Mondial* chair**, 1957
Made by Gispen (Netherlands)
Metal, polyester
Museum Boijmans Van Beuningen, Rotterdam

OPPOSITE
656. **Poster for Expo '58 in Brussels**
Bibliothèque Forney, Paris

EXPO '58

The 1958 World's Fair in Brussels, known as Expo '58, contributed greatly to the popularity of modern interior design in Belgium and the Netherlands. Belgium had already established itself as a leading organizer of international exhibitions in the first half of the twentieth century, hosting World's Fairs in Liège (1905, 1939), Brussels (1910, 1935), Ghent (1913), and Antwerp and Liège together (1930). At Expo '58, the first postwar World's Fair, the exhibits were centered around the futuristic Atomium, shaped like an iron crystal magnified 165 billion times and constructed of steel, with nine globes clad in a shiny new building material—aluminum. Inside the globes was an exhibition called *Between Utopia and Reality*.

Forty million people from all over the world visited this event in the summer of 1958, at the height of the Cold War. About 60 percent of the exhibitors were from Belgium. The Pavilion of Honor was designed by Emiel Veranneman and the colorful logo and poster were created by Belgian graphic designer Jacques Racket (plate 656). In addition to the strikingly large pavilions of Marie Thumas, a producer of canned vegetables, there was the Cote d'Or exhibit, which was conceived as a self-service restaurant with a fountain that spouted the company's chocolate. Of course, the utilities, the trades, and the shipping companies were all represented, along with the leading Belgian manufacturers of furniture, textiles, and ceramics.

The Dutch section was also relatively large and progressive. Designed by J. W. C. Boks, the firm Van den Broek & Bakema, and Gerrit Rietveld, it presented an interesting set of low, terraced pavilions with water features (including a basin with real waves), a lighthouse, and bridges. Rietveld was also responsible for the display of modern home furnishings, along with his son Jan. All the famous new designs could be admired there, such as the *Revolt* chair by Friso Kramer or the *Mondial* chair by Gerrit Rietveld and another son, Wim (plate 655). Auping exhibited a bed by André Cordemeijer; Artifort, a padded armchair by Gerrit Rietveld; and Gispen, the first Dutch chair with a fiberglass shell, by Cordemeijer and Wim Rietveld. Linoleum Krommenie, the Gero metalworks, the carpet maker KVT, and the wallpaper manufacturer Rath & Doodeheefver also mounted displays; the Leerdam glassworks even had its own pavilion. All the graphics for these displays, including the stylized dolphin that served as the logo of the Dutch delegation, were designed by Jan Bons.

Part of the Dutch section, though completely detached from the rest, was the pavilion of the electronics manufacturer Philips, designed by Le Corbusier in collaboration with Iannis Xenakis. This complicated tentlike construction displayed a multimedia show that included a new electronic music composition by Edgar Varese, entitled *Poème Electronique*. The presence of this stand-alone pavilion led to some animosity between the Dutch architects and the famous Corbusier.

58
BRUXELLES
AVRIL 1958 OCTOBRE
EXPOSITION UNIVERSELLE
marfurt

OPPOSITE, TOP LEFT
657. **HA 2728 fan**, 1955
Made by Philips (Netherlands)
Melamine, low-density polyethylene, polystyrene, chromium-plated brass
Jean-Bernard Hebey collection, Paris

OPPOSITE, RIGHT
658. **HA 2380 travel iron**, 1959
Made by Philips (Netherlands)
Chrome-plated steel, Bakelite, cast aluminum
Jean-Bernard Hebey collection, Paris

OPPOSITE, BOTTOM LEFT
659. **UV lamp**, 1951
Made by Philips (Netherlands)
Lacquered steel, Bakelite, glass
Jean-Bernard Hebey collection, Paris

INDUSTRIAL PROSPERITY IN THE 1960S AND '70S

The Belgian furniture and textile industries flourished in the 1960s, becoming a significant economic force, with a large number of companies whose products were distributed worldwide. During this period of growth and prosperity, the first Biënnale Interieur was held at Kortrijk in 1968. The renowned Belgian furniture manufacturer Van De Coene played a significant role in the event, for which the name "biennale," with its cultural connotations, was deliberately chosen. This was meant to be more than an ordinary trade fair—and, indeed, it was almost an instant success. Moreover, Italian companies participated from the outset, which partially explains the strong influence of Italian design in Belgium in subsequent years.

One designer who was inspired by Italy at an early stage was Pieter De Bruyne. Through his teaching at Sint-Lukas College in Brussels and his numerous trips to Italy, he gradually began to think differently about furniture design. He decided to go to work afresh, more simply and on a smaller scale, with more attention to aesthetics, symbolism, and emotion as opposed to functionality and efficiency. One milestone was his hybrid *Chantilly* cabinet of 1975, which may be considered his first piece of postmodern furniture. Emiel Veranneman, who was a collector and gallery owner as well as a designer, also abandoned mass production in the 1960s and turned, as De Bruyne did, to smaller editions, bringing together a variety of influences in his abstract furniture concepts.

The Netherlands was also experiencing strong growth: production doubled from 1948 to 1962, with new businesses springing up everywhere. At the same time, the field of industrial design became more widely known and accepted. In the furniture industry, Pastoe, 't Spectrum, and Artifort led the way as far as design was concerned. After the war, Pastoe was best known for its simple storage systems in light birchwood, designed by Cees Braakman. At 't Spectrum, Martin Visser was the central figure between 1954 and 1974. Some of his sleek minimalist pieces—such as his *BR02.7* sofa bed of 1960 and his *SZ02* armchair of 1965—are still being produced today. In the 1970s, Visser introduced the Belgian design duo Claire Bataille and Paul Ibens to 't Spectrum, and they produced a number of successful pieces for the firm. In the 1960s, Artifort, based in Maastricht, became the most internationally oriented furniture manufacturer in the Netherlands, with designs by Theo Ruth and especially Kho Liang Ie, whose promising career was cut short by his untimely death in 1975. An important design prize, awarded annually between 1979 and 2002, was named in his honor.

In the rapidly growing market for office furniture, Ahrend and De Cirkel became the new leaders, although Gispen remained an important player as well. Entirely in the spirit of the times, Friso Kramer designed the *MEHES* office system (whose name stood for "mobility, efficiency, humanization, environment, and standardization") for Ahrend; it consisted of modular desks that could be arranged at the user's discretion to create a flexible open workspace. Meanwhile, in the field of office equipment, the Dutch company Océ van der Grinten became famous for its beautifully designed and ergonomic photocopiers.

In this period, industrial design also conquered the home appliances sector. Philips, founded in 1891, has become one of the world's largest industrial conglomerates, specializing in electrical and electronic consumer goods, particularly television sets (plates 657–59). Philips has always been at the forefront of sound and imaging technology, creating the CD and DVD formats in association with the Sony Corporation of Japan. Plastics also experienced major growth: The Dutch firms Mepal, Tiger Plastic, and Curver collaborated with professional designers, while Meurop began producing a broad range of plastic furniture in Belgium. In 1960, the Tupperware company opened a large factory in the Belgian city of Aalst.

THE DESIGN AGENCIES

A new phenomenon in the Netherlands in the 1960s and '70s, once again following in the footsteps of the U.S., was the creation of large, versatile design agencies. Among the leading agencies were Tel Design and Total Design. The former was founded in 1962 by the industrial designers Emile Truijen, Rob Parry, and Jan Lucassen, who had the ambition of being active in all fields of design. In 1957, Truijen and Parry had designed the efficient twin mailbox, made partly from plastic, that was used in the Netherlands for decades. The graphic designer Gert Dumbar joined the firm in 1967, when it was chosen to give Dutch Railways a new corporate identity.

Total Design (TD) was founded in 1963 by Wim Crouwel, Benno Wissing, Friso Kramer, and the brothers Paul and Dick Schwartz. At TD, there was a somewhat

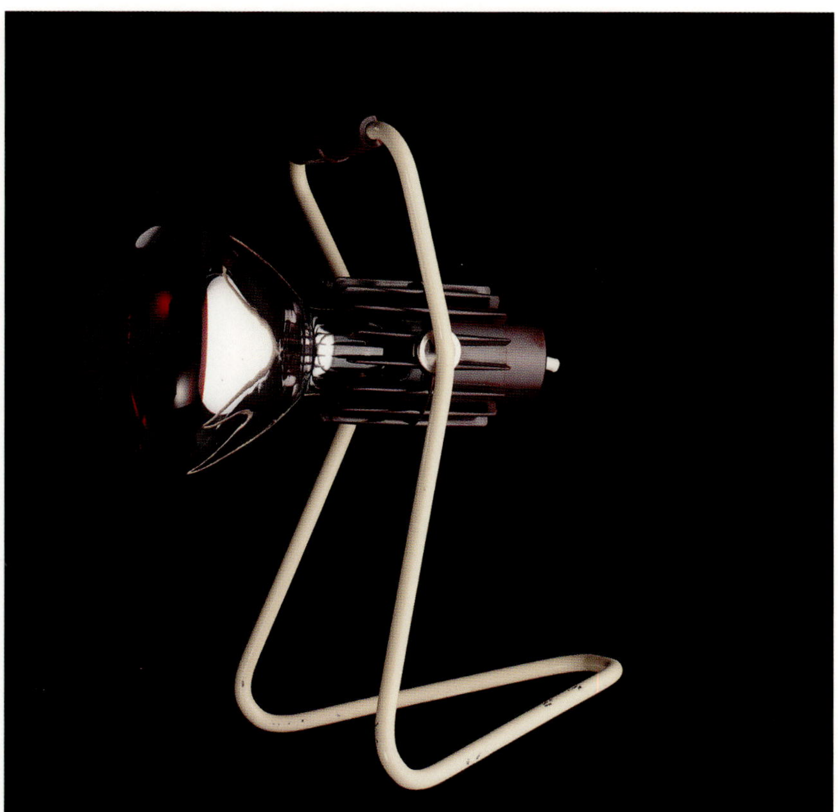

PHILIPS
COTON

BELOW
660. Wim Crouwel (b. 1928)
Poster for the *Vijftig Jaar Zitten* exhibition, Stedelijk Museum, Amsterdam, 1966
Stedelijk Museum, Amsterdam

OPPOSITE
661. Kho Liang Ie (1927–1975)
***K46* floor lamp**, 1947
Made by Artifort (Netherlands)
Metal, plastic
Museum Boijmans Van Beuningen, Rotterdam

greater focus on graphic design. Crouwel, for years the face of TD, had begun his design career with a short but mutually inspiring collaboration with Kho Liang Ie, in which their work was typified by a tight, crisp minimalism (plate 661). Crouwel was a convinced modernist, and his graphic work was strongly influenced by Swiss typography (plate 660).

Wissing was an idealistic graphic designer enormously inspired by the constructivists El Lissitzky and Moholy-Nagy, as well as by his fellow Dutchmen Piet Zwart and Paul Schuitema. One of Wissing's most important commissions was to create new signage for Amsterdam's Schiphol Airport when it was reconstructed in 1967. His restrained and carefully considered design set the standard for all similar projects. (The airport's interiors were designed by Kho Liang Ie.)

The roots of the "corporate identity" phenomenon may be traced to the groundbreaking work done in the Netherlands in the 1930s—by Jac Jongert for Van Nelle, Paul Schuitema for Van Berkels Patent, and Piet Zwart for Nederlandse Kabelfabriek. In the 1960s, Total Design and Tel Design broadened the concept of house style with their designs for a wide range of corporations and cultural institutions. For example, the identity that Ben Bos of Total Design created for the temp agency Randstad in 1967 is still in use today.

The third-largest design agency in the Netherlands was Premsela Vonk, which focused on textiles for interiors. Together with Marijke de Ley, the firm's charismatic principal Benno Premsela (1920–1997) designed the innovative cotton loop carpet for the firm of Van Besouw. Premsela, who began his career as a window dresser at the De Bijenkorf department store, was a major force in Dutch culture in the last quarter of the twentieth century, holding numerous executive positions and even being dubbed an "art pope." In 1982, Premsela designed the *Lotek* lamp, which is still popular today.

Of course, corporate identities and logos were also designed in Belgium in the 1960s, and the Brussels Design Center even dedicated an exhibition to the theme in 1966. Paul Ibou, also an inventive book designer and publisher, was the principal graphic designer in this field. However, in contrast to the Netherlands, Belgium did not have any large design agencies focusing on corporate identities, only individual designers, a circumstance that made collaborations with major corporations more difficult.

In these years, Dutch design attracted increasing international interest, as well as official and semiofficial involvement. The large commissions for Dutch Railways and Schiphol Airport have already been mentioned. Equally important was the work done for the national post office and telephone company, the PTT, whose engagement with design dates to the beginning of the twentieth century, when J. F. van Royen, then secretary of the board, set out to improve the quality of the organization's printed matter. In the decades that followed, this led to dozens of commissions for stamps, forms, signs, post office interiors, postmen's caps, and the lettering on postal vehicles.

Although he died in a German prison camp in World War II, Van Royen's progressive policies lived on in the PTT's design office, which has had a series of inspired directors, including Ootje Oxenaar, who served from 1976 to 1994. Oxenaar was already well known for modernizing the design of the Dutch banknotes in the 1960s. In the 1980s, this was followed by his much-talked-about, colorful, and completely original 50, 100, and 250 guilder notes, depicting a sunflower, a snipe, and a lighthouse. This attractive currency helped give foreign visitors the impression that the Netherlands itself was carefully and responsibly designed.

OPPOSITE, TOP
662. Gijs Bakker (b. 1942)
Fruit bowl, 2000
Made by Royal VKB (Netherlands)
Stainless steel
Philadelphia Museum of Art

OPPOSITE, BOTTOM
663. Gijs Bakker (b. 1942)
Dish, 2004
Made by Salviati (Italy)
Distributed by Droog (Netherlands)
Glass
Centre National des Arts Plastiques/Fonds National d'Art Contemporain, Paris

POSTMODERNISM

Many changes occurred in the mid-1970s, in both Belgium and the Netherlands. The economic slump resulting from the oil crisis led to the closure of many factories, while others shifted their production abroad, where wages were lower. The number of commissions for professionally trained industrial designers fell sharply. At the same time, the climate in the design world was changing. First, modernism was increasingly criticized: Designers were no longer convinced that there was only one ideal, functional style that they all should pursue. Rather, they found that the rigid uniformity of modernism restricted their creativity. Meanwhile, individualism had taken root in society as a whole, resulting in the emergence of a separate youth culture and a growing segmentation of the entire commercial market.

At the same time, environmental threats, consumerism, and the unbridled growth of world markets were coming under scrutiny, as was the role of design in these phenomena. The Dutch critic Simon Mari Pruys raised some of these issues as early as 1972 in his book *Dingen Vormen Mensen* (Things Form People), a study of production, consumption, and culture. Good design was no longer synonymous with modernism, and designers were asked to think about the social implications of their activities. Pruys introduced Jean Baudrillard's *Le Système des Objets* (The System of Objects) into the realm of design, opening the way for the consideration of design as a sociological phenomenon, and of its sometimes undesirable influence on the modern consumer society.

Even in the prosperous field of graphic design, criticism was now being heard. The tight and uniform visual approach that the Netherlands had applied to so many daily objects, from traffic signals to postage stamps and phone books, was now the subject of general disapproval, with the term Nieuwe Lelijkheid (New Ugliness) even being used. People had become bored with what they considered to be anonymous simplicity, and an emerging movement of young designers began to resist the hegemony of Wim Crouwel and "his" TD. They were buoyed by a rising tide of technical developments in the graphic arts: rub-on lettering, the photocopier, and cheaper offset printing made experiments with alternative graphics easy to design and disseminate.

Two magazines, *Hard Werken* (Hard Work) in Rotterdam (edited by Gerard Hadders, Rick Vermeulen, and Henk Elenga) and *Wild Plakken* (Wild Paste) in Amsterdam (edited by Lies Ros and Rob Schröder), set a totally new course, characterized as "postmodern," that others followed. It is typical of the open-mindedness of Dutch society in those years that these revolutionary initiatives were accepted and adopted by the government, including the state printing office, as well as by the leading cultural institutions. Among the most important graphic innovators of the time were Anthon Beeke and Jan van Toorn.

After the war, those working in traditional crafts—potters, weavers, glassblowers, and goldsmiths—gradually gained greater visibility. Their handmade and individual work offered a welcome alternative to the mass-produced objects that were increasingly seen as dull and anonymous. For example, the studio ceramics of Hans de Jong, Johan van Loon, Jan van der Vaart, Jan de Roode, Johnny Rolf, and Lies Cosijn were acquired by the Stedelijk Museum in Amsterdam and the Museum Boijmans Van Beuningen in Rotterdam. The same was true of the unique pieces crafted by the glass designers Andries Copier, Floris Meydam, and Willem Heesen, and the textile works of Ria van Eyk, Margot Rolf, and Herman and Desirée Scholten.

Industrial designers, craftspeople, and artists have traveled back and forth over the same ground. Artists have made craft products, while craftspeople have produced art, and occasionally even industrial design, as in the case of Jan van der Vaart. Particularly influential were the experiments of some jewelry makers. Gijs Bakker and his wife Emmy van Leersum, Françoise van den Bosch, Maria van Hees, and Hans Appenzeller, to name but a few, worked in what seemed to be an austere mode of modernist geometric abstraction, but their pieces were closer to sculpture than wearable jewelry, and were not always made from the traditional precious metals (plates 662, 663). Their jewelry was presented in an important group exhibition in Amsterdam in 1969, called *Objects to Wear*.

OPPOSITE
664. Tejo Remy (b. 1960)
***You Cannot Lay Down Your Memory* chest of drawers**, 1991
Made by Droog (Netherlands)
Drawers, maple, belt
Museum of Modern Art, New York

RIGHT
665. Piet Hein Eek (b. 1967)
***Scrapwood* dresser**, 1990
Scrap wood

BELOW
666. Tejo Remy (b. 1960)
Milk Bottle Lamp, 1991
Made by DMD (Netherlands)
Frosted glass milk bottles, stainless steel

SELF-PRODUCING DESIGNERS AND DROOG DESIGN

The shortage of commissions continued into the 1970s, leading an increasing number of young designers to handle their own production and sales. The "self-producing" designer was a new and important phenomenon in the Netherlands. At first, their handmade objects retained an industrial appearance, such as in the pendant watch and minimalist tube lamp by Ninaber van Eyben. Besides Van Eyben, the explorers of this new terrain included Frans van Nieuwenborg, Martijn Wegman, Herman Hermsen, the Vormgeversassociatie Arnhem (Design Association Arnhem: Hans Ebbing, Ton Haas, and Paul Schudel), and the architect Mart van Schijndel. The magazine *Items*, first published in 1982, became a showcase for all these new experiments, which would gradually succeed in changing the face of Dutch design.

The work of some of these self-producing designers was presented in the 1981 exhibition *Design from the Netherlands*, sponsored by the Dutch government's Office for Visual Art. Curated by Gijs Bakker, this show traveled worldwide for more than six years. In addition to his work as a designer, Bakker also taught at the Institute of the Arts in Arnhem, where he encouraged his young students to use their own initiative and be more creative, original, and artistic in their work. Bakker felt that design should once again be seen as a part of the arts, and from 1987 he promoted this new mindset in another teaching post, at the Design Academy in Eindhoven.

It was in this artistic climate that Bakker founded Droog in 1992, with Renny Ramakers, a design historian and editor of the magazine *Industrieel Ontwerpen* (Industrial Design). Bringing together those Dutch designers who were exploring radical new directions, Droog organized exhibitions of products that were not always functional, nor even attractive or artistic, but which carried a message. Moreover, after the exhibitions, Droog made considerable efforts to produce and market these items. The success of Droog's first presentation outside the Netherlands, at the Kortrijk biennial in 1992, led to a high-profile entry in the Triennale di Milano several months later.

The name *Droog*, or "dry," is meant to evoke a sober, honest, no-nonsense mode of design. At first this conceptual approach referred mainly to the conventions of

LEFT
667. Marcel Wanders (b. 1963)
Knotted Chair, 1996
Made by Droog (Netherlands)
Knotted carbon fiber, epoxy resin
Museum of Modern Art, New York

BELOW
668. Richard Hutten (b. 1967)
***Sexy Relaxy* chair**, 2002
Made by Richard Hutten Studio (Netherlands)
Wood (walnut, cherry, ash, maple, or oak), 2007 version
Edition of eight
Museum Boijmans Van Beuningen, Rotterdam

OPPOSITE
669. Marcel Wanders (b. 1963)
***Sinusitis* vase**, 2001
Made by Wanders Wonders (Netherlands)
Sintered polyamide
Centre National des Arts Plastiques/Fonds National d'Art Contemporain, Paris

design itself. Early examples include Tejo Remy's chest of drawers, poetically titled *You Cannot Lay Down Your Memory* and consisting of a random collection of old desk drawers piled together and tightly secured with a strap (plate 664); Piet Hein Eek's cabinet made from scrap wood with peeling paint (plate 665); and Rody Graumans's lamp, a bundle of twenty light bulbs on individual wires.

Droog became a mentality, a refreshing way of thinking about and doing design. Ideas were exchanged, from reflections on the design profession and proposals for recycling materials, to the quest for the simplest and purest form. In many cases, the result was a kind of ironic anti-design. In 1999, Jurgen Bey presented his *Tree Trunk Bench*, consisting of three bronze chair backs attached to a tree trunk (plate 673). Richard Hutten called one of his series *No Sign of Design*, claiming that it represented the elemental form of the table and chair. The irony in his *Sexy Relaxy* chair of 2002 is unmistakable (plate 668).

Experimentation with innovative materials was another theme, exemplified by Marcel Wanders's *Knotted Chair* of 1996, in which the corny home craft technique of macramé is combined with an advanced type of rope used in the modern aircraft industry (plate 667). In the early Droog period, Hella Jongerius, the most successful Dutch designer of the last two decades, produced a sink made of rubber and ceramics covered in embroidery. In her more recent and spectacular *Frog* table, a giant wooden frog and a simple table are merged into one (plate 670). Droog's tremendous success and the attention it received for its unorthodox, against-the-grain design were partly due to the funding opportunities available in the Netherlands at the end of the twentieth century. The government encouraged and generously financed exhibitions and publications.

Among the new generation of successful designers, many of them Eindhoven graduates, we should mention Maarten Baas (plate 671), Bertjan Pot (plate 672), Wieki Somers (plates 674, 677), Job Smeets, Christien Meindertsma, Demakersvan (plate 675), and the duo Scholten & Baijings (plate 676). Baas displays a

BELOW
670. Hella Jongerius (b. 1963)
***Frog* table**, 2009
Made by Galerie Kreo (France)
Walnut, resin
Museum Boijmans Van Beuningen, Rotterdam

OPPOSITE
671. **View of the exhibition *Maarten Baas, les curiosités d'un Designer*, Musée des Arts Décoratifs, Paris**, 2011
Left: *Sculpt* wardrobe, 2007; right: *Grandfather Clock*, 2009; on the floor: *Animal Skin* rug, 2008–9

groundbreaking creativity with his furniture in charred and blackened wood coated in epoxy resin (including a scorched Rietveld chair), and in clay. Also highly original are his video-based *Real Time* clocks: for example, in the *Sweepers Clock*, filmed high above a city square, two street sweepers push around piles of street clutter that form the hands of a clock. Smeets and his partner Nynke Tynagel are the directors of Studio Job, which makes the most unlikely pieces, as impressive as they are kitschy, and exhibits them in the studio's own gallery in Antwerp. Christien Meindertsma is concerned with elucidating the production process and the origins of her raw materials: each of the sweaters in her *One Sheep Sweater* project was knitted from the wool of a single sheep and embroidered with that animal's number.

It is interesting to see how the radical designers of recent years have found their way back to industry: Jongerius, Wanders, Hutten, and Bey now work for dozens of famous brands around the world. At the same time, it should not be forgotten that other designers have continued to work for industry all along. Delft University of Technology has had a Department of Industrial Design Studies since the mid-1960s. Without many artistic pretensions, but with great technical expertise, its graduates address the less glamorous areas of design, such as electrical appliances, medical equipment, and the transport sector. Examples of their work well known to the general public include the *Maxi Cosi* car seat for babies, first designed by Huibert Groenendijk in 1981, and the forward-sloping *Senseo* coffee maker, designed by the Waacs agency for Douwe Egberts and Philips in 2001. The economic importance of this sometimes prosaic type of design is illustrated by Jeroen Verbrugge's new system for transporting Coca-Cola bottles.

LEFT
672. Bertjan Pot (b. 1975) and
Marcel Wanders (b. 1963)
Carbon Chair, 2004
Made by Moooi (Netherlands)
Epoxy resin, carbon fiber

BELOW
673. Jurgen Bey (b. 1965) for
Makking & Bey
Tree Trunk Bench, 1999
Made by Droog (Netherlands)
Chair backs in bronze,
tree trunk

OPPOSITE, TOP
674. Wieki Somers (b. 1976)
High tea pot, 2004
Porcelain, coypu fur, leather,
stainless steel
Fonds National d'Art Contemporain, Paris, on loan to the
Musée des Arts Décoratifs,
Paris

OPPOSITE, BOTTOM
675. Demakersvan
(est. 2005)
***Cinderella* table**, 2010
Carrara marble
Carpenters Workshop Gallery,
Paris and London

ABOVE
676. Stefan Scholten (b. 1972) and Carole Baijings (b. 1973)
Colour Porcelain service, 2012
Made by 1616 (Arita, Japan)
Museum Boijmans Van Beuningen, Rotterdam

OPPOSITE
677. Wieki Somers (b. 1976)
Big Frozen Vase, 2010
Made by Galerie Kreo (France)
Resin, metal, silk, gel coat
Fonds National d'Art Contemporain, Paris, loan to the Musée des Arts Décoratifs, Paris

BELOW
678. Maarten Van Severen (1965–2005)
MVS chaise longue, 2000
Made by Vitra (Switzerland)
Polyurethane, stainless steel
Centre National des Arts Plastiques/Fonds National d'Art Contemporain, Paris

OPPOSITE
679. Maarten Van Severen (1965–2005)
***LCP* chaise longue**, 2000
Made by Kartell (Italy)
Methacrylate
Museum of Modern Art, New York

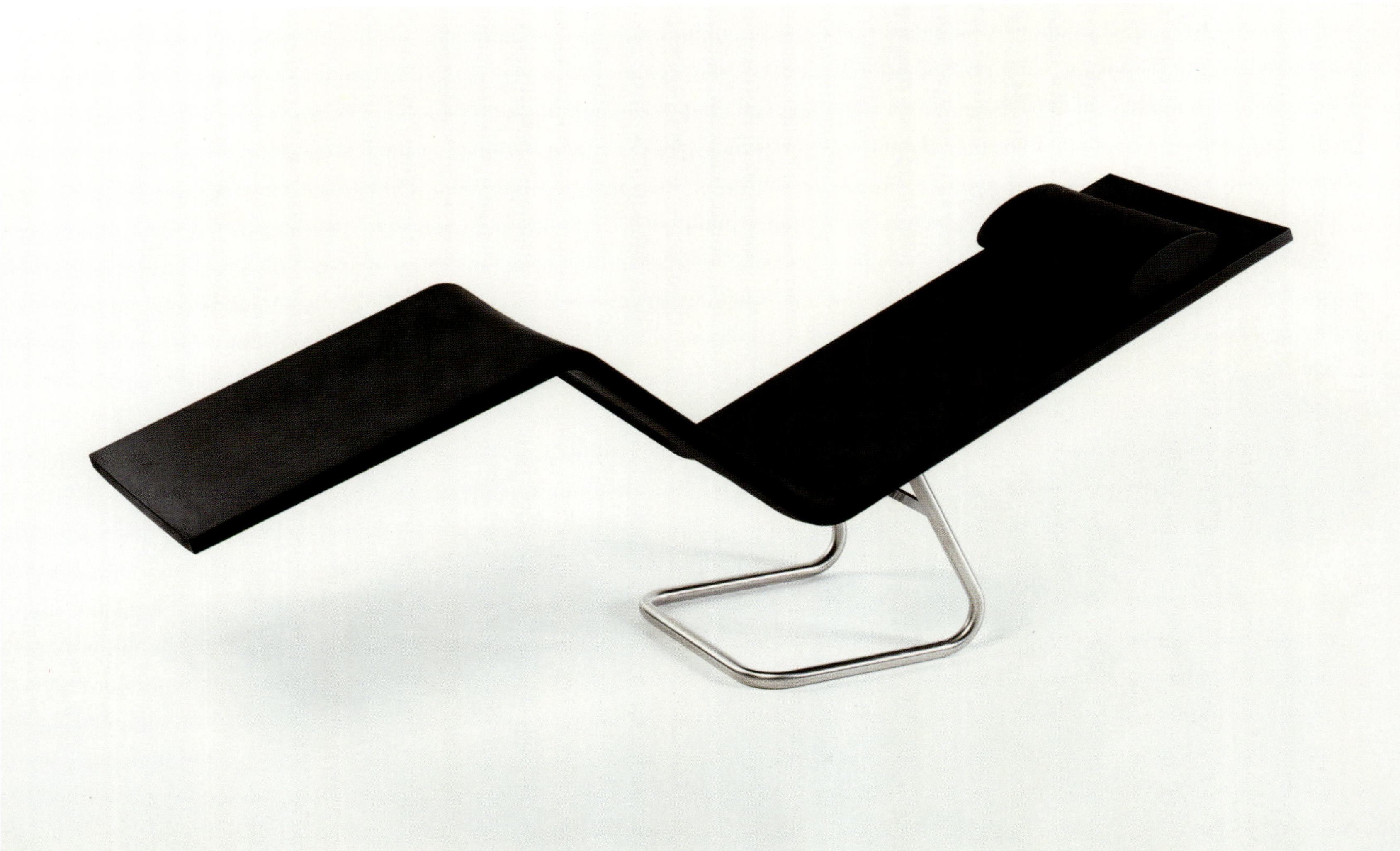

NEW TENDENCIES IN BELGIUM

Developments in Belgium were not quite parallel to those in the Netherlands. The economic blow to the furniture and textile industries in the late 1970s affected Belgium somewhat less than the Netherlands, which meant that more opportunities remained for designers in these sectors. Nevertheless, Belgian design has not acquired the same international reputation as Dutch design. The Netherlands, a trading nation by tradition, still knows how to sell things—in this case, a new image of Dutch design.

This does not mean that the quality of Belgian design lags behind that of its northern neighbor, only that Belgian designers have been less inclined to present themselves as a group, perhaps in part because of the country's language and political divisions. However, Design Vlaanderen (Design Flanders; formerly VIZO) diligently promotes new developments in the field with the support of the Flemish government of Belgium. Design Flanders has held the Triennial for Design since 1995, concentrating on the most current themes, and has awarded the Henry van de Velde design prize since 1994. It also publishes the magazine *Kwintessens*.

As in the Netherlands, small-scale, artistic design began to regain importance in Belgium around 1980. We have already seen how Pieter De Bruyne and Emiel Veranneman laid the foundation for this trend in the

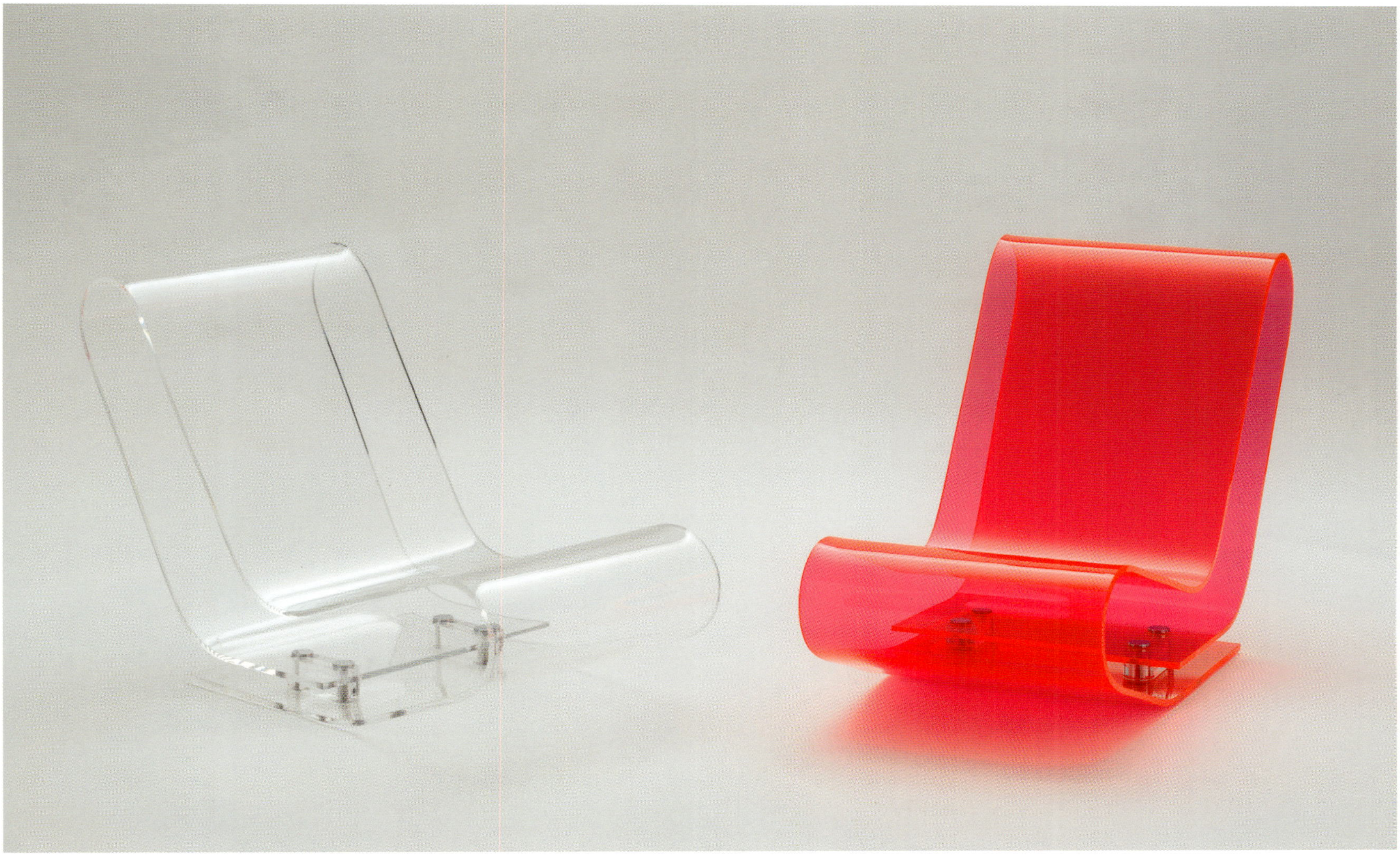

BELOW
680. Xavier Lust (b. 1969)
Credenza, 2003
Made by DePadova (Italy)
Lacquered anodized aluminum
Fonds National d'Art Contemporain, Paris, on loan to the Musée des Arts Décoratifs, Paris

OPPOSITE
681. Bram Boo (b. 1971)
***Overdose* desk**, 2007–9
Made by Bulo (Belgium)
Lacquered MDF, wood

1970s. There was also a growing interest in craft-based design and in combining craft, art, and design. The goldsmith Siegfried De Buck was a pioneer in the field of jewelry design, reviving it with the use of new materials. In the 1990s, he also created larger pieces, such as tableware, that attracted the attention of progressive Italian designers. Piet(er) Stockmans played a similar role in ceramics. He was a designer for the Mosa manufactory in Maastricht from 1966 to 1989, and the stackable porcelain coffee cup he created in 1976, called *Sonja*, was produced in the tens of millions. But gradually he also began to work more freely, sculpturally, and conceptually in porcelain. Meanwhile, Andries Verroken opened new frontiers in furniture with his search for pure, mathematical, sculpted forms.

Maarten Van Severen, who died prematurely in 2005, earned a special place in Belgian design with his minimalist creations that sought out the very essence of form. He collaborated closely with the celebrated Dutch architect Rem Koolhaas, notably on the interior of the Lemoine house in Floirrac, near Bordeaux. In 1998, Van Severen designed his famous *.03* chair for Vitra, which combines polyurethane, steel, and aluminum in a very elegant way. He was also occupied for some years with the concept of the chaise longue (plates 678, 679), of which he presented his first prototype in 1995. In 2006, the Dutch firm Pastoe produced Van Severen's last design, the *Leather Lounge Chair LL04*. Van Severen's younger brother Fabiaan and his sons Hannes and David are also leading designers.

The last ten to fifteen years have seen the emergence of many more innovative young Belgian designers, including Xavier Lust (plate 680), Bram Boo (plate 681), Sylvain Willenz, and Danny Venlet. At the same time, industrial designers and design agencies with a more classical approach have remained very active. Axel Enthoven, for example, who also taught for many years at Eindhoven, was responsible for the interior of the high-speed *Thalys* train.

A new phenomenon in the design world, both in Belgium and in the Netherlands, has been a greater attention to fashion. Here Belgium took the lead. The Antwerp Six, including Ann De Meulemeester, Dries Van Noten, and Walter van Beirendonck, have earned much attention since 1988. The Dutch followed in 1992 with the group show *Le Cri Néerlandais*, in which Viktor & Rolf and Alexander van Slobbe participated. Antwerp's Mode Museum (MoMu) opened in 2002.

In the twenty-first century, Belgian and Dutch designers have come to realize that their creative abilities might contribute to the solution of major world problems. The phenomenon of "social design" has thus made its appearance. Since 2010, Amsterdam has hosted the annual international conference What Design Can Do. And for its 2010–11 triennial, Design Flanders chose the theme "Belgium Is Design. Design for Mankind," another indication that the social aspect of design is now paramount.

BULO
BUOVE

THE LANDSCAPE OF DESIGN TODAY

CONSTANCE RUBINI

PRECEDING PAGE
682. Pucci de Rossi (1947–2013)
***Mobile Metafisico* writing case, an homage to Giorgio de Chirico**, 1990
See plate 688.

BELOW
683. Jaime Hayon (b. 1974)
***Pixel Ballet* exhibition, presented at the Fondazione Bisazza, Montecchio Maggiore, Italy**, 2007

OPPOSITE
684. Jaime Hayon (b. 1974)
***Conversation Vase I-B* from the *Fantasy* collection**, 2008
Made by Lladrò (Spain)
Enameled porcelain

The start of the twenty-first century saw a paradigm shift. The age of design as the orchestration of form, function, and cost for the purpose of industrial production was over. That vision had been sustained by an old utopian ideal—upheld in France, for example, since the creation of the UAM (Union des Artistes Modernes) in 1929. Politically inflected design, which played a role in the postwar period, especially in Italy, at a time when the whole world had to be reconstructed anew, had also lost its appeal. In fact, it was in Italy that the doors were flung open to change. Since the 1980s, other dimensions beyond function and use—those of emotion and culture—have been explored. The radical Italian designers Andrea Branzi and Ettore Sottsass invented a world of feeling, and opened the way to a mode of design freed from the demands of homogeneity. Current practices, reflecting the proliferation of cultural identities, express this new diversity.

LIFE AS COMMUNICATION

"To live effectively is to live with adequate information," wrote Norbert Wiener, the inventor of cybernetics and apostle of a new vision of the world whose fundamental features are information and communication. The progressive unfolding of this vision has influenced design, leading to the birth of "information design," which treats objects as a bundle of messages whose vectors are forms, ornament, and functions. Today we lead our lives entirely within this new society. In the 1960s, Marshall McLuhan's provocative aphorism "The medium is the message" summed up just how far technological innovation was going to rock modern civilization. As the container claimed precedence over the content, the door was opened to a type of ephemeral creation that favored process over product and the theatricality of the presentation over the character of the

ABOVE
685. Studio Job
***Taj Mahal* table**, 2012
Made by Carpenters Workshop Gallery (U.K. and France)
Bronze, gilt bronze

OPPOSITE
686–87. Front
Sketch Furniture, 2005
The designers' movements are recorded in 3-D and used as a blueprint for furniture, here executed in polyamide.

object. The object in turn lost its desire for permanence. Today what designers produce is judged by the media, and hence on the creativity and novelty of the concepts they conjure up rather than on criteria of functionality and use. A product is no longer designed to last, but to improve the competitiveness of the businesses involved, by contributing to the ever more rapid turnover of what they sell. Businesses bring in designers to stage a new product and create a buzz around it. They prefer the kind of design whose codes flirt with those of art: the Spanish designer Jaime Hayon and Studio

Job (Job Smeets and Nynke Tynagel in the Netherlands) are among the most skilled at these maneuvers. The extravagant installations they each produced for the Italian mosaic manufacturer Bisazza at the Milan furniture fair in 2007 are striking examples of this. Studio Job displayed tableware covered with silver mosaic, including an outsize spoon hanging on the wall and a giant teapot and salad bowl set on a classical pedestal in sculpted white wood; these were objects that played with the conventions of the decorative arts to surprise or sometimes even astonish. Jaime Hayon used white, black, and gold mosaic to adorn furniture, accessories, and figurines, among them a giant seated figure inspired by Pinocchio, which occupied the center of the space like a supernatural being (plate 683).

Design galleries also exhibit this kind of whimsical production, with its potential for extreme flamboyancy, as they seek to reposition the decorative arts in the twenty-first century. Thus Carpenters Workshop Gallery has exhibited the *Taj Mahal* table (2012) by Studio Job (plate 685), whose work has also been shown at the Victoria and Albert Museum in London, in the exhibition *Telling Tales* (2009).

The market for art and design is riding this trend: for example, the young Dutch designer Maarten Baas, the Swedish design collective Front (Sofia Lagerkvist, Charlotte von der Lancken, and Anna Lindgren), and the Spanish designer Martino Gamper all engaged in real-time performances at Design Miami/Basel 2007, in response to the theme of "Performance/Process" chosen by Ambra Medda, the fair's director. The aim of these performances was to create products that were unique and hence more valuable. As ephemeral as they were, the performances of Gamper, who appropriated and reconceived furniture originally designed by Gio Ponti in 1960 for the Parco dei Principi Hotel in Sorrento, or of Front, who used motion-capture technology to design their furniture (plates 686, 687), helped to shape the new landscape of design. The image of Front's performance, using technological tools that seemed almost magical, has also gone around the world, becoming the symbol of this collective, who enjoy telling stories about their design processes.

688. Pucci de Rossi (1947–2013)
***Mobile Metafisico* writing case, an homage to Giorgio de Chirico**, 1990
Beech, plywood, nails
Musée des Arts Décoratifs, Paris

HUMOR, GESTURE, AND ANECDOTE

What designers have to do now is provide a kind of narrative that did not previously exist, but now represents a form of communication with the public. The 1980s saw the start of this trend. Memphis, the group founded by Ettore Sottsass in 1981, rejected the conventions articulated by the Bauhaus and introduced freedom, feeling, and humor into design. In France, there developed an approach to design that emphasized uniqueness, from the *Siège Impérial* by Garouste & Bonetti to Pucci de Rossi's *Mobile Metafisico* (plate 688)—an outsize writing case intended for penning a love letter—or Philippe Starck's *Juicy Salif* citrus squeezer, whose shape, while scarcely suitable for its ostensible purpose, has aroused much commentary. Each of these creations tells a one-of-a-kind story. Starck likes to claim that the *Juicy Salif* "starts a conversation in the kitchen." The object has become a means of communication, and its semantic richness contributes significantly to its aesthetic quality. The next generation of designers took this link between design and communication to a new level with humorous objects such as the furniture of the Campana brothers from Brazil, who do not hesitate to upholster an armchair with stuffed animals, using narrative associations to evoke a fantastic image of comfort (plate 689).

Hayon has designed the *Fantasy* collection for the Spanish porcelain manufacturer Lladrò. As its name suggests, this china service plays on the comical character of its decoration, which has discarded the symbolic and explanatory purposes formerly assigned to ornament to become simply a source of originality, one ingredient in the crafting of a personal vision. The impish figurine enthroned on the top of his *Conversation Vase I-B*, for instance, is a fanciful image whose function matters less than its humorous effect (plate 684).

This same humorous approach was a hit with the audience at the 2007 design festival at the Villa Noailles in Hyères, France, where the young Spaniard Nacho Carbonell exhibited *Pump It Up* (plate 690). Sitting down on this foam chair compresses the air inside, which is then driven along an umbilical cord attached to animal figurines, seeming to breathe life into them. Like a living organism, the chair creates an emotional presence, although the effect is first of all strange and suprising. The surprise, however, can only be experienced once, the first time you sit in the chair; after that it becomes a story. Odd, amusing narratives of this sort now form part of design.

The point of these stories is often to make people smile. We may think of the 1998 exhibition by Radi Designers (a French collective active 1993–2003, including Laurent Massaloux, Florence Doléac, Robert Stadler, and Olivier Sidet, who now work individually)

RIGHT
689. Fernando Campana (b. 1953) and Humberto Campana (b. 1961)
***Banquete* chair**, 2002
Limited edition
Stuffed animals, upholstery, stainless steel

BELOW
690. Nacho Carbonell (b. 1980)
***Pump It Up* armchair**, 2007

LEFT
691. Radi Designers
***Coffee Drop Splash* cookie**, 1994
Made by Radi Designers (France)
Chocolate cookie fitted to a coffee cup

BELOW
692. Radi Designers
Whippet Bench, 1998
Made by Radi Designers (France)
Wood, polyurethane foam, printed fabric

RIGHT
693. Jasper Morrison (b. 1959)
***Plywood* chair**, 1988
Made by Vitra (Switzerland)
Birch, plywood

BELOW
694. Jasper Morrison (b. 1959)
Cork Chair, 2007
Made by Vitra Edition
(Switzerland)
Cork

BELOW
695. IDEO
LifePort Kidney Transporter,
2003

OPPOSITE
696–97. Ronan Bouroullec
(b. 1971) and Erwan Bouroullec
(b. 1976)
***Joyn* office furniture**, 2002
Made by Vitra (Switzerland)

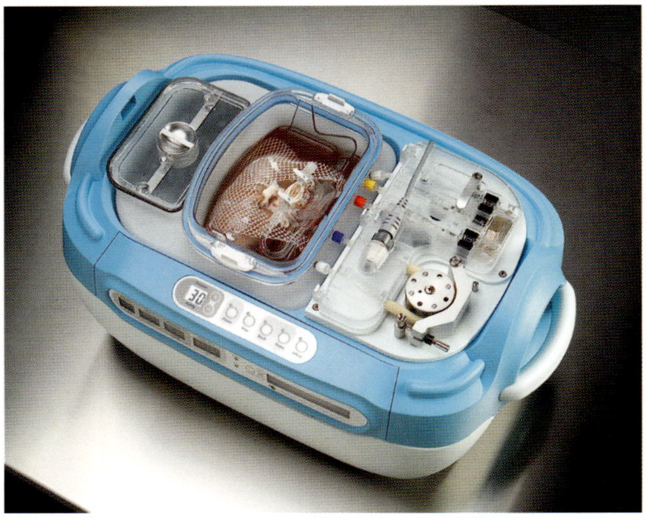

at the Galerie Perrotin in Paris, where they displayed their *Sleeping Cat* carpet, *Miroir Vibrant*, *Coffee Drop Splash* (plate 691), and *Whippet Bench* (plate 692). Stories can also be a way of creating displacement or ambiguity, propelling us into another world. *Winter Arrives*, designed by the Singaporean-Argentine-Spanish collective Outofstock, is a chair with a foam seat upholstered in white so that it looks like a sled covered with six inches of snow, creating the absurd union of two conflicting worlds.

Designers who work in this playful mode have lately found a receptive audience. This has not escaped the notice of businesses, who rush to commission ephemeral products from them. The ice-cream maker Häagen-Dazs, for example, commissioned Doshi Levien (Nipa Doshi from India and Jonathan Levien from Britain) to design a unique ice-cream cake for Christmas 2012, asking for something "attractive, cheery, and ephemeral." Doshi Levien responded by treating the ice cream as they would any other material and imagined a cake in the shape of the moon, which required a mold just as complex as one for an industrial design, although of course much less expensive.

These event-driven creations are being produced in a world with a new relation to time, in which events move faster and faster, and everything happens simultaneously in a gravity-free environment.

A WONDERFULLY CONCRETE WORLD

But in parallel with this realm of fun there exists a kind of everyday design, restrained and scarcely visible, whose sole purpose is to quietly create a world that is wonderfully concrete, filled with well-designed objects. This is the "super-normal" design that Jasper Morrison of Great Britain and Naoto Fukasawa of Japan speak of, each in their own way: the design of a concrete reality that continues to exist despite all the new alternatives. Even as science and technology propel us into a virtual world, sociologists are agreed that online networks do not result in a dematerialized society: on the contrary, they help to increase the number of real relations between people. We know that in the very near future it may be technically possible to have chips implanted in our bodies to enhance our human capacities, but we are still not ready to swap our bowls and spoons for food injections. We will continue to need everyday objects for a very long time.

Since the late 1980s, Morrison has been exploring these everyday things—in particular their mysterious aspects, the aura that emanates from them and affects the surrounding atmosphere. There is no need for originality in order to accomplish this, quite the reverse. Rather than breaking with tradition, Morrison seeks to design "super-normal" objects whose form results from the slow changes that time imprints on things—objects that are contemporary without breaking with the past (plates 693, 694). Modernity does not necessarily spring from a tabula rasa.

Fukasawa shares this sensitivity and attention to the atmosphere that objects produce. He has been design director of ±0 (Plus Minus Zero), a line of household goods, since its creation by a leading Japanese toy manufacturer in 2003. The name of the brand seems to echo his design credo: variations around a neutral center, in minuscule gradations from the smallest to the largest. Details that are almost trifling can arouse specific sensations, make objects seem more friendly, generate pleasure, produce a feeling of involvement. Ideas can affect our shared sensibility and thus become part of our behavior in a barely perceptible way. This is

RIGHT
698. Naoto Fukasawa (b. 1956)
***Haptic* fruit juice packaging**, 2004
Prototype

OPPOSITE
699. Konstantin Grcic (b. 1965)
***MayDay* lamp**, 1998
Made by Flos (Italy)
Plastic, polycarbonate
Museum of Modern Art, New York

a type of design that privileges the senses. Fukasawa's objects possess the intelligence and precision of modern functional design, but they also convey something more. Reduced to its essence, its simplest manifestation, form encourages a direct relationship with the object, in which wit and humor have their place. Take the packaging he designed for a banana-flavored milk drink: the solid yellow of the octagonal shape, with traces of brown along the edges, suggests the fruit even without its characteristic curve (plate 698). To put the finishing touches on the concept, opening the package is like peeling a banana. Neither style nor form is ultimately essential; beyond the object itself, what matters is the link it constructs with its surroundings.

Fukasawa paved the way for a generation that has acknowledged his formative influence and been inspired by his design philosophy. At the Ziba agency (Portland, Oregon), Mera Middleton hung pictures of his objects on the studio walls, like yardsticks. Tim Brown (CEO of Ideo, an American design agency with branches worldwide) learned from Fukasawa that design is at once the product of an idea and of a form. When Ideo designed the *LifePort Kidney Transporter* (2003), they came up with an object that looks like an egg, expressing the fragility of the kidney being transported (plate 695). All are deeply conscious of the way Fukasawa gets to the essence of the object in order to reinterpret it with wit and intelligence.

Design is naturally tied to its context. Morrison and Fukasawa are the elders in a line of designers who all strive in their own way to create manufactured objects—often products for household use—that will come to life in our everyday world. Whether they choose to work in industrial design or, less often, a more artisanal mode, they focus on the reproducibility of the object. Among this group are the French brothers Ronan and Erwan Bouroullec, who try to achieve a degree of diversity even though their work is industrially produced and subject to the constraints of standardization, cost management, packaging, transport, and many other highly complex criteria. They are careful to give each of their objects a character that is entirely its own. They seek a type of difference that astonishes us and brings to light the variety of contemporary lifestyles (plates 696, 697). Their *Algues*, made by Vitra, address the question of the separation and definition of interior spaces in a quite unprecedented way, offering users the option of making their own walls.

The furniture designs of the German Konstantin Grcic are also industrially produced. His startling forms permanently call into question the fundamental principles of design: comfort, structure, use, the relation to the body. *Chair_ONE*, made by Magis, and the *MayDay* lamp (plate 699), made by Flos, are absolutely characteristic of Grcic's approach: these are objects whose unique personalities quickly earned them a place among the supreme icons of design.

RIGHT
700. Anthony Dunne (b. 1964) and Fiona Raby (b. 1963)
Designs for an Overpopulated Planet: Foragers, 2009
Photograph conceived for the exhibition *Between Reality and the Impossible* at the Saint-Étienne Biennale Internationale de Design, 2010

OPPOSITE, TOP
701. Susana Soares (b. 1977)
Bee's: Fertility Object, 2007
Prototype produced by Vilabo (Portugal)
Project developed with Fiona Raby and Anthony Dunne for the Design Interaction program at the Royal College of Art, London
Borosilicate glass

THE INVENTION OF NEW CATEGORIES

The conspicuous overproduction of objects—especially chairs—has led a new generation of designers to turn away and think of something else. While for decades the first challenge assigned to any designer was to create a chair, today's students react with "Another chair—what's the point?" Now design itself is being redefined. The fields of design are proliferating, echoing the fragmentation of contemporary society and culture, in which each group is distinguished by its own codes and language. Far from the utopian objectivity of the rationalistic modern movement, products are launched in the name of some specific cultural value. They are often a response to the problems of our time: environmental challenges, overpopulation and food insecurity, the pursuit of well-being, ever-greater demands for mobility, and so on.

In France, Mathieu Lehanneur began his career by focusing on objects in the fields of medicine and wellness. After a brilliant diploma project that reimagined the packaging of pharmaceuticals, he made his public debut with *Andrea*, a household air purification device (plate 702). With *Bee's*, Portuguese designer Susana Soares explores how we might coexist with natural systems and use them to increase our own perceptual abilities. This 2007 project, developed with designers Fiona Raby and Anthony Dunne for the Design Interactions program at the Royal College of Art in London, grew out of research into the extraordinary olfactory capacity of bees. A set of alternative diagnostic tools was designed that would use bees for medical monitoring. Each object, made of glass, is composed of two compartments: a small one, which functions as the diagnostic zone, and a larger one, in which the bees are kept during the short time they need in order to react. The patient blows into the smaller compartment, and the bees fly into it only if they detect the odor they have been trained to recognize in the patient's breath. The *Face* and *Precise* objects, for example, are intended to accurately diagnose a wide variety of diseases at an early stage, including lung cancer, tuberculosis, and diabetes. The *Fertility Object* can diagnose pregnancy by recognizing specific molecular markers (plate 701).

Dunne and Raby have also developed a project on alternative methods of nutrition that could help to address the food security problems of an exponentially increasing global population. The objects they designed—such as artificial stomachs that would permit us to consume foods we cannot digest naturally—belong to larger scenarios intended to help audiences imagine possible biofutures (plate 700). The goal was to stimulate democratic debate on a more human use of new technologies. In New Zealand, designer Grant Ryan has developed a folding electric bike, the *YikeBike* (2010), to improve mobility in urban settings (plate 703). People can carry this compact, lightweight object onto a bus, then mount it after getting off in order to travel the last few hundred yards. We could cite many more examples, each more inventive than the next, of a liberated approach to design that draws on new technologies to reveal or generate our current needs.

ABOVE
702. Mathieu Lehanneur
(b. 1974)
***Andrea* air purifier**, 2009
Made by Le Laboratoire
(France), in collaboration
with David Edwards, Harvard
University
Electric fan, polycarbonate,
plant

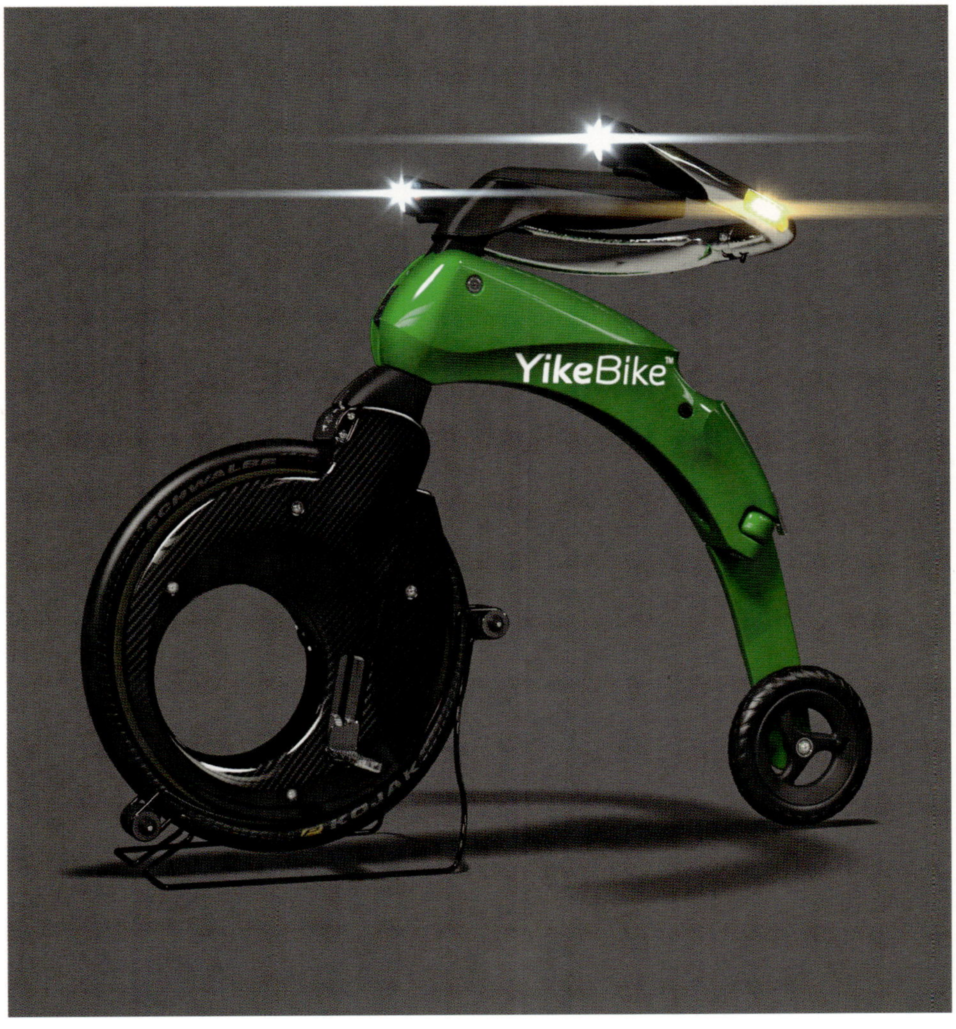

RIGHT
703. Grant Ryan
***YikeBike* folding electric bicycle**, 2010
Made by YikeBike
(New Zealand)
Carbon, electrical components

704–6. Marc Sarrazin (b. 1982)
***Falling Asleep* alarm clock,**
2010
Prototype

THE INTERACTIVE SOCIETY

As society became saturated with consumer goods, a market for intangible services and experiences developed. The desire to acquire a particular object was supplanted by the desire to have a particular experience; in the words of philosopher André Gorz in *L'Immatériel* (2003), we are seeing "the transformation of material capitalism into a capitalism of the immaterial." In the 1990s, traffic on the information superhighway consisted mostly of text and images; in the 2010s, the Internet is getting ready to connect billions of objects beyond computers and smartphones.

At home, automation is emerging from its first phase—remote-controlled devices—and advancing toward the smart house, which uses automatically recorded data to anticipate actions that need to be taken. For example, homeowners in France can now use the networked *Linky* electricity meter (made by ERDF), which frees them from having to stay home when the meter is read and the bill calculated. Similarly, Samsung's *RF 4289* refrigerator is equipped with an LCD touch screen that connects it to the Internet—allowing the user to find recipes, listen to streaming music, or manage e-mail. These interactive objects often go beyond current needs. The most interesting are those that rethink existing objects in order to provide a form of interactivity that is simple and easily accessible but offers emotionally satisfying content. The *Falling Asleep* alarm clock designed by Marc Sarrazin induces the user to let go, to surrender to a dream world of words, and so be sent to sleep (plates 704–6): in 2012, it won the Audience Award at the Design Parade festival in Hyères. The physical form of products like these is merely the vector or packaging of the immaterial content that design makes visible.

In 1999, Ruth Kedar, a Brazilian living in the U.S. who specializes in Web and user experience design, created the Google home page, one of the best design innovations of the last fifteen years. This clean, empty space—its only image the Google logo, a simple combination of letters and colors—aptly embodies the promise of the unimaginably vast content available through an invisible search engine that handles billions of queries per day.

707. Hella Jongerius (b. 1963)
View of *Jongeriuslab*, the designer's studio in Rotterdam, 2007

THE HOMEMADE

New technologies are an integral part of the production process. Many young designers who had hoped to work in collaboration with industry are now rethinking the production methods appropriate to their work. In a quest for greater freedom, they are moving toward self-production—emulating the Dutch, who revolutionized design in the 1990s and 2000s by restoring an artistic dimension to the studio. The Jongeriuslab, the Rotterdam studio of designer Hella Jongerius, is home to a team that assists her not only in the design but also in the manufacture of her objects (plate 707). In the studio, one designer paints embroidery patterns on the china of the *B-Set Repeat Series* (2002), and another uses colored Scotch tape to join the glass and china sections of the *Long Neck and Groove Bottles* (2000), while a third irons the linen *Embroidered Tablecloth* (2000). *Designers' Universe*, the 2007 film directed by Marie Farman and Benjamin Loyauté, shows us Maarten Baas's team at work on the various stages of the production of his furniture. The same film also lets us into the apartment of Gitta Gschwendtner, cluttered with all her objects stored there between exhibitions.

Designers today are working actively to invent their own tools in their personal studios. They are using new technologies to develop manufacturing systems that open up the possibilities of diversified production. Some designers are attracted by the idea of a new technological world in which they can invent a comprehensive system of design, production, and distribution. Reed Kram and Clemens Weisshaar's *Breeding Tables* project (2005) is perhaps a case in point. The idea was to create a process rather than a product. Observing that computer-aided design had existed for more than twenty years, but that the various programs used for design, communication, production, and inventory management have been developed in isolation, they decided to maximize the utility of digital tools by inventing a comprehensive program that would include all stages of design and production, with the goal of producing a series of different tables. "What we propose here," Weisshaar explained, "is to apply existing infrastructures in such a way that the company can go beyond the concept of mass production." He added that the potential for a new level of custom production, combined with the efficiency of the assembly line, exists locally, but is underused: "All we need to do is get it going." In the long term, Kram suggested, this will be the only way the West can hope to compete with the productive capacity of China or India.

Many designers are developing a unique connection between artisanal and industrial production. Sick and tired of being turned down by industry, which is only profitable at a mass-production level, the English designer Max Lamb bypasses its requirements. He came up with the idea of casting his *Pewter Stool* (2006) on a beach in Cornwall. Instead of ordering an impossibly expensive mold, he dug a hole in the sand and poured the liquid metal into it. The result is a seat with a unique appearance, combining the sheen of the metal with the irregularities of a surface imprinted by the coarse sand. In 2011, in his London studio, he created the *Woodware* series in collaboration with Gallery Fumi. This furniture, which references traditional English craftsmanship and has a beautiful physical presence, is constructed entirely from standard wooden dowels, in order to be easily reproducible if demand should grow to exceed Lamb's own production capacity.

No more sitting in front of the computer, no more design studio that looks like the office of an investment firm: the body comes back into play, and physical movement takes on a role in the creative process. This is how Sandeep Sangaru, an Indian designer based in Bangalore, describes the centrality of craftsmanship in his projects. In the Sangaru Design Studio, craftsmanship is at the foundation of every project and necessarily has

BELOW
708. Thomas Maincent
Spider Farm, 2011

OPPOSITE
709. Anna Loskiewicz (b. 1980)
Cork Beehive, 2012
In collaboration with Corticeira Amorim (Portugal)

its place alongside technology and the most innovative materials. Manual and digital artisans now play their part in the same studios, in the service of designers searching for ways to manufacture their own designs. The industrial revolution was made by engineers. Today, it is the designers who are building machines, by making ever better use of new technologies such as 3-D printers, initiating a new revolution that is emerging from "fab labs" worldwide. These young designers are capable of conceiving and producing their own objects and materials, creating an alternative to traditional production systems and moving toward a sustainable economy. In one instance, the Belgian designer Thomas Maincent has used his research on new biomaterials to design a *Spider Farm* that produces an ultra-resistant fabric (plate 708).

We are seeing the birth of a new kind of reciprocity between craftsmanship and industry, encompassing both global and local perspectives. Business is following in the footsteps of individual initiatives. The Portuguese company Amorim, primarily a manufacturer of bottle corks, turned to design to find other ways of using its raw material. In 2001, it launched its Materia line, for which a dozen or so international designers created objects whose forms took advantage of the physical and mechanical characteristics of cork. Then, in the summer of 2012, Amorim collaborated with Alexander von Vegesack, the head of the Domaine du Boisbuchet design center, in organizing an experimental workshop led by the Portuguese agency Pedrita. Fifteen students worked away in complete freedom to imagine new uses and possibilities for cork. The winning project was a model of a modern beehive by Polish designer Anna Loskiewicz (plate 709). Drawing on one of the most ancient uses of the material—the cork beehive being, in fact, a Portuguese tradition—and in complete accord with nature, the Loskiewicz's design is part of the trend toward a sustainable urban landscape in which beekeeping is increasingly popular. It exemplifies the reconciliation of the urban and the natural, promoted by an industrial business that integrates the values of craftsmanship.

Another company that stands out in this way is the Italian furniture maker Mattiazzi. This small family business has developed a hybrid work process that combines the traditional manual skills of the cabinetmaker—precise knowledge of the materials, how they feel, and how they make us feel—with the sophisticated digital tools they were wise enough to invest in. Mattiazzi has developed a commercially viable system by mechanizing where necessary and relying on the human hand to continue to craft high-quality items. Some of the most famous designers on the international scene—Konstantin Grcic, Nitzan Cohen, Sam Hecht, and Ronan and Erwan Bouroullec—have worked with Mattiazzi, a sign that their hybrid model points to future trends. Today it coexists with the more conventional mass production carried out by companies like Apple. On very different scales, both models rely on design to generate objects that will seem close to handmade.

DESIGN AS A MASS OCCUPATION?

The number of designers—and of schools, universities, fairs, exhibitions, and museums of design—is expanding rapidly, including in developing countries. Is design about to become a new engine of worldwide economic growth? Its purpose has indeed been transformed: design is no longer intended merely to find a correct and hence permanent solution to some problem, but rather to invent new products, new markets, and new economies. The goal is no longer to produce finished objects, but to imagine strategies of innovation with which to compete in the marketplace. The great challenge is novelty.

Design offers new possibilities, new services that enable industries to address the globalization of markets and develop new business models such as online production and publishing. The Kickstarter platform now makes it possible to collectively fund projects from design to technology to film.

While industrial progress has been the primary factor of emancipation since the Enlightenment, its value is now being challenged, and it is gradually being supplanted by artistic creation. Designers have taken over—especially as they now possess a unique tool, the Internet, which enables them to gain visibility through a system of personal communication that begins, at virtually no cost, on their own Web sites.

Deyan Sudjic, the director of the Design Museum in London, writes in *The Language of Things* that the overuse of the term "designer" has "rubbed it clean of earlier meanings, or left it as a synonym for the cynical and the manipulative." But if we are to remain calm in the face of change, we should stop thinking that every aspect of design was defined once and for all in the twentieth century. Since then the world has mutated into a space full of uncertainties. There is a new appreciation of the values of sharing, collective commitment, and sociability—which were given a place of honor, for example, in British designer Jerszy Seymour's First Supper, organized in 2008 at the Museum für Angewandte Kunst in Vienna (plates 710, 711).

What is work today? How do we organize our lives, our leisure time, our food? How can we be responsible to the world as a whole and promote sustainable

development? These are the questions that interest young designers now. In our postindustrial society, fragmented into numerous groups, each with its own vernacular, products are successful because of their cultural value. The signs produced are no longer universal, as they were in the day of Le Corbusier and European modernism; rather, they are visually less powerful, more diffuse, less recognizable, yet more widely shared through the Internet—a tool that enables communication without the need for robust and recognizable codes.

OPPOSITE AND ABOVE
710–11. Jerszy Seymour
(b. 1968)
***First Supper* installation**, 2008
Museum für Angewandte
Kunst, Vienna

NOTES

SELECTED BIBLIOGRAPHY

INDEX OF NAMES

PAGE 575
712. Frank O. Gehry (b. 1929)
Wiggle Side Chair, 1972
Made by Jack Brogan
See plate 111.

NOTES

Old Continent and New World: The Emergence of Design

1. Adolf Loos, *Ornement et Crime*, Petite Bibliothèque (Paris: Rivages, 2003), p. 244.
2. Walter Gropius, *Architecture et société*, preface by Lionel Richard (Paris: Éditions du Linteau, 1995).

Germany and Switzerland

1. Claude Lichtenstein, *Playfully Rigid: Swiss Architecture, Graphic Design, Product Design 1950–2006* (Zurich, 2007), pp. 7–12.
2. Artur Ruegg, ed., *Swiss Furniture and Interiors in the 20th Century* (Basel, 2002), pp. 166–67, and Lichtenstein, *Playfully Rigid*, p. 48.
3. Gwendolyn Ristant, in *Deutsches Design 1950–1990: Designed in Germany*, ed. Michael Erlhoff (Munich, 1990), pp. 101–3.
4. Quoted in Annemarie Jaeggi, ed., *Egon Eiermann (1904–1970), Architect and Designer: The Continuity of Modernism* (Ostfildern-Ruit, 2004), p. 93. See also Christopher Oestereich, *'Gute Form' im Wiederaufbau: Geschichte der Produktgestaltung in Westdeutschland nach 1945* (Berlin, 2000).
5. Gerd Fleischmann, Hans Rudolf Bosshard, and Christoph Bignens, *Max Bill: Typographie, Reklame, Buchgestaltung (Typography, Advertising, Book Design)* (Sulgen, 1998).
6. Sophie Lovell, Klaus Kemp, and Jonathan Ive, eds., *Dieter Rams: As Little Design as Possible* (London, 2011).
7. Beate Mansk, ed., *Wie Wohnen: Von Lust und Qual der richtigen Wahl, Aesthetische Bildung in der Alltagskultur des 20. Jahrhunderts* (Ostfildern-Ruit, 2005).
8. Renate Ulmer, "Anspruchsvolles Wohndesign," in *Im Designerpark/Leben in künstlichen Welt*, exh. cat., ed. Kai Bucholz and Klaus Wolbert (Darmstadt, 2004), pp. 644–65.
9. Greg Castillo, "The Bauhaus in Cold War Germany," in *Bauhaus Culture: From Weimar to the Cold War*, ed. Kathleen James-Chaktraborty (Minneapolis, 2006).
10. Tibor Kalman, *Chairman: Rolf Fehlbaum* (Zurich, 1997).
11. Hans Wichmann, *Die Realisation eines neuen Museumtyps: Die Neue Sammlung: Bilanz 1980/90*, with essays by Florian Hufnagl and Corinna Rösner (Basel, 1990); Florian Hufnagl, ed., *A Century of Design: Insights/Outlook on a Museum of Tomorrow* (Stuttgart, 1996). For one approach taken by the Museum für Gestaltung, Zürich, see Christian Brändle, ed., *Every Thing Design: The Collections of the Museum für Gestaltung Zürich*, exh. cat. (Ostfildern-Ruit, 2009).
12. Michael J. Hogan, *The Marshall Plan: America, Britain, and the Reconstruction of Western Europe, 1947–1952* (Cambridge, 1987); Volker Berghahn, *The Americanisation of West German Industry* (Leamington Spa, 1986).
13. *Bundespreis Produktdesign 1996: Katalog der Prämierten Produkte, Rat für Formgebung*, edited by Hans Höger (Frankfurt am Main, 1996). See also Jeremy Aynsley, *Designing Modern Germany* (London, 2009), pp. 192–95.
14. David Childs, "The Economy: A Kind of Miracle?" in *The GDR: Moscow's German Ally* (London, 1988), pp. 140–64.
15. Heinz Hirdina, *Gestalten für die Serie in der DDR, 1949–1985* (Dresden, 1988); Deutscher Werkbund, Regine Halter, ed., *Vom Bauhaus bis Bitterfeld: 41 Jahre DDR-Design* (Giessen, 1991); Eli Rubin, *Synthetic Socialism: Plastics and Dictatorship in the German Democratic Republic* (Chapel Hill, NC, 2008).
16. Hirdina, *Gestalten für die Serie in der DDR*, p.78; Siegfried Gronert and Elke Beilfuss, eds., *Horst Michel: Formgestalter in Weimar*, exh. cat. (Weimar, 2004).
17. Joachim Petsch, *Geschichte des Auto-Design* (Cologne, 1982), p. 225.
18. Ulrich Kubisch, *Aller Welts Wagen, die Geschichte eines automobilen Wirtschaftswunders* (Berlin, 1986).
19. Peter Schirmbeck, *Fabrikstadt Opel: 130 Jahre Industriearchitektur von Weltrang* (Tübingen, 2001); http://en.wikipedia.org/wiki/Opel_Kadett (accessed November 11, 2012).
20. Ulrich von Mende, *The VW Golf* (Frankfurt, 1999).
21. *100 Jahre Schweizer Grafik, Museum für Gestaltung Zürich*, exhibition (February 10, 2012–June 3, 2012); http://www.museum-gestaltung.ch/de/ausstellungen/ausstellungen-2012/100-jahre-schweizer-grafik/ (accessed November 11, 2012); Christian Brändle, Karin Gimmi, Barbara Junod, and Bettine Richter, eds., *Hundert Jahre Schweitzer Grafik* (Zurich, 2013).
22. Jens Müller and Karen Welland, eds., *Lufthansa and Graphic Design: Visual History of an Airline* (Zurich, 2011).
23. Ruegg, *Swiss Furniture and Interiors*, pp. 174–75, and Lichtenstein, *Playfully Rigid*, pp. 40–47.
24. http://edenspiekermann.com/projects/deutsche-bahn (accessed November 11, 2012).
25. Ruegg, *Swiss Furniture and Interiors*, pp. 192–95.
26. Ruegg, *Swiss Furniture and Interiors*, pp. 172–73.
27. Paul Sigel, *Exponiert, Deutsche Pavillons auf Weltausstellungen* (Berlin, 2000); Annemarie Jaeggi, *Egon Eiermann*.
28. Winfried Nerdinger, *Frei Otto: Das Gestamtkunstwerk; Leicht bauen, natürlich gestalten* (Munich, 2005).
29. Wilhelm Siemen, *Suomi: Die Rosenthal Studio-Linie; Form & Varianten* (Hohenberg an der Eger, 2000).
30. Siemen, *Suomi*; Edith Behnke, *Limitierte Kunstobjekte in Porzellan: 1968–2003* (Bonn, 2011).
31. Lotte Schilder Bär and Norbert Wild, *Designland Schweiz: Gebrauchsgüterkultur im 20. Jahrhundert* (Zurich, 2001), pp. 148–64; http://www.swatch.com/zz_en/about/history.html (accessed November 11, 2012).
32. Tony Lewin, *Smart Thinking: The Little Car That Made It Big* (Minneapolis, 2004); http://www.moma.org/collection/object.php?object_id=85995 (accessed November 11, 2012).
33. Michael Webb, *Richard Sapper* (San Francisco, 2002); Ute Brandes, *Richard Sapper: Tools for Life* (Göttingen, 1993); Hans Höger, *Tizio Light by Richard Sapper*, Design Classics (Hamburg, 1997).
34. Helmut Bauer, ed., *Ingo Maurer: Making Light*, exh. cat. (Munich, 1992); Alexander von Wegesack, ed., *Ingo Maurer: Light; Reaching for the Moon*, exh. cat. (Weil am Rhein, Germany, 2004).
35. Florian Böhm, ed., *Konstantin Grcic Industrial Design* (London, 2006).
36. Grcic quoted in Alice Rawsthorn, Style section, *New York Times*, July 12, 2007.
37. *Authentics Artipresent GmbH* (Gütersloh, 1999); http://www.authentics.de/ (accessed November 11, 2012).
38. Renate Menzi, *Freitag: Out of the Bag*, exh. cat. (Zurich, 2012).

Italy

1. A. Pansera and M. Chirico, eds., *1923–1930: Monza verso l'unità delle arti; Oggetti d'eccezione dalle Esposizioni internazionali di arti decorative* (Milan: Silvana Editoriale, 2004).
2. He served as general editor of the six-volume *Enciclopedia delle moderne arti decorative* (Milan: Ceschina, 1925–28).
3. See below, pp. 266, 329.
4. See below, p. 336.
5. A. Pansera, *Flessibili splendori: I mobili in tubolare metallico; Il Caso Columbus* (Milan: Electa, 1998).
6. Gillo Dorfles, "Il disegno industriale in Italia," *Civiltà delle macchine* (January–February 1957).
7. For more information on Ponti, see F. Irace, *La casa all'italiana* (Milan: Electa, 1988); U. La Pietra, *Gio Ponti: L'arte si innamora dell'industria* (Rome: Coliseum, 1988; Milan: Rizzoli, 2009); L. Licitra Ponti, *Gio Ponti: L'opera* (Milan: Leonardo, 1990); Graziella Roccella, *Gio Ponti: Master of Lightness* (Cologne: Taschen, 2009).
8. See below, p. 328.
9. A. Saggio, *L'opera di Giuseppe Pagano tra politica e architettura* (Bari: Dedalo, 1984); A. Bassi and L. Castagno, *Giuseppe Pagano* (Bari: Laterza, 1994).
10. A. Bassi, "L'archivio storico Breda e la storia del design aereonautico," in P. Ferrari, *L'aeronautica italiana: Una storia del Novecento* (Milan: Franco Angeli, 2004).
11. Pallavicino would also design the first *Lambretta* scooter.
12. A. T. Anselmi, *Carrozzeria italiana: Cultura e progetto* (Milan: Alfieri, 1978).
13. L. Dova, *Dante Giacosa, l'ingegno e il mito: Idee, progetti e vetture targate FIAT* (Boves: Araba Fenice, 2008).
14. Several articles and books (particularly exhibition catalogs) have been published on Albini and his studio. The most recent is

G. Bosoni and F. Bucci, *Il design e gli interni di Franco Albini* (Milan: Electa, 2009). A foundation was created in his name in 2008.
15. On the subject of women designers in Italy, see A. Pansera and T. Occleppo, *Dal merletto alla motocicletta: Artigiane/artiste e designer nell'Italia del Novecento* (Milan: Silvana Editoriale, 2002); A. Pansera, ed., *D come design: La mano, la mente, il cuore* (Biella: Eventi & Progetti Editore, 2008); A. Pansera, *Nientedimeno: Nothing less; La forza del design feminile* (Turin: Allemandi, 2011). The association DcomeDesign was founded in 2008 to highlight women's creativity (www.dcomedesign.org).
16. F. Dal Co and G. Mazzariol, *Carlo Scarpa 1906–1978* (Milan: Electa, 1984); S. Los, *Scarpa* (Cologne: Taschen, 2009).
17. Most works on Carlo Mollino appeared after his death, including *Carlo Mollino: Arabeschi*, published in Turin in 2006 for a major exhibition at the Galleria Civica d'Arte Moderna e Contemporanea. The largest retrospective dedicated to him was held at the Centre Pompidou (1989).
18. M. Nizzoli, *Marcello Nizzoli*, intro. by A. C. Quintavalle (Milan: Electa, 1990).
19. Several works address the different facets of Munari's career. The most recent is F. Montrasio, B. Finessi, and M. Meneguzzo, eds., *Bruno Munari* (Milan: Silvana Editoriale, 2007).
20. See below, p. 275.
21. Beginning in 1952, Munari was the artistic director of Pigomma (a division of the Pirelli Group), in charge of producing foam rubber toys with a wire core.
22. Sold in 1992 to the French company Strafor-Facom, Danese was purchased by Carlotta De Bevilacqua in 1999 and is now part of the Artemide Group. In 1998, Bruno Danese and Jacqueline Vodoz founded an organization bearing their names in Paris to promote young designers.
23. The Museo Scooter & Lambretta opened in Rodano (Milan) in 2001; nearly eighty Italian or foreign scooters and many archival documents are held there.
24. That company, founded by Giovan Battista Pirelli in 1873 and world-famous for its tires, has contributed a great deal to industrial design, producing both finished and semifinished products in elastic rubber: sporting goods (from soccer balls to tennis balls), nautical gear (inflatable boats, flippers, masks, wet suits), clothing (raincoats), industrial supplies (cable), and items for the bathroom, to name a few. In 1898, Pirelli purchased the Società del Linoleum, whose products were used by the rationalist architects, especially in the 1930s, to infuse a little color into homes.
25. See above, p. 270.
26. C. Avogadro, ed., *Cini Boeri, architetto e designer* (Milan: Silvana Editoriale, 2004).
27. F. Burkhardt, *Marco Zanuso. Design* (Milan: Federico Motta, 1994).
28. H. W. Holzwarth, *Kartell* (Cologne: Taschen, 2011). The collections of the Museo Kartell, which opened its doors at the company's headquarters in Noviglio (near Milan) in 1999, include products, prototypes, drawings, photographs, and printed matter.
29. C. Morozzi, *Anna Castelli Ferrieri* (Bari: Laterza, 1993).
30. Arteluce would be sold to Flos in 1973.
31. *Gino Sarfatti: Il design della luce*, Triennale Design Museum, September 2012; M. Romanelli and S. Severi, *Gino Sarfatti: Opere scelte 1938–1973* (Milan: Silvana Editoriale, 2012).
32. Sarfatti received a Compasso d'Oro in 1954 for the *Mod. 559* table lamp, and another in 1955 for the *Mod. 1055* adjustable lamp.
33. A public enterprise in charge of managing and selling off (to finance reconstruction) the colossal inventories of military vehicles and war materiel confiscated from the enemy or abandoned on Italian soil by Allied troops.
34. In particular, the company would be awarded two Compasso d'Oro awards (one in 1962 for the *Doney 14* television set by Zanuso and Sapper, and the other in 1970 for its activities as a whole).
35. The theme of the thirteenth Triennale, in 1964, was "Free Time"; in this spirit, the architect-designer Gae Aulenti created the installation *Arrival at the Sea* for the Italian pavilion.
36. E. Frateili, "Insofferenza e dinamica del design italiano 1960–1990," in *Mobili italiani 1961–1991: Le Varie Età dei linguaggi* (Milan: Cosmit, 1991).
37. For a more detailed chronology of the Compasso d'Oro prize, see the catalogs for the various years.
38. Now the Museo Nazionale della Scienza e della Tecnologia Leonardo da Vinci.
39. A. Pansera and A. Grassi, *L'Italia del design* (Casale Monferrato: Marietti, 1986); R. De Fusco, *Una storia dell'ADI* (Milan: Franco Angeli, 2010).
40. See below, p. 337.
41. Magistretti's offices on via Conservatorio in Milan (which he inherited from his father, Pier Giulio Magistretti) have been turned into the Fondazione Vico Magistretti, with an archival collection accessible to researchers. See F. Irace and V. Pasca, *Vico Magistretti: Architetto e designer* (Milan: Electa, 1999); B. Finessi, *Vico Magistretti* (Mantua: Corraini, 2003).
42. See below, p. 310.
43. From 1976 to 1991, *Abitare* was edited by Franca Santi Gualtieri, who produced many monographic issues, particularly for the Milan furniture fair. He was succeeded by Italo Lupi, Stefano Boeri (in 2007), and Matteo Piazza (since 2011).
44. *Ottagono*, edited until 1988 by Sergio Mazza in collaboration with Giuliana Gramigna, is now published in Bologna by Compositori, under the editorial direction of A. Colonetti and A. Lolli.
45. Achille Castiglioni has been the subject of many publications. Among the most recent: G. Cavaglià, *Di Achille Castiglioni* (Mantua: Corraini, 2006); Y. Taki, *Achille Castiglioni: Design as a Quest for Freedom* (Tokyo: Axis, 2007) (in Japanese). The Studio Museo Achille Castiglioni, located in his former offices on the piazza Castello in Milan, serves as the headquarters for the Fondazione Achille Castiglioni.
46. S. Limonta, ed., *Jonathan De Pas, Donato D'Urbino, Paolo Lomazzi* (Corsico, Milan: Ricerche Design Editrice, 2011); V. Pasca, ed., *Il gioco e le regole* (Milan: Triennale Design Museum, 2012). The DDL archives are held at the Centro di Alti Studi sulle Arti Visive in Milan.
47. After many vicissitudes, Gufram was purchased by the Poltrona Frau Group and then sold to Sandra Vezza, who has begun to reissue some classic pieces, such as the *Bocca* divan designed by Studio 65 (a design agency founded in 1965 and active until 1972).
48. V. Vercelloni, *L'avventura del design: Gavina* (Milan: Jaca Book, 1987).
49. B. Radice, *Ettore Sottsass* (Milan: Electa, 1993).
50. I. Favata, *Joe Colombo Designer 1930–1971* (Milan: Idea Books, 1988); M. Kries, ed., *Joe Colombo: Inventare il futuro* (Milan: Skira, 2005).
51. A. Pansera, *Antonia Campi: Creatività, forma e funzione*, catalogue raisonné (Milan: Silvana Editoriale, 2008).
52. Two major exhibits focusing on Gaetano Pesce were held in Paris: the first, titled *Le futur est peut-être passé* (The Future May Be Past), at the Musée des Arts Décoratifs in 1975; the second, *Le temps des questions* (The Time of Questions), at the Centre Pompidou in 1996. In 2005, the Triennale di Milano devoted a retrospective to him, *Gaetano Pesce: Il rumore del tempo* (Gaetano Pesce: The Sound of Time); the catalog was copublished by Charta Editore.
53. R. Pedio, *Enzo Mari, designer* (Bari: Dedalo, 1980). Many catalogs have been devoted to Enzo Mari's work. Among his own publications is *Progetto e passione* (Turin: Bollati Boringhieri, 2001).
54. A. Barrese and A. Marangoni, *MID, Alle origini della multimedialità: Dall'arte programmata all'arte interattiva* (Milan: Silvana Editoriale, 2007).
55. M. Costanzo, *Leonardo Ricci e l'idea di spazio comunitario* (Macerata: Quodlibet, 2010).
56. M. Nocchi, *Leonardo Savioli: Allestire, arredare, abitare* (Florence: Alinea, 2008).
57. All the documentation on Archizoom is held at the Centro Studi e Archivio per la Comunicazione at the University of Parma, a study center founded by C. A. Quintavalle in 1980, which collects and preserves materials relating to the culture of design (including plans by Munari, Mari, Sambonet, Sottsass, Nizzoli, Rosselli, and others). On Archizoom, see also R. Gargiani, *Archizoom Associati, 1966–1974: Dall'onda pop alla superficie neutra* (Milan: Mondadori Electa, 2007).
58. At the time, Cassina's artistic director was Franco Binfarè, who was very much attuned to trends in contemporary design.
59. In addition to Lapo Binazzi, who wrote an artist's book giving an account of the group's experiments (*La ricostruzione radicale dell'universo*, published in Florence in 1997), Gruppo UFO included Riccardo Foresi, Titti Maschietto, Carlo Bachi, and Patrizia Cammeo.
60. Giorgio Birelli, Carlo Caldini, Fabrizio Fiumi, and Paolo Galli.
61. The Ziggurat group, founded in Florence in 1969 by Alberto Breschi, Gigi Gavino, and Roberto Pecchioli, took a special interest in speculative architecture.
62. R. Orsini, ed., *Ultrarazionale ultramobile di Dino Gavina* (Bologna: Compositori, 1998).
63. B. Radice, *Memphis* (Milan: Electa, 1984).
64. Alighiero Boetti, Pier Paolo Calzolari, Sandro Chia, Joseph Kosuth, Maurizio Mochetti, Mimmo Paladino, Michelangelo

Pistoletto, Susana Solano, Lawrence Weiner, and Franz West.
65. P. Scarzella, *Il bel metallo: Storia dei casalinghi Alessi* (Milan: Arcadia, 1985).
66. A. Alessi, *La fabbrica dei sogni: Il design italiano nella produzione Alessi* (Milan: Electa/Alessi, 1998; 2003); C. Alessi, ed., *Oggetti e Progetti: Alessi, storia e futuro di una fabbrica del design italiano* (Milan: Electa, 2010).
67. The coauthor, with B. Orlandini, of *Architettura radicale* (Segrate, Milan: Casabella, 1974).
68. R. Poletti, *Zanotta: Design per passione* (Milan: Electa, 2004).
69. See below, pp. 334–35.
70. F. Dessì, G. Canestrini, and M. Pagano, *Storia illustrata dell'auto italiana* (Milan: G. di Piero de Martini, 1961–62).
71. W. Oude Weernink, *La Lancia* (Milan: Giorgio Nada, 1994).
72. A. Prunet, *Pininfarina: Arte e industria, 1930/2000* (Milan: Giorgio Nada, 2000).
73. The consortium was composed of the leading Italian locomotive builders: Ansaldo, Breda, Fiat Ferroviaria, ABB Tecnomasio, and Firema Trasporti.
74. In the meantime, the second generation has joined Giugiaro: Fabrizio Giugiaro has been responsible for styling since 1996. In 2012, Volkswagen AG purchased 90.1 percent of the shares of Italdesign Giugiaro.
75. The important Centro di Ricerca sulle Strutture Ambientali in Verrucchio (Rimini) is named for Pio Manzù. He was one of the founding members of this organization, which holds seminars and colloquia every year on the "economic and scientific aspects of the relationship between man and his environment."
76. See above, p. 282.
77. See above, p. 282.
78. See above, p. 282.
79. See above, p. 282.
80. Meda and Rizzatto won the Compasso d'Oro in 2008 for the *Mix* lamp (Luceplan).
81. See above, p. 256.
82. M. Canfailla, A. Lee, E. Martera, and P. Perra, eds., *Architettura del Mare: La progettazione nella nautica da diporto in Italia* (Florence: Alinea, 1994).
83. It was recently purchased by the Azimut-Benetti Group, active since 1969.
84. In 2001, a guided tour was set up to introduce visitors to the extraordinary architecture and urban planning of Ivrea.
85. T. Faravelli Giacobone, P. Guidi, and A. Pansera, *Dalla casa elettrica alla casa elettronica: Storia e significati degli elettrodomestici* (Milan: Arcadia, 1989).
86. See A. Pansera and A. Grassi, *Sportdesign, progettare la competizione*, exh. cat. (Ivrea, 2008; Biella: Eventi&Progetti, 2008–9).
87. Gilera is now owned by the Piaggio group.
88. See above, p. 312.
89. At the 1972 MoMA exhibition *Italy: The New Domestic Landscape*, Hosoe displayed a mobile home codesigned with Rosselli.
90. Starck also worked with the motorcycle maker Aprilia, designing the *Lama* (1993), an innovative prototype scooter, and the *Moto 6.5*, a city motorcycle that was more a cult object than a commercial success. (Slightly fewer than four thousand were produced between 1995 and 2000.)
91. E. Commessatti, *Cavalicco Friuli Mondo* (Milan: Il Sole 24 Ore, 2008).
92. C. T. Parmegiani, *Spirito Mobile: Piero Gervasoni, Gervasoni* (Udine, 2009).
93. It was this realization that inspired the creation of Museimpresa in 2001, with the objective of forming a network of leading manufacturers and offering, among other things, public tours of their museums and archives. A guidebook on this theme, *Turismo Industriale in Italia*, was published by the Touring Club Italiano in 2003 and updated in 2008.

France

1. Arlette Barré-Despond, *Jourdain* (Paris: Éditions du Regard, 1988), p. 328.
2. Charlotte Perriand, *Une vie de création* (Paris: Éditions Odile Jacob, 1998), p. 278.
3. Patrick Favardin, *Les Décorateurs des années 1950* (Paris: Editions Norma, 2002), p. 166.
4. During the 1950s, in France as in other countries, industrial design was promoted by the award of quality labels. The Beauté France designation, inaugurated in 1955 by the Institut d'Esthétique Industrielle, may not have achieved the prestige of the Italian Compasso d'Oro, but its intent was the same: to honor manufacturers and attract buyers.
5. In 1961, the Salon des Arts Ménagers moved to the larger, though less opulent, CNIT exhibition space in La Défense.
6. Marc Berthier, interview with Dominique Forest, October 7, 2008.
7. Catherine Millet, "Roger Tallon designer," in *Roger Tallon: Itinéraires d'un designer industriel*, exh. cat. (Paris: Centre Pompidou, October 20, 1993–January 10, 1994), p. 23.
8. Established in 1810, Peugeot made a variety of small metal goods over the course of the nineteenth century, including saw blades, gears, and its already legendary coffee mills. Following a family dispute in 1896, the still-embryonic automobile division split off from the rest of the firm. There were now two distinct companies: the toolmaker Peugeot Frères and the carmaker Société Anonyme des Automobiles Peugeot.
9. Roger Tallon, interview with Dominique Forest, December 18, 2008.
10. Before creating the design department at the École Nationale Supérieure des Arts Décoratifs (where he stayed until 1994), Tallon had taught at the École des Arts Appliqués, in a course on the "industrial aesthetic" organized by Jacques Viénot.
11. "Les sièges en liberté. Avez-vous essayé?" *Meubles et Décors* 870, February 1970, p. 52.
12. Roland Topor, "Le style lisse," *Le Monde*, December 9, 1979, supplement to no. 10843, p. I.
13. Franco Bertoni, *Philippe Starck: L'architecture* (Liège: Éditions Mardaga, 1994), p. 12.
14. *Starck Explications*, exh. cat. (Paris: Centre Pompidou, February 26–May 12, 2003), p. 119.
15. Jasper Morrison and Marc Berthier have designed appliances for Rowenta; Radi Designers, Jerszy Seymour, Sebastian Bergne, and George Sowden for Moulinex; Delo Lindo and Marc Newson for Tefal; and Konstantin Grcic for Krups.

SELECTED BIBLIOGRAPHY

General

Années pop, 1956–1968, Les. Exh. cat. Paris: Centre Pompidou, March 15–June 18, 2001.

Antonelli, P. *Objects of Design from the Museum of Modern Art*. New York: MoMA, 2003.

Assises du siège contemporain, Les. Exh. cat. Paris: Union Centrale des Arts Décoratifs, May 3–July 29, 1968.

Barré-Despond, A., ed. *Dictionnaire international des arts appliqués et du design*. Paris: Éditions du Regard, 1996.

Beauffet, J., ed. *Design: Collection du musée d'Art moderne de Saint-Étienne Métropole*. Saint-Étienne: Cité du Design Éditions, 2008.

Bons Génies de la vie domestique, Les. Exh. cat. Paris: Centre Pompidou, October 11, 2000–January 22, 2001.

Bony, A. *Les Années 40*. Paris: Éditions du Regard, 1985.

———. *Les Années 50*. Paris: Éditions du Regard, 1982.

———. *Les Années 60*. Paris: Éditions du Regard, 1983.

———. *Les Années 70*. Paris: Éditions du Regard, 1993.

Bradley, Q. *Design Futures*. London, New York: Merrell, 2011.

Branzi, A., and M. Pederbelli. *Qu'est-ce que le design?* Paris: Gründ, 2010.

Cent chefs-d'oeuvre de la collection du Vitra Design Museum. Exh. cat. Weil am Rhein: Vitra Design Museum, July 10, 1995–January 21, 1996.

Colin, C. *Design aujourd'hui*. Paris: Flammarion, 1988.

———. *Questions de design*. Paris: Flammarion, 2010.

D Day: Le design aujourd'hui. Exh. cat. Paris: Centre Pompidou, June 29–October 17, 2005.

Decelle, P. *L'Utopie du tout plastique, 1960–1973*. Paris: Éditions Norma, 1994.

Design, miroir du siècle. Exh. cat. Paris: Grand Palais, May 19–July 25, 1993.

Design Since 1945. Exh. cat. Philadelphia: Philadelphia Museum of Art, 1983.

Fiell, C. and P. *Design of the 20th Century*. Cologne: Taschen, 2000.

Guidot, R., ed. *Design, techniques et matériaux*. Paris: Flammarion, 2006.

———. *Histoire du design, 1940–1990*. Paris: Hazan, 1994.

Hebey, J.-B., with A. Ménard. *Esthétique domestique: Les arts ménagers 1920–1970*. Milan: 5 Continents, 2002.

Jousset, M.-L., and M. Moinot, M., eds. *La Collection de design du Centre Pompidou*. Paris: Éditions du Centre Pompidou, 2001.

Made 4 You/Design für den Wandel/Design for Change. Exh. cat. Vienna: MAK, June 6–October 7, 2012.

Mécanisation au pouvoir, La. Paris: Centre Pompidou/Centre de Création Industrielle, 1980.

Midal, A. *Design: Introduction à l'histoire d'une discipline*. Agora series. Paris: Pocket, 2010.

Miller, C. R., P. Sparke, and C. McDermott. *European Design since 1945: Shaping the New Century*. London: Merrell, 2009.

Moins et plus: 620 pièces de la collection design du Fonds national d'art contemporain 1980–2002. Exh. cat. Saint-Étienne: Musée d'Art Moderne de Saint-Étienne Métropole, November 16, 2002–February 10, 2003.

Munari, B. *L'Art du design*. Paris: Pyramid, 2012.

Noblet, J. de. *Design: Le geste et le compas*. Paris: Somogy, 1988.

Noblet, J. de, and C. Bressy. *Design: Introduction à l'histoire de l'évolution des formes industrielles de 1820 à aujourd'hui*. Paris: Stock-Chêne, 1974.

Pevsner, N. *The Sources of Modern Architecture and Design*. London: Thames and Hudson, 1968.

Phaidon Design Classics. 3 vols. London: Phaidon, 2007.

Quarante, D. *Éléments de design industriel*. 3rd ed. Paris: Économica, 2001.

Qu'est-ce que le design? Joe Colombo, Charles Eames, Fritz Eichler, Verner Panton, Roger Tallon. Exh. cat. Paris: Union Centrale des Arts Décoratifs/Centre de Création Industrielle, October 24–December 31, 1969.

Sparke, P. *An Introduction to Design and Culture*. London: Routledge, 2004.

Structures gonflables. Exh. cat. Paris: Musée d'Art Moderne de la Ville de Paris, March 1968.

Sudjic, D. *Le Langage des objets*. Paris: Pyramid, 2012.

The United States

Aldersey-Williams, H. *Cranbrook Design: The New Discourse*. New York: Rizzoli, 1991.

Bayley, S. *Coca-Cola 1886–1986: Designing a Megabrand*. London: The Boilerhouse, 1986.

———. *Harley Earl*. Design Heroes. London: Trefoil, 1990.

Beadle, T., and I. Penberthy. *The American Automobile*. New York: Salamander, 2000.

Behar, Y. *Yves Behar + Fuseproject: Concept/Commerce*. Germany: Mudac/Birkhauser, 2004.

Berry, J. *Herman Miller: The Purpose of Design*. New York: Rizzoli, 2004.

Brown, T., and J. Wyatt. "Design Thinking for Social Innovation." *Stanford Social Innovation Review*, Winter 2010.

Bruce, G. *Eliot Noyes* London: Phaidon, 2006.

Clarke, A. J. *Tupperware: The Promise of Plastic in 1950s America*. Washington, D.C.: Smithsonian, 2001.

Dreyfuss, H. *Designing for People*. New York: Simon and Schuster, 1955.

DuPont: The Autobiography of an American Enterprise. Wilmington, Delaware: E. I. Du Pont de Nemours & Company, 1952.

Eidelberg, M. *Eva Zeisel: Design for Industry*. Montreal: Musée des Arts Décoratifs, 1984.

Eizenbrand, J., and A. Von Vegesack. *George Nelson: Architect, Writer, Designer, Teacher*. Weil am Rhein: Vitra Design Museum, 2013.

Ewen, Stuart. *All Consuming Images: The Politics of Style in Contemporary Culture*. New York: Basic Books, 1988.

Fehlbahm, R., and D. Eames. *The Furniture of Charles and Ray Eames*. Weil am Rhein: Vitra Design Museum, 2007.

Flinchum, R. *Henry Dreyfuss, Industrial Designer: The Man in the Brown Suit*. New York: Rizzoli, 1997.

Ganzer, G. *Harry Bertoia: It All Started with a Chair*. Milan: Silvana Editoriale, 2009.

Genat, R. *American Cars of the '50s*. New York: Motorbooks International, 2007.

Handley, S. *Nylon: The Manmade Fashion Revolution*. London: Bloomsbury, 1999.

Hennessey, E. *Russel Wright*. New York: MIT Press, 1983.

Hine, T. *Populuxe: The Look and Life of America in the '50s and '60s, from Tailfins and TV Dinners to Barbie Dolls and Fallout Shelters*. New York: Alfred A. Knopf, 1986.

Hirshmann, K. *Jonathan Ive: Designer of the iPod*. New York: Greenhaven Press, 2007.

Ive, J. "The Apple Bites Back." *Design* 1 (Autumn 1998).

Kaufmann Jr., E. *Introductions to Modern Design*. New York: MoMA, 1950.

Kirkham, P. *Charles and Ray Eames: Designers of the Twentieth Century*. New York: MIT Press, 1998.

Long, C. *Paul T. Frankl and Modern American Design*. New Haven: Yale University Press, 2007.

Lutz, B. *Eero Saarinen: Furniture for Everyman*. New York: Pointed Leaf Press, 2012.

———. *Knoll: A Modernist Universe*. New York: Rizzoli, 2010.

Meikle, J. L. *American Design in the Twentieth Century*. London: Oxford University Press, 2005.

———. *Industrial Design in America, 1925–1939*. Philadelphia: Temple University Press, 2001.

Moggridge, B. *Designing Interactions*. New York: MIT Press, 2006.

Nelson, G., and H. N. Wright. *Tomorrow's House: How to Plan Your Post-war Home Now*. New York: Simon and Schuster, 1945.

O'Brian, T. *American Modern*. New York: Harry N. Abrams, 2010.

Packard, V. *The Hidden Persuaders*. New York: David McKay, 1957.

———. *The Status Seekers*. New York: David McKay, 1959.

———. *The Waste Makers*. New York: David McKay, 1960.

Papanek, V. *Design for the Real World: Human Ecology and Social Change*. New York: Pantheon, 1971.

"Power of Design, The." *Business Week*, May 17, 2004.

Pulos, A. J. *American Design Ethic*. New York: MIT Press, 1986.

Schultze, S., and I. Grätz, I., eds. *Apple Design*. Germany: Hatje Cantz, 2011.

Smith, M. S., and C. Pittel. *Michael S. Smith: House and Home; Creating a House*. New York: Rizzoli, 2013.

Teague, W. D. *Design This Day: The Technique of Order in the*

Machine Age. New York: Harcourt Brace, 1940.
Trétiack, P. *Raymond Loewy*. Paris: Editions Assouline, 2006.
Venturi, R. *Complexity and Contradiction in Architecture*. New York: MoMA, 1966 (1984).
Votolato, G. *American Design in the Twentieth Century*. Manchester, UK: Manchester University Press, 1998.

Scandinavia

Aav, M., and N. Stritzler-Levine, eds. *Finnish Modern Design: Utopian Ideals and Everyday Realities, 1930–1997*. New Haven: Bard Graduate Center for Studies in the Decorative Arts and Yale University Press, 1998.
Beer, E. H. *Scandinavian Design: Objects of a Life Style*. New York: American-Scandinavian Foundation, 1975.
Bengtsson, S. *Norway Calling: Touch-Down in a New World of Norwegian Design*. Stockholm: Arvinius Förlag, 2008.
Boman, M. *Estrid Ericson: Founder of Svenskt Tenn*. Stockholm: Carlson Bokförlag, 1989.
Bruno Mathsson, Architect and Designer. Exh. cat. Malmö: Bokförlaget Arena, 2006.
Cottard-Olsson, P. *Angles suédois*. Exh. cat. Paris: Centre Culturel Suédois, 2000.
Creagh, L., H. Kåberg, and B. Miller Lane, eds. *Modern Swedish Design: Three Founding Texts*. New York: MoMA, 2008.
Design démocratique: Un livre sur la forme, la fonction et le prix, les troisdimensions du design IKEA. Ikea France, 2001.
Fallan, K., ed. *Scandinavian Design: Alternative Histories*. London: Berg, 2012.
Fiell, C. and P. *Scandinavian Design*. Cologne: Taschen, 2002.
Gura, Judith. *Sourcebook of Scandinavian Furniture: Designs for the Twenty-First Century*. New York: W.W. Norton, 2007.
Halén, W., and K. Wickman, eds. *Scandinavian Design Beyond the Myth: Fifty Years of Design from the Nordic Countries*. Stockholm: Arvinius Förlag/Form Förlag, 2003.
Hans J. Wegner: Om Design. Copenhagen: Danish Design Centre, 1994.
Helgeson, S., and K. Nyberg. *Swedish Design: The Best in Swedish Design Today*. London: Octopus, 2002.
Hiort, E. *Finn Juhl: Furniture, Architecture, Applied Art*. Copenhagen: Danish Architectural Press, 1990.
Keinänen, T., P. Korvenmaa, K. Mikonranta, and A. Ólafsdóttir. *Alvar Aalto Designer*. Paris: Gallimard, 2003.
Klemp, K., and K. M. Wagner, eds. *On the Cutting Edge: Design in Iceland*. Berlin: Die Gestalten Verlag, 2011.
Lunning Prize, The. Stockholm: Nationalmuseum, 1986.
Møller, H. S. *Motion and Beauty: The Book of Nanna Ditzel*. Copenhagen: Rhodos, 1998.
Polster, B. *Dictionnaire du design scandinave*. Paris: Éditions du Seuil, 1999.
Poul Kjærholm. Copenhagen: Arkitektens Forlag, 1999.
Revere McFadden, D., ed. *Scandinavian Modern Design 1880–1980*. New York: Harry N. Abrams, 1982.
Sommar, I. *Scandinavian Style: Classic and Modern Scandinavian Style and Its Influence on the World*. London: Carlton, 2003.
Tapio Wirkkala. Exh. cat. Helsinki: Finnish Society of Crafts and Design, 1981.
Thau, C., and K. Vindum. *Arne Jacobsen*. Copenhagen: Danish Architectural Press, 2001.
Transforme: Design islandais; Nouvelle génération. Exh. cat. Reykjavik: Icelandic Design Council, 2004.
Vegesack, A. von, and M. Remmele, eds. *Verner Panton: Das Gesamtwerk*. Exh. cat. Weil am Rhein: Vitra Design Museum, 2000.

Germany and Switzerland

Aynsley, J. *Designing Modern Germany*. London: Reaktion, 2009.
Böhm, F., ed. *Konstantin Grcic Industrial Design*. London: Phaidon, 2006.
Brändle, C., ed. *Every Thing Design: The Collections of the Museum für Gestaltung Zürich*. Ostfildern-Ruit: Hatje Cantz, 2009.
Bucholz, K., and K. Wolbert. *Im Designerpark/Leben in künstlichen Welt*. Darmstadt: Institut Mathildenhöhe, 2004.
Droste, M. *Bauhaus 1919–1933*. Cologne: Taschen, 2002.
Erlhoff, M., ed. *Deutsches Design 1950–1990: Designed in Germany*. Munich: Prestel Verlag, 1990.
Fleischmann, G., H. R. Bosshard, and C. Bignens. *Max Bill: Typographie, Reklame, Buchgestaltung (Typography, Advertising, Book Design)*. Sulgen: Niggli, 1998.
Hirdina, H. *Gestalten für die Serie in der DDR, 1949–1985*. Dresden: VEB Verlag der Kunst, 1988.
Jaeggi, A., ed. *Egon Eiermann (1904–1970), Architect and Designer: The Continuity of Modernism*. Ostfildern-Ruit: Hatje Cantz, 2004.
Less and More: The Design Ethos of Dieter Rams. Exh. cat. Osaka: Suntory Museum, November 15, 2008–January 25, 2009; Tokyo: Fuchu Art Museum, May 23–July 20, 2009; London: Design Museum, November 18–March 7, 2010; Frankfurt am Main: Museum für Angewandte Kunst, May 22–September 5, 2010.
Lichtenstein, C. *Playfully Rigid: Swiss Architecture, Graphic Design, Product Design 1950–2006*. Zurich: Lars Müller, 2007.
Lovell, S., K. Kemp, and J. Ive, eds. *Dieter Rams: As Little Design as Possible*. London: Phaidon, 2011.
Mansk, B., ed. *Wie Wohnen: Von Lust und Qual der richtigen Wahl, Aesthetische Bildung in der Alltagskultur des 20. Jahrhunderts*. Ostfildern-Ruit: Hatje Cantz, 2005.
Mende, U. von. *The VW Golf*. Frankfurt: Verlag Form, 1999.
Menzi, R. *Freitag: Out of the Bag*. Zurich: Lars Müller, 2007.
Ruegg, A., ed. *Swiss Furniture and Interiors in the 20th Century*. Basel: Birkhäuser, 2002.
Schilder Bär, L., and N. Wild. *Designland Schweiz: Gebrauchsgüterkultur im 20. Jahrhundert*. Zurich: Pro Helvetia, 2001.
Webb, M. *Richard Sapper*. San Francisco: Chronicle Books, 2002.
Wegesack, A. von, ed. *Ingo Maurer: Light; Reaching for the Moon*. Weil am Rhein: Vitra Design Museum, 2004.

Italy

Alessi, A. *La fabbrica dei sogni: Il design italiano nella produzione Alessi*. Milan: Electa/Alessi, 1998 (2003).
Alessi, C., ed. *Oggetti e Progetti: Alessi; Storia e futuro di una fabbrica del design italiano*. Milan: Electa, 2010.
Anselmi, A. T. *Carrozzeria italiana: Cultura e progetto*. Milan: Alfieri Ed., 1978.
Antonelli, P., and S. Guarnaccia. *Achille Castiglioni*. Mantua: Corraini, 2008.
Avogadro, C., ed. *Cini Boeri, architetto e designer*. Milan: Silvana Editoriale, 2004.
Barrese, A., and A. Marangoni. *MID, Alle origini della multimedialità: Dall'arte programmata all'arte interattiva*. Milan: Silvana Editoriale, 2007.
Bassi, A. *Design anonimo in Italia: Oggetti comuni e progetto incognito*. Milan: Electa, 2007.
———. *La luce italiana: Il design della lampade 1945–2000*. Milan, Electa, 2003.
———. (with M. Mulazzani). *Le macchine volanti di Corradino D'Ascanio*. Milan: Electa, 1999.
Bassi, A., and L. Castagno. *Giuseppe Pagano*. Bari: Laterza, 1994.
Binazzi, L. *La ricostruzione radicale dell'universo*. Florence, 1997.
Bosoni, G., ed. *Made in Cassina*. Milan: Skira, 2008.
Bosoni, G., and F. Bucci. *Il design e gli interni di Franco Albini*. Milan: Electa, 2009.
Branzi, A. *Introduzione al design italiano: Una modernità incompleta*. Milan: Baldini & Castoldi Dalai, 2008.
———. *La Casa calda*. Paris: L'Équerre, 1985.
Burkhardt, F. *Marco Zanuso. Design*. Milan: Federico Motta, 1994.
Canfailla, M., A. Lee, E. Martera, and P. Perra, eds. *Architettura del Mare: La progettazione nella nautica da diporto in Italia*. Florence: Alinea, 1994.
Carlo Mollino: Arabeschi. Exh. cat. Turin: Galleria Civica d'Arte Moderna e Contemporanea, September 20, 2006–January 7, 2007; Milan: Electa, 2006.
Cavaglià, G. *Di Achille Castiglioni*. Mantua: Corraini, 2006.
Costanzo, M. *Leonardo Ricci e l'idea di spazio comunitario*. Macerata: Quodlibet, 2010.
Cotoranu, A. *Gaetano Pesce: Ideas and Innovation*. Paper, Simon Fraser University, 2008.
Dal Co, F. and G., *Mazzariol: Carlo Scarpa 1906–1978*. Milan: Electa, 1984.
Dessì, F., G. Canestrini, and M. Pagano. *Storia illustrata dell'auto italiana*. Milan: Giumar di Pietro De Martino, 1961–62.
Dorfles, G. "Il disegno industriale in Italia." *Civiltà delle macchine*, January–February 1957.
Dova, L. *Dante Giacosa, l'ingegno e il mito: Idee, progetti e vetture targate FIAT*. Boves: Araba Fenice, 2008.
Faravelli Giacobone, T., P. Guidi, and A. Pansera. *Dalla casa elettrica alla casa elettronica: Storia e significati degli elettrodomestici*. Milan: Arcadia, 1989.
Favata, I. *Joe Colombo Designer 1930–1971*. Milan: Idea Books, 1988.
Ferrari, P. *L'aeronautica italiana: Una storia del Novecento*. Milan: Franco Angeli, 2004.
Finessi, B. *Fabio Novembre*. Milan: Skira, 2008.
———. *Vico Magistretti*. Mantua: Corraini, 2003.

Gargiani, R. *Archizoom Associati, 1966–1974: Dall'onda pop alla superficie neutra*. Milan: Mondadori Electa, 2007.
Gramigna, G. *Repertorio 1950/1980: Immagini et contributi per una storia dell' arredo italiano*. Milan: Mondadori, 2001.
Grassi, A., and A. Pansera. *Atlante del disegno italiano: 1940–1980*. Milan: Fabbri, 1980.
Gregotti, V. *Il disegno del prodotto industriale: Italia 1860–1980*. Milan: Electa Mondadori, 1986.
Holzwarth, H. W. *Kartell*. Cologne: Taschen, 2011.
Irace, F. *La casa all'italiana*. Milan: Electa, 1988.
Irace, F., and V. Pasca. *Vico Magistretti: Architetto e designer*. Milan: Electa 1999.
Italy: The New Domestic Landscape. Exh. cat. New York: MoMA, 1972.
Joe Colombo, l'invention du futur. Exh. cat. Paris: Musée des Arts Décoratifs, March 28–August 19, 2007.
Kries, M., ed. *Joe Colombo: Inventare il futuro*. Milan: Skira, 2005.
La Donation Kartell. Exh. cat. Paris: Centre Pompidou, September 12, 2000–January 1, 2001.
La Pietra, U. *Giò Ponti: L'arte si innamora dell'industria*. Rome: Coliseum, 1988 (Milan: Rizzoli, 2009).
Licitra Ponti, L. *Gio Ponti: L'opera*. Milan: Leonardo Ed., 1990.
Limonta, S., ed. *Jonathan De Pas, Donato D'Urbino, Paolo Lomazzi*. Corsico (Milan): Ricerche Design Editrice, 2011.
Los, S. *Scarpa*. Cologne: Taschen, 2009.
Marangoni, G. *Enciclopedia delle moderne arti decorative*. Milan: Ceschina, 1925–28.
Marcello Nizzoli. Milan: Electa, 1990.
Mari, E. *Progetto e passione*. Turin: Bollati Boringhieri, 2001.
Mitomacchina: Il design dell'automobile; Storia, tecnologia e futuro. Milan: Skira, 2006.
Montrasio, F., B. Finessi, and M. Meneguzzo, eds. *Bruno Munari*. Milan: Silvana Editoriale, 2007.
Morozzi, C. *Anna Castelli Ferrieri*. Bari: Laterza, 1993.
Nizzoli, M. *Marcello Nizzoli*. Introduction by M. C. Quintavalle. Milan: Electa, 1990.
Nocchi, M. *Leonardo Savioli allestire arredare abitare*. Florence: Alinea, 2008.
Orsini, R., ed. *Ultrarazionale ultramobile di Dino Gavina*. Bologna: Compositori, 1998.
Oude Weernink, W. *La Lancia*. Milan: Giorgio Nada, 1994.
Pansera, A. *Antonia Campi: Creatività, forma e funzione*. Catalogue raisonné. Milan: Silvana Editoriale, 2008.
———. *Bracciodiferro: Gaetano Pece–Alessandro Mendini, 1971–1975*. Milan: Silvana Editoriale, 2013.
———, ed. *D come design: La mano, la mente, il cuore*. Biella: Eventi & Progetti Editore, 2008.
———. *Il design del mobile italiano dal 1946 a oggi*. Bari: Laterza, 1990.
———. *Flessibili Splendori: I mobili in tubolare metallico; Il Caso Columbus*. Milan: Electa, 1998.
———. *Nientedimeno: Nothingless; La forza del design feminile*. Turin: Allemandi Ed., 2011.
———. *Storia del disegno industriale italiano*. Bari: Laterza, 1993.
Pansera, A., and M. Chirico, eds. *1923–1930: Monza verso l'unità delle arti; Oggetti d'eccezione dalle Esposizioni internazionali di arti decorative*. Milan: Silvana Editoriale, 2004.
Pansera, A., and A. Grassi. *L'Italia del design*. Casale Monferrato: Marietti, 1986.
———. *Sportdesign, progettare la competizione*. Exh. cat. Ivrea, 2008; Biella: Eventi&Progetti, 2008–9.
Pansera, A., and T. Occleppo. *Dal merletto alla motocicletta Artigiane: Artiste e designer nell'Italia del Novecento*. Milan: Silvana Editoriale, 2002.
Parmegiani, C. T. *Spirito Mobile: Piero Gervasoni*. Udine: Gervasoni, 2009.
Pasca, V. *Gae Aulenti*. Milan: Triennale Design Museum, 2013.
———, ed. *Il gioco e le regole*. Milan: Triennale Design Museum, 2012.
Pedio, R. *Enzo Mari, designer*. Bari: Dedalo, 1980.
Poletti, R. *Zanotta: Design per passione*. Milan: Electa, 2004.
Prunet, A. *Pininfarina: Arte e industria, 1930/2000*. Milan: Giorgio Nada, 2000.
Radice, B. *Ettore Sottsass*. Milan: Electa, 1993.
———. *Memphis*. Milan: Electa, 1984.
Roccella, G. *Gio Ponti: Maestro della leggerezza*. Cologne: Taschen, 2009.
Romanelli, M., and S. Severi. *Gino Sarfatti: Opere scelte 1938–1973*. Milan: Silvana Editoriale, 2012.
Saggio, A. *L'opera di Giuseppe Pagano tra politica e architettura*. Bari: Dedalo, 1984.
Scarzella, P. *Il bel metallo: Storia dei casalinghi Alessi*. Milan: Arcadia, 1985.
Sparke, P. *Italian Design, from 1870 to the Present*. London: Thames and Hudson, 1988.
Taki, Y. *Achille Castiglioni: Design as a Quest for Freedom*. Tokyo: Axis, 2007.
Vanlaethem, F. *Gaetano Pesce Architetturadesignarte*. Milan: Idea Books, 1989.
Vercelloni, M. *Breve storia del design italiano*. Rome: Carocci, 2008.
Vercelloni, V. *L'avventura del design: Gavina*. Milan: Jaca Book, 1987.
Vitta, M. *Dell'abitare: Corpi spazi oggetti immagini*. Turin: Einaudi, 2008.
———. *Il disegno delle cose*. Naples: Liguori, 1996.
———. *Il progetto della bellezza*. Turin: Einaudi, 2001.

Great Britain

Aldersley-Williams, H. *British Design*. New York: MoMA, 2010.
Banham, M., and B. Hillier. *A Tonic for the Nation: The Festival of Britain 1951*. London: V&A, 1976.
Buckley, C. *Designing Modern Britain*. London: Reaktion, 2007.
Conran, T. *Terence Conran on Design*. London: Conran Octopus, 2001.
Conway, H. *Ernest Race*. London: Design Council, 1982.
Dixon, T. *The Interior World of Tom Dixon*. London: Conran Octopus, 2008.
Dyson, J. *James Dyson: Against the Odds; An Autobiography*. London: Orion, 1997.
Frayling, C., and C. Catterall, eds. *Design of the Times: One Hundred Years of the Royal College of Art*. London: Richard Dennis, 1996.
Hebdige, D. *Subculture: The Meaning of Style*. London: Routledge, 1979.
Hewison, R. *The Heritage Industry: Britain in a Climate of Decline*. London: Methuen, 1987.
Hicks, A. *David Hicks*. New York: Rizzoli, 2009.
Hughes, H. *John Fowler: The Inventor of the Country House Style*. Dorset: Donhead Publishing, 2005.
Huygen. F. *British Design: Image and Identity*. London: Thames and Hudson, 1989.
Jackson, L. *Robin and Lucienne Day: Pioneers of Contemporary Design*. London: Mitchell Beazley, 2001.
McCarthy, F. *British Design since 1880*. London: Allen and Unwin, 1982.
McDermott, C. *Street Style: British Design in the 1980s*. London: Design Council, 1987.
Myersen, J. *Gordon Russell: Designer of Furniture, 1892–1992*. London: Design Council, 1992.
Sparke, P. *Did Britain Make It? British Design in Context, 1946–1986*. London: Design Council, 1986.
Sudjic, D. *Ron Arad (Design)*. London: Laurence King, 1999.
———, et al. *Kenneth Grange: Making Britain Modern*. London: Black Dog, 2011.
Thackera, J. *New British Design*. London: Thames and Hudson, 1986.
Whiteley, N. *Pop Design: From Modernism to Mod*. London: Design Council, 1987.
Williams, R. *Culture and Society*. London: Chatto and Windus, 1958.

France

Alain Richard: Luminaires et mobilier 1950–1970. Exh. cat. Saint-Ouen: Galerie Pascal Cuisinier, October 9–November 16, 2009.
Amic, Y. *Le Mobilier français 1945–1964*. Intérieurs series. Paris: Via/Éditions du Regard, 1983.
Barsac, J. *Charlotte Perriand: Un art d'habiter*. Paris: Éditions Norma, 2005.
Bony, A. *Prisunic et le design, une aventure unique*. Paris: Éditions Alternatives, 2008.
Brunhammer, Y. *Sylvain Dubuisson, la face cachée de l'utile*. Paris: Éditions Norma, 2006.
Brunhammer, Y., and M.-L. Perrin. *Le Mobilier français, 1960–1998*. Paris: Éditions Massin, 1998.
Bure, G. de. *Le Mobilier français 1965–1979*. Intérieurs series. Paris: Éditions du Regard, 1983.
Calloway, S., F. Baudot, and G.-G. Lemaire. *Élisabeth Garouste et Mattia Bonetti*. Paris: Michel Aveline Éditeur, 1990.
Charlotte Perriand. Exh. cat. Paris: Centre Pompidou, December 7, 2005–March 27, 2006.
Charlotte Perriand, un art de vivre. Exh. cat. Paris: Union Centrale des Arts Décoratifs, February 6–April 1, 1985.
Colin, C. *Starck*. Brussels: Édition Mardaga, 1989.
Design français. Exh. cat. Paris: Union Centrale des Arts Décoratifs/Centre de Création Industrielle, October 22–December 21, 1971.
Design français 1960–1990: Trois décennies. Exh. cat. Paris: APCI/Centre Pompidou, June 22–September 26, 1988.
Favardin, P. *Les Décorateurs des années 50*. Paris: Éditions Norma, 2002.
———. *Steiner et l'aventure du design*. Paris: Éditions Norma, 2007.

Fayolle, C. *Le Design*. Paris: Scala, 1998.

Formes utiles: Les arts ménagers. Exh. cat. Saint-Étienne: Musée d'Art Moderne de Saint-Étienne Métropole, October 9, 2004–January 16, 2005.

Gareth, W. (preface). *Matali Crasset*. Paris: Pyramid, 2003.

Geel, C., ed. *Pierre Paulin designer*. Paris: Éditions Archibooks, 2008.

Germain, É. *Marc Held: Du design à l'architecture*. Paris: Éditions Norma, 2009.

Jean Prouvé constructeur. Exh. cat. Paris: Centre Pompidou/Centre de Création Industrielle, October 24, 1990–January 28, 1991.

Jollant Kneebone, F., C. Braunstein, and A. Atelier. *Rencontre de l'art et de l'objet*. Paris: Éditions Norma, 2003.

Jousse, P., ed. *Mathieu Matégot*. Paris: Jousse Entreprise Éditions, 2003.

Koivu, A. *Ronan and Erwan Bouroullec: Works*. London: Phaidon, 2012.

Lebovici, E. *Martin Szekely*. Zurich: JPR-Ringier, 2010.

Les Années UAM, 1929–1958. Exh. cat. Paris: Union Centrale des Arts Décoratifs, September 27, 1988–January 29, 1989.

Leymonerie, C. "Des formes à consommer. Pensées et pratiques du design industriel en France (1945–1980)." Diss., Paris, École des Hautes Études en Sciences Sociales, 2010.

Marc Held. Exh. cat. Nantes: Musée des Arts Décoratifs, Château des Ducs de Bretagne, November 10, 1973–January 7, 1974.

Mathey, F., ed. *Au bonheur des formes: Design français, 1945–1992*. Paris: Éditions du Regard, 1992.

Meubles TV: Éditeur d'avantgarde, 1952–1959. Exh. cat. Saint-Ouen: Galerie Pascal Cuisinier, April 17–May 31, 2010.

Mobi Boom: L'explosion du design en France, 1945–1975. Exh. cat. Paris: Musée des Arts Décoratifs, 2010–2011.

Olivier Mourgue. Exh. cat. Nantes: Musée des Arts Décoratifs, Château des Ducs de Bretagne, March 12–April 19, 1976.

Olivier Mourgue: Peintures, design, jardins imaginaires et petits théâtres. Exh. cat. Brest: Musée des Beaux-Arts, November 20, 2004–February 20, 2005.

Perriand, C. *Une vie de création*. Paris: Éditions Odile Jacob, 1998.

Pralus, P.-E. *Serge Mouille, un classique français*. Saint-Cyrau-Mont-d'Or: Les Éditions du Mont Thou, 2006.

Ragot, G., and M. Dion. *Le Corbusier en France: Projets et réalisations*. Paris: Le Moniteur, 1997.

Raymond Loewy (1893–1986), un pionnier du design industriel. Exh. cat. Paris: Centre Pompidou, June 27–September 24, 1990.

Rémy, C., ed. *Création en France: Arts décoratifs 1945–1965*. Paris: Éditions Gourcuff Gradenigo, 2009.

Renous, P. *Portraits de décorateurs: Entretiens publiés par la Revue de l'ameublement*. Paris: Éditions Vial, 1969.

Roger Tallon: Itinéraires d'un designer industriel. Exh. cat. Paris: Centre Pompidou, October 20, 1993–January 10, 1994.

Rouaud, J. *Soixante ans d'arts ménagers: 1948–1983, la consommation*. Vol. 2. Paris: Syros-Alternatives, 1989.

Starck Explications. Exh. cat. Paris: Centre Pompidou, February 26–May 12, 2003.

Sulzer, P. *Jean Prouvé: Oeuvres complètes/Complete Works*. Vol. 1, *1917–1933*. Tübingen: Wasmuth/Axel Menges, 1995.

———. *Jean Prouvé: Oeuvres complètes/Complete Works*. Vol. 2, *1934–1944*. Basel: Birkhäuser, 2000.

———. *Jean Prouvé: Oeuvres complètes/Complete Works*. Vol. 3, *1944–1954*. Basel: Birkhäuser, 2005.

Védrenne, É., and A.-M. Fèvre. *Pierre Paulin*. Design series. Paris: Éditions Dis Voir, 2001.

Japan

Andreini, L. *Arata Isozaki*. Milan: Motta Architettura, 2008.

Bayley, S. *Sony Design*. London: V&A, 1982.

Design japonais 1950–1995. Exh. cat. Paris: Centre Pompidou, February 14–April 29, 1996.

Dietz, M. *Japanese Design*. Cologne: Taschen, 1992.

Earle, J. *New Bamboo: Contemporary Japanese Masters*. New Haven: Yale University Press, 2008.

Ekuan, K. *The Aesthetics of the Japanese Lunchbox*. Tokyo: MIT, 1998.

Evans, S. *Contemporary Japanese Design*. Tokyo: Collins and Brown, 1991.

Freeman, M. *Space: Japanese Design Solutions*. New York: Rizzoli, 2004.

Frolet, E., N. Pearce, and S. Yanagi. *Mingei: The Living Tradition in Japanese Arts*. Tokyo: Japan Folk Crafts Museum, 1991.

Fukusawa, N. *Naoto Fukusawa*. London: Phaidon, 2007.

Galindo, M. *Japanese Interior Design*. New York: Braun, 2011.

Irvine, G., and J. Earle, ed. *Japanese Art and Design*. London: V&A, 2009.

Japon des avantgardes, 1910–1970. Exh. cat. Paris: Centre Pompidou, 1986.

Lim, S. *Japanese Style*. New York: Gibbs Smith, 2007.

Lonsdale, S. *Japanese Design*. London: Carlton, 2002.

Mitsuhashi, I. *Shigeru Uchida Interior Design*. Tokyo: Richuyo-Sha Publishing, 2004.

Morrison, J., N. Fukusawa, and K. Hara. *Muji*. New York: Rizzoli, 2010.

Nendo Works 2007–2010. Tokyo: Ad Publishing, 2011.

Slesin, S., and D. Rozensztroch. *Japanese Style*. New York: Thames and Hudson, 1998.

Sparke, P. *Japanese Design*. London: Michael Joseph, 1988.

———. *Japanese Design*. New York: MoMA, 2009.

Sudjic, D. *Shiro Kuramata*. London: Phaidon, 2013.

The Netherlands

Antonella, P., ed. *Wander Wonders: Design for a New Age*.'s-Hertogenbosch: Museum of Contemporary Art, 1999.

Bakker, G., and E. Rodrigo. *Design from the Netherlands/Design aus den Niederlände*. Amsterdam: Bureau Beeldende Kunst Buitenland, Ministerie CRM, 1980.

Bakker, W. *Droom van helderheid, ontwerpbureaus en modernisme in Nederland: 1960–1975*. Rotterdam, 2011.

Berg, L. van den. *Benno Premsela: Een vlucht naar voren*. Utrecht: Centraal Museum, 1996.

Bergvelt, E. "The Decorative Arts in Amsterdam 1890–1930." In *Designing Modernity: The Arts of Reform and Persuasion 1885–1945*. Miami Beach: Wolfsonian, 1995.

Bergvelt, E., F. van Burkom, and K. Gaillard, eds. *From Neorenaissance to Postmodernism: A Hundred and Twenty-Five Years of Dutch Interiors (1870–1995)*. Rotterdam, 1996.

Betsky, A., and A. Eeuwens. *False Flat: Why Dutch Design Is So Good*. London, 2004.

Blotkamp, C., et al. *De Stijl: The Formative Years 1917–1921*. Utrecht, 1982.

Bolten, J. *Het Nederlandse bankbiljet en zijn vormgeving*. Amsterdam, 1987.

Bullhorst, R., and R. Eggink. *Friso Kramer: Industrieel ontwerper*. Rotterdam, 1991.

Clarijs, J. *'t Spectrum: Moderne meubelvormgeving en naoorlogs idealisme*. Rotterdam, 2002.

Dosi Delfini, L. *The Furniture Collection of the Stedelijk Museum Amsterdam 1850–2000: From Michael Thonet to Marcel Wanders*. Amsterdam, 2004.

Duits, T., ed. *The Origin of Things*. Rotterdam: Museum Boijmans Van Beuningen, 2002.

Eek, P. H. *Piet Hein Eek 1990–2006*. Eindhoven, 2006.

Eliëns, T. M., M. Groot, and F. Leidelmeijer. *Dutch Decorative Arts (1880–1940)*. Bussum, 1997.

Ginneke, I. van. *Kho Liang Ie: Interieurarchitect; industrieel ontwerper*. Rotterdam, 1986.

Groot, M. *Vrouwen in de vormgeving 1880–1940*. Rotterdam, 2007.

Hinte, E. van. *Huibert Groenendijk: Down to Design*. Rotterdam, 2004.

———. *Wim Rietveld: Industrieel ontwerper*. Rotterdam, 1996.

Huygen, F. *Visies op Vormgeving: Het Nederlandse ontwerpen in teksten*. 2 vols. Amsterdam, 2007–8.

Huygen, F., and H. Boekraad. *Wim Crouwel: Mode en Module*. Rotterdam, 1997.

Industry & Design in the Netherlands 1850–1950, Exh. cat. Amsterdam: Stedelijk Museum, 1985.

Joris, Y. *Jewels of Mind and Mentality: Dutch Jewelry Design 1950–2000*. Exh. cat.'s-Hertogenbosch: Museum Het Kruithuis, 2000.

Junte, J. *Hands On: Dutch Design in de 21ste eeuw/in the 21st Century*. Zwolle, 2011.

Koch, A., ed. *Dichtbij klopt het hart der wereld: Nederland op de Expo 58*. Schiedam, 2008.

———. *W.H. Gispen: A Pioneer of Dutch Design*. Rotterdam, 1998.

Kras, R. *Nederlands Fabrikaat*. Utrecht, 1997.

Kuper, M. *De Stoel van Rietveld/Rietveld's Chair*. Rotterdam, 2011.

———. *Gerrit Th. Rietveld: L'oeuvre complet*. Utrecht, 1993.

Lambert, S., and M. Chanaud. *Droog*. Paris, 2008.

Langendijk, E., and M. Simon Thomas. *Dutch Art Nouveau and Art Deco Ceramics 1880–1940*. Rotterdam: Museum Boijmans Van Beuningen, 2001.

Lauwen, T. *Holland in beeld 1895–2008*. Bussum, 2007.

Liefkes, R. *Andries Copier: Glass Designer, Glass Artist*. Zwolle, 2002.

Lootsma, B. *Mentalitäten: Niederländisch Design; Prämierte Arbeiten des Designpreises Rotterdam 1993–1996/Mentalities: Dutch Design; Nominated*

Products of the Design Prize Rotterdam. Amsterdam, 1996.
Middendorp, J. *Dutch Type*. Rotterdam, 2004.
Molenkamp, M. *The Style of the State: The Visual Identity of the Dutch Government*. The Hague, 2010.
Oldewarris, H. *The Covers of Wendingen*. Rotterdam, 1995.
Powilleit, I. *How They Work: The Hidden World of Dutch Design*. Rotterdam, 2010.
Pruys, S. M. *Dingen vormen mensen: Een studie over produktie, consumptie en cultuur*. Bilthoven, 1972.
Rijk, T. de, ed. *Designers in Nederland: Een eeuw productvormgeving*. Amsterdam/Ghent, 2003.
Rock, M. "Mad Dutch Disease: The Strange Case of Dutch Design and Other Contemporary Contagions." In *Jaarboek Nederlandse Vormgeving 03/04*. Rotterdam, 2004.
Schouwenberg, L. *Hella Jongerius, Misfit*. London, 2010.
Schouwenberg, L., and G. Staal. *House of Concepts: Design Academy Eindhoven*. Amsterdam, 2008.
Simon Thomas, M. *Dutch Design: A History*. London, 2009.
———. *Jaap Gidding: Art Deco in Nederland*. Rotterdam: Museum Boijmans Van Beuningen, 2006.
Staal, G., and H. Wolters. *Holland in Vorm: Dutch Design 1945–1987*. The Hague, 1987.
Staal, G., and A. van der Zwaag. *Pastoe: 100 jaar vernieuwing in Vormgeving*, 2013.
Temmink, J. *Complete Copier: The Oeuvre of A.D. Copier 1901–1991*. Rotterdam, 2011.
Teunen, J. *Bruno Ninaber van Eyben: With Compliments*. Rotterdam, 2002.
Unger, M. *Het Nederlandse sieraad in de 20ste eeuw*. Bussum/Utrecht, 2004.
Wanders, M. *Interiors: Marcel Wanders*. London, 2011.

Belgium

Bouchez, H. "Desperately Seeking Identity: A Case-study on the Construction of Flemish Identity in Design." http://www.monthofdesign.com/2010/01/30.
Breuer, G., ed. *Vormgeving in Vlaanderen 1980–1995*. Ghent: Museum voor Sierkunst, 1995.
Buismeubelen in België-Tijdens het interbellum. Ghent: Museum voor Sierkunst, 1987.
Coirier, L. *Schoonheid: Enkelvoud-meervoud/Beauté: Singulier-pluriel/Beauty: Singular-plural*. Brussels: Koninklijke Musea voor Kunst en Geschiedenis, 2007.
Daenens, L., ed. *Forms from Flandres: From Henry van de Velde to Maarten van Severen 1900–2000*. Ghent/Amsterdam, 2001.
De Caigne, S. *Bouwen aan een nieuwe thuis: Wooncultuur in Vlaanderen tijdens het interbellum*. Leuven, 2010.
Defour, F. *Belgische meubelkunst in de XXe eeuw*. Tielt, 1979.
Design Book: Made in Belgium. Brussels: Maison des Métiers d'Art et du Patrimoine (MMAP), 2001.
Dubois, M. *Albert Van Huffel (1877–1935)*. Ghent, 1983.
Floré, F. *Lessen in goed wonen*. Leuven 2010.
Föhl, T., and S. Walter. *Leidenschaft Funktion und Schönheit: Henry van de Velde und sein Beitrag zur Europäischen Moderne*. Weimar: Neuen Museum, 2013.
Henry van de Velde Awards & Labels. Brussels: Design Vlaanderen, 1994–2012.
Jansen, E. "Creating Textiles in Belgium During the 20th Century: A Large Weaving Manufacturer, a Small Textile Studio, and a Contemporary Textile Designer." *Textile Society of America Symposium Proceedings*. Paper 178, 1998. http://digitalcommons.unl.edu/tsaconf/178.
Kooning, M., and F. Floré. *Alfred Hendrickx en het fifties-meubel in België*. Mechelen: Stedelijk Musea, 2000.
Lampaert, H., and C. Lapinne. *In Koeien van letters: 50 jaar grafische vormgeving in Vlaanderen*. Ghent: Museum voor Sierkunst en Vormgeving, 1997.
Leblanc, C. *Art nouveau & design: Les arts décoratifs de 1830 à l'Expo 58*. Brussels: Koninklijke Musea voor Kunst en Geschiedenis, 2005.
Middendorp, J. "Twee werelden. Grafisch ontwerpen in Vlaanderen en Nederland na 1945." In *Magistraal: Meesters van het Plantin Genootschap 1951–1974*. Antwerp: Museum Plantin Moretus/Prentenkabinet, 2007.
Perren, J. van der, *Huib Hoste: Architekt en meubeldesigner*. Ghent: Museum voor Sierkunst, 1980.
Severen, M. van. *Maarten van Severen Werk: Work*. Ghent, 2004.
Smets, M. *Propagateur d'une architecture renouvelée*. Brussels, 1972.
Verroken, A. *André Verroken*. Ghent, 2006.
"Vormgeving in Vlaanderen. Aspecten van Meubeldesign." Special issue, *Vlaanderen: Tweemaandelijks Tijdschrift* 50, no. 1 (2001).

INDEX OF NAMES

Numbers in *italics* refer to plates. In the main text, both designers and manufacturers are indexed. In the captions, only the designers are indexed.

Photography Credits

Numbers refer to pages. Page numbers in bold type indicate images that also appear on the slipcase.

Photo Les Arts Décoratifs, Paris: front cover, pp. 12, 90t, 91, 97t, 105, 128, 130, 133, 134, 152, 153, 401t, 406, 419b, 420, 421t, 439, 447, 454. **Photo Les Arts Décoratifs, Paris/Jean Tholance:** pp. 4, 5, 39tr, 131tr, 135r, 203, 208, 231, 234b, 240, 297tl, 298, 299, 313, 317, 320tl, 364h, 372, 379, **383**, 392tl, 396, 397, 399b, 401b, 402, 403, 410t, 411, 414–15, 417, 442, 443, 448, 485, 486, 493, 509, 519, 545t, 550, 553, 558. **Photo Les Arts Décoratifs/Béatrice Hatala:** pp. 13, 279t, 392tr, 412, 416, 484. **Photo Les Arts Décoratifs, Paris/Laurent Sully-Jaulmes:** pp. 136, 138, 257t, 329, 424t, 446. **Photo Les Arts Décoratifs/Luc Boegly:** p. 543. **Julius Shulman Photography Archive, Research Library at the Getty Research Institute, © J. Paul Getty Trust:** pp. 2, 73, 87. **Scala Archives, Florence:** pp. 8, 10, 21, 22, 25, 32, 34, 36, 47, 53, 62, 67, 69b, 75, 80, 81, 82, 83, **86b**, 99l, **102**, 103r, 106r, 109, 113, 120, 126r, 129, 137, 151b, 154, 156, 157, 159tr, **159b**, 167c, 194, 209, 211, 216, **235**, 238t, 239, 261t, 263b, 270t, 271, 272, 274t, 276, 280c, 280b, 287, 292, 293, 294, 296, 297tr, 299bl, 299br, 307br, 311, 318, 332, 344l, 374, 407, 413tr, 440, 441tr, 441b, 454, 459, 460l, 467b, 468c, 473, **478l**, 478r, 479l, 487, 489, 492, 494, 496, 497, 499, 501, 506, 507, 516, 520, 538, 540t, 549, 565, 575, back cover (Museum of Modern Art, New York); 14, 26, 58l, 60, 121, 139, 155, 161, 192l, 236, 315tl, 319, 340, 345, 348, **349tl**, 349tr, 349b, 350, 351, 352b, 353, 360, 362, 363b, 364b, 365, 366, 367, 370, 371, **467t**, 491, 500l, 515 (V&A Images/Victoria and Albert Museum, London), 16, 205 (Cooper-Hewitt, National Design Museum, Smithsonian Institution/Art Resource, NY), 31, 247 (DeAgostini Picture Library), 33t (Namur Archive), 43l, 190 (Cooper-Hewitt, National Design Museum, Smithsonian Institution/Art Resource, NY. Photo: Matt Flynn), 49 (Cooper-Hewitt, National Design Museum, Smithsonian Institution/Art Resource, NY. Photo Andrew Garn © Smithsonian Institution), 50b (Cooper-Hewitt, National Design Museum, Smithsonian Institution/Art Resource, NY. Photo Ellen McDermott © Smithsonian Institution), 59 (Yale University Art Gallery/Art Resource, NY), 103l, 483b, 537t (Philadelphia Museum of Art/Art Resource), 218b, 219, 223, 422 (Heritage Images), 463 (Museum of Modern Art, New York/© Sony), 518 (Museum Associates/LACMA/Art Resource, NY). **Photo agency of the Réunion des Musées Nationaux–Grand Palais, Paris:** pp. 6, 9, 11, 23, 28l, 37, 48, 89, 99d, 227, 233t, 270b, 280t, **289**, 309, 320br, 321, 333, 369, 387, 389, **392b**, 434t, 436, 437, 449b (Centre Pompidou, MNAM-CCI, Dist. RMN-Grand Palais/Jean Claude Planchet), 38, **284b** (Centre Pompidou, MNAM-CCI, Dist. RMN-Grand Palais/Jacques Faujour), 39tl, 39b, 90b, 308, 466 (Centre Pompidou, MNAM-CCI, Dist. RMN-Grand Palais/Bertrand Prévost), 69t, 258t, **316** (Metropolitan Museum of Art, Dist. RMN-Grand Palais), 93, **376**, 419t, 441tr, 490(Centre Pompidou, MNAM-CCI, Dist. RMN-Grand Palais/Philippe Migeat), 119 (Centre Pompidou, MNAM-CCI, Dist. RMN-Grand Palais/Georges Meguerditchian/Philippe Migeat), 210t, 284t, 286, 295, 312, 326, 373, 377, 431, 451, 468b (Centre Pompidou, MNAM-CCI, Dist. RMN-Grand Palais/Georges Meguerditchian), 221 (BPK, Berlin. Dist. RMN-Grand Palais), 423 (Ministère de la Culture–Médiathèque du Patrimoine, Dist. RMN-Grand Palais/ René-Jacques). **Musée d'Art Moderne de Saint-Etienne-Métropole/Yves Bresson:** pp. 7, 96, 118, 131tl, 151t, **210b**, 305, 395, 410b, 413tl, 413bl, **413br**, 472. **YMER&MALTA:** p. 15, 452t, 453. **Corbis Images, Paris:** p. 17, 79 (© Bruce Benedict/Transtock), 42, 55 (© Schenectady Museum; Hall of Electrical History Foundation), 56b (© Underwood & Underwood), 58r, 64l, 160 (© Condé Nast Archive), 71 (© Geneviève Naylor), 78t (© Car Culture), 78b (© Glenn Paulina/TRANSTOCK Inc/Transtock), 106l (© Adrianna Williams), 163, 195t, 228, 356, 359, 425t (© Bettmann), 220b (© Fuchs/dpa), 222 (© David LeBon/Transtock), 324, 384t (© Martyn Goddard), 357b (© Hulton-Deutsch Collection), 445 (© Adrian Wilson/Beateworks). **Rue des Archives, Paris:** pp. 27l, 77t, (© Mary Evans), 46, 65, 66, 68, 70, 74r, 100, 107, 234t, 346, 347, 464 (© The Advertising Archives), 357t, 358b, 424b (AGIP), 470 (Spaarnestad). **Leemage, Paris:** pp. 27r, **74l**, 476 (© SSPL), 57 (© Selva), 249, 322r (© Costa), 262 (© Luisa Ricciarini), 264 (© Imagebroker), 274b (© Photo-Re-Pubblic), 323 (© Farabola), 325b (© Marcello Mencarini), 444 (© Stephane Couturier/Philippe Starck/Artedia). **AKG-Images, Paris:** pp. 28r, 218t, 220t, 224, 29 (Bildarchiv Monheim), 149 (Archives CDA/St-Genès), 213 (ddrbildarchiv.de), **259** (De Agostini Pict. Lib.), 265h (MPortofolio/Electa). **Bildarchiv Foto Marburg:** p. 30. **Albers Foundation/Art Resource, NY. Photo Tim Nighswander:** p. 33b. **© 2013 Charlotte Perriand/ADAGP:** pp. 35, 394, 461. **Musée des Beaux-Arts de Montréal:** pp. 40 (Collection Liliane and David M. Stewart, Don Eric Brill/photo Denis Farley), 266 (Collection Liliane and David M. Stewart/photo Christine Guest). **Library of Congress, Washington:** p. 41. **Gabrielle Cadringher:** p. 43r. **The Bridgeman Art Library, Paris:** pp. 44 (Peter Newark American Pictures), 51, 61 (Lake County Discovery Museum/UIG), 94 (© The Advertising Archives), 115 (© Nationalmuseum, Stockholm, Sweden), 159tl, 165, 192r, 193, 229t, 229 br, 233b, 269, 315tr, 344r, 361, 363t (Photo © Christie's Images), 191 (Saint Louis Art Museum, Missouri/Museum Shop Fund), 398, 460r (Archives Charmet). **General Motors:** pp. 45, 50t, 76. **Getty Images:** pp. 56t (Three Lions), 63 (Time & Life Pictures), 197br (MTC/Kansas City Star), 217, 471, 475t (Bloomberg), 368 (Peter Macdiarmid), 465 (SSPL), 474, 477 (SIM). **Vitra Design Museum, Weil am Rhein:** pp. 64r, 84–85, 92, 140, 141, 282 (Archives Vitra Design Museum, Weil am Rhein, Germany), 95, 214, 215, 238b, 297b, 315b (photo: Thomas Dix), 230 (Collection Alexander von Vegesack, photo: Andreas Sütterlin), 512 (Collection Vitra Design Museum, Weil am Rhein, Germany). **Kharbine-Tapabor:** pp. 77b (Collection NB), 225b (Collection IM), 425b. **© 2013 Eames Office, LLC (eamesoffice.com):** p. 86t. **Knoll Inc:** pp. 88, 97b. **Roger-Viollet, Paris:** pp. 98, 409, 531 (© Bibliothèque Forney), 131b (© TopFoto), 263t (© Alinari). **Bibliothèque des Arts Décoratifs, Paris/photo Suzanne Naguy:** pp. 101, 391, 432, 433. **Centre National des Arts Plastiques/Fond National d'Art Contemporain, Paris:** pp. 104 (©Michael Graves/CNAP/photo Tobia Décoration), 162 (© D.R./CNAP/photo Adelta International), 164t (© D.R./CNAP/photo Jean Tholance, Les Arts Décoratifs, Paris), 167t (© D.R./CNAP/photo Forum Diffusion), 237, 288 (© Richard Sapper and Marco Zanuso/CNAP/photo Y. Chenot, Paris), 273 (© D.R./CNAP/photo Danese, Milan), 330 (© Archivio Achille Castiglioni, Pier Giacomo Castiglioni, Pio Manzu/CNAP/photo Bruno Scotti, Paris), 380t, 502 (© D.R./CNAP/photo Y. Chenot, Paris), **380b** (© Jerszy Seymour/CNAP/photo Y. Chenot, Paris), 450t (© ADAGP/CNAP), 504 (© D.R./CNAP), 511, 548 (© Marteen Van Severen/CNAP/photo Y. Chenot, Paris), 537b (© ADAGP, Paris/CNAP/photo Y. Chenot, Paris), **541** (© D.R./CNAP/photo: TOOLS Galerie). **Fuseproject:** pp. 108, 110, 111. **Wright:** pp. 116, 117, 158, 277. **© Archivio Fotografico La Triennale di Milano:** pp. 122, 255. **Brooklyn Museum Library Collection, New York:** pp. 123. **© Iben Kaufmann. Eyespice.dk:** pp. 124, 125, 126l. **Dansk Moebelkunst, Paris:** p. 127. **Radisson Blu Royal Hotel, Copenhagen:** p. 132. **Design Museum, Copenhagen/photo Pernille Klemp:** 135l. **Photo K. Helmer-Petersen:** p. 142. **Photo Kryn Tagonis:** p. 143. **© Bang & Olufsen:** p. 144. **Hans Sandgren Jakobsen:** p. 145. **© Lightyears:** p. 146l. **© Gubi.com:** p. 146r. **Fritz Hansen/photo Ditte Isager:** p. 147. **Luxproduction.com:** pp. 148, 150. **Fiskars/photo Cyril Villain:** p. **164b**. **Nokia:** p. 166. **© Terhi Tuominen:** p. 167b. **Saas Instrumentti Oy/photo Chikako Harada:** p. 168. **Secto Design:** p. 169. **Photo Ragnar Th. Sigurdsson:** p. 170. **© Andrés Kolbeinsson/Reykjavik Museum of Photography:** pp. 171, 174b. **Photo Vigfús Birgisson:** pp. 172, 173. **Einar þorsteinn Ásgeirsson:** p. 174t. **Photo Gudmundur Ingolfsson/Ímynd:** p. 175. **PBL Design, Reykjavik:** p. 176. **Kusch + Co:** p. 177. **© Dögg Guðmundsdóttir:** p. 178. **Katrin Olina:** p. 179. **Northernlighting/photo Damian Heinisch & Simen Skyer, Norway:** p. 180. **Nasjonalmuseet for Kunst, Arkitektur og Design, Oslo:** pp. 181, 183bl, 183r, 187 (photo Anne Hansteen Jarre). **Norks Folkemuseum, Oslo/photo Anne-Lise Reinsfelt:** p. 182. **Private collections:** pp. 183tl, 185br, 241, 245, 248, 250, 251, 252, 253, 260, 261b, 267, 268, **275**, 283, 290, 291, 306, 307t, 307bl, 310, 314, 322l, 325t, 327, 328, 331, 393, 527r. **Stokke:** pp. 185tl, 185cr. **Norsk Møbelfaglig Senter, Norway/photo Dag Brøther:** pp. 185cl, 185tr. **© Andreas Engesvik:** p. 185bl. **Johan Verde:** p. 186. **Line Böhmer Lökken, Oslo:** p. 188. **Kalle Sanner & Daniel Rybakken:** p. 189. **Volvo cars:** p. 195b. **Veryday:** p. 196. **Inter IKEA Systems B.V. 2013:** pp. 197l, 197tr. **Tacchini:** p. 198. **TEA:** p. 199t. **Asplund:** p. 199b. **Källemo AB:** p. 200. **© Iittala:** p. 201t. **Booo/photo Par Olofsson:** p. 201b. **Museum für Gestaltung Zürich, ©ZHdK:** pp. 204, 206, 207, **225t**, 226 (photo Alexander von Steiger), 232. **© Vitra (www.vitra.com):** pp. 229bl (photo Andreas Sütterlin), 561b (photo Nicole Bachmann), 563. **Nils Holger Moormann GmbH:** p. 242t. **Photo "Moustache" Stéphane Arriubergé:** p. 242b. **Photo Roland Tännler:** p. 243. **Gio Ponti Archives:** pp. 256, 257b. **Hotel Parco del Principi, Sorrento:** p. 258b. **© Tajan:** p. 265b. **Studio arch. Cini Boeri:** pp. 278, 279b. **Archives Kartell:** p. 281. **© Fabrice Gousset/Galerie Kreo:** pp. 285, 404t, 547. **Ignazia Favata/Studio Joe Colombo, Milan:** p. 300–301. **Office of Gaetano Pesce:** pp. 302, 303, 304 (photo Bruno Falchi and Liderno Salvador). **Alessi Museum:** pp. 334, 335, 337 (photo Alessandro Milani). **FLOS:** p. 336. **Imperial War Museums, London:** p. 341. **Design Council/University of Brighton Design Archives, www.brighton.ac.uk/designarchives:** pp. 342, 343. **Galerie de Casson/photo Maxime Champion:** p. 352t. **John Maltby/RIBA Library Photographs Collection:** p. 354. **Archives Conran:** p. 355. **© Xavier de Nombel:** pp. 358t, 426, 427, 429l. **Tom Dixon Studio:** p. 375. **Jasper Morrison Ltd/photo Marc Eggiman:** p. 378. **Ross Lovegrove/photo John Ross:** p. 381t. **Established & Sons/photo Peter Guenzel:** p. 381b. **Paul Cocksedge Studio (www.paulcocksedgestudio.com)/photo Richard Brine (www.richardbrine.co.uk):** p. 382t. **Philippe Joner, blacksquare.ch:** p. 382b. **©Miro Zagnoli/Ustra:** p. 384b. **Anglepoise®:** p. 385. **Ville du Havre/photo Jean Collas:** p. 390. **Jousse Entreprise/© Marc Domage:** p. 399t. **Demish Danant/photo Thierry Depagne:** p. 399c. **Archives Step Simon/Galerie Downtown François Laffanour (photo Marie Clérin):** p. 404b. **Galerie Pascal Cuisinier/Xavier Béjot:** p. 405. **Sumo/Hidac (photo Christophe Filliaux):** pp. 408, 533. **Photo Jean Pierre Leloir:** p. 418. **SNCF Médiathèque/Jean-Marc Fabbro:** p. 421b. **ELLE/SCOOP:** p. 428. **© Petit-Roulet:** p. 429r. **© Dominique Genet:** p. 430. **Photo Pierre Berdoy:** p. 434b. **Artifort, www.artifort.com:** p. 435. **Ligne Roset:** p. 438. **Photo © Fabrice Gousset:** p. 449t. **Masaki Okumura:** p. 450b. **Photo Michel Bonvin:** p. 452b. **Studio Massaud, with the scientific cooperation of ONERA:** pp. 456, 457. **© Sony:** pp. 462, 469, 503. **© Kiyoshi Soeno:** p. 475t. **Naoto Fukasawa Design Ltd:** pp. 479r, **564**. **GK-Design:** pp. 480, 481. **Noguchi Museum, NY/photo Nomadic Studio:** p. 482. **Memphis s.r.l./photo Aldo Ballo, Guido Cegani, Peter Ogilvie:** p. 500r. **© Muji:** p. 505. **Gemeentemuseum, The Hague:** p. 513. **Centraal Museum, Utrecht/photo Ernst Moritz:** p. 517. **Museum Boijmans Van Beunigen, Rotterdam:** pp. **526**, 528, **535** (photo Studio Tromp, Rotterdam), 527l, 529, 530 (photo Tom Haartsen, Ouderkerk a/d Amstel), 540b (photo Bob Goedewaagen, Rotterdam), 542. **Design Museum, Ghent:** p. 521. **Galerie Catherine Houard:** pp. 522, 523, 524, 525. **Collection Stedelijk Museum, Amsterdam:** p. 534. **Tejo Remy & René Veenhuizen/photo Bob Goedewagen:** p. 539l. **Piet Hein Eek/photo Nob Ruijgrok:** p. 539r. **Moooi:** p. 544t. **Jurgen Bey/Droog Design for Oranienbaum/photo Marsel Loermans (www.studiomakkinkbey.nl):** p. 544b. **Carpenters Workshop Gallery:** pp. 545b, 556. **Scheltens & Abbenes:** p. 546. **Photo Kris Druant:** p. 551. **Hayon Studio:** p. 554 (photo Nienke Klunder), 555. **Front Design:** p. 557. **Campana/photo © Luis Calazans:** p. 559t. **Nacho Carbonell/photo © Lilian van Rooij:** p. 559b. **Photo Patrick Gries:** p. **560t**. **Radi Designers:** p. 560b. **Photo Studio Frei:** p. 561t. **IDEO (www.ideo.com):** p. 562. **Dunne & Raby/photo Jason Evans:** p. 566. **Susana Soares:** p. 567t. **Mathieu Lehanneur/photos © Véronique Huyghe:** p. 567bl. **YikeBike:** p. 567r. **Marc Sarazin/photos Claire Lavabre:** p. 568. **Photo Roel van Tour:** p. 569. **Thomas Maincent:** p. 570. **Photo © Mauricio Freyre for Boisbuchet:** p. **571**. **Studio Jerszy Seymour/photo Peter Kainz:** pp. 572, 573.

© 2013 Adagp, Paris, for Isamu Noguchi (pp. 4, 482, 484, 485), Harry Bertoia (pp. 12, 90, 91, 97), Cédric Ragot (pp. 15, 452), Wilhelm Wagenfeld (pp. 16, 21, 205), Thomas Gerrit Rietveld (pp. 25, 37, 515, 516, 517, 530), Joost Schmidt (p. 31), Herbert Bayer (pp. 32, 33), Walter Gropius (pp. 32, 232), Pierre Jeanneret (pp. 34, 35), Charlotte Perriand (pp. 34, 35, 394, 395, 396, 461), Alvar Aalto (pp. 37, 120, 121, 148, 149, 150, 151t), Tapio Wirkkala (pp. 122, 123, 151b, 152, 153), David Lewis (p. 144), Ingeborg Lundin (p. 192r), Sixten Sason (p. 195t), Jonas Bohlin (p. 200), Anna Lindgren (pp. 201b, 557), Max Bill (p. 208), Stiletto (p. 238b), Giacomo Balla (p. 247), Fortunato Depero (pp. 248, 249), Marcello Dudovich (p. 262), Jean Prouvé (pp. 392t, **392b**, 393), Fernand Léger (p. 394), Mathieu Matégot (p. 399tr), Serge Mouille (p. 404b), Jacques Nathan-Garamond (p. 407), Roger Tallon (pp. 416, 417, 418, 419, 420, 421), Danielle Quarante (p. 432), Mattia Bonetti (pp. 446, 447), Elizabeth Garouste (pp. 446, 447), Matali Crasset (p. 450), Henry Van de Velde (p. 512), Giks Bakker (p. 537), Pucci De Rossi (pp. 553, 558), Fernando and Umberto Campana (p. 559t). **© 2013 The Josef and Anni Albers Foundation/Adagp, Paris:** p. 33b. **© 2013 FLC/Adagp, Paris:** pp. 34, 35, 396. **© 2013 Stig Lindberg/Adagp, Paris:** p. 192l.

© Verner Panton Design: front cover, pp. 14, 138, 139, 140, 141. **© Eames Office (eamesoffice.com):** pp. 53, 62, 80, 82, 83, 84, 85, 86t, **86b**, 87. **© Louis Poulsen Lighting:** pp. 119, 137. **© Stelton:** p. 134. **© 2012 Estate of Sveinn Kjarval/Myndstef:** p. 170. **© Bruno Munari/courtesy Maurizio Corraini s.r.l.:** pp. 260, 272, 273, 306. **© Ignazia Favata/Studio Joe Colombo, Milano:** pp. 280b, 298, 299, 300. **© Castelli Spa (www.castelli1877.com):** p. 281. **© Archives Pio Manzù, Bergamo, Italy:** pp. 306, 325t, 328l. **© The Robin and Lucienne Day Foundation:** pp. 350, 351, 352, 353. **© Ron Arad Associates:** pp. 339, 368, 369, 370, 371, 372, 373. **© Collection Aerospace 1968/Velvet Galerie:** p. 430. **© Philippe Starck:** pp. 440, 441, 442, 443, 444, 445, back cover. **© Maarten van Severen Foundation:** pp. 511, 548, 549.

Every effort has been made to properly credit the rights holders of the images reproduced in this book. We apologize for any unintentional errors or omissions, which will be corrected in future printings.

Author Biographies

Dominique Forest is chief curator of the modern and contemporary department at the Musée des Arts Décoratifs in Paris. She has organized numerous exhibitions on contemporary design, including *Small Talk: Konstantin Grcic dialogue avec le musée des Arts décoratifs* (2007); *Jasper Morrison: Take a Seat!* (2009); *Normal Studio: Design élémentaire* (2010); *Mobi Boom: L'explosion du design en France, 1945–1975* (2010); *Maarten Baas: Les curiosités d'un designer* (2011); *Les Frères Campana: Barroco Rococó* (2012); and *Ronan et Erwan Bouroullec: Momentané* (2013). She is the author of many books and articles on ceramics and design.

Jeremy Aynsley is professor of design history at the University of Brighton, specializing in European design, particularly Swiss and German. He is the author of *Nationalism and Internationalism in Design* (1994), *Graphic Design in Germany, 1890–1945* (2000), and *Designing Modern Germany* (2010), and has contributed to several international exhibitions of design.

Ásdís Ólafsdóttir, who holds a doctorate in the history of art, is a specialist in twentieth-century design and a noted authority on the furniture of Alvar Aalto. An exhibition curator, author, and conference organizer active in France and abroad, Ólafsdóttir is director of the Maison Louis Carré in Yvelines (designed by Alvar Aalto) and editor of the journal *ARTnord*.

Anty Pansera, a historian of Italian art and design, is a professor at the Accademia di Belle Arti in Brera and founder and president of the Associazione D come Design. Pansera has been president of the faculty of the Istituto Superiore per le Industrie Artistiche in Faenza since 2010, and a board member of the Fondazione Museo del Design of the Triennale di Milano since 2012. She is the author of many publications, including *L'Italia del design* (1986) and *Il design del mobile italiano dal 1946 a oggi* (1990).

Constance Rubini, a historian of design, teaches at the École des Arts Décoratifs in Paris. Currently director of the Musée des Arts Décoratifs in Bordeaux, Rubini has organized numerous exhibitions, including the 2010 Biennale Internationale Design Saint-Étienne, and was editor of the journal *Azimuts* from 2004 to 2010. She is the author of, most notably, *Ronan et Erwan Bouroullec, entretien* (2012) and *La Ville mobile* (2012).

Penny Sparke is professor of the history of design at Kingston University in London, where she directs the Modern Interiors Research Centre. She is the author of numerous publications on design, including *An Introduction to Design and Culture, 1900 to the Present* (1986, revised 2004), *Japanese Design* (1986), *Design in Context* (1987), and *Italian Design* (1988).

Mienke Simon Thomas, who holds a doctorate in the history of design from the Vrije Universiteit Amsterdam, is curator of decorative arts at the Museum Boijmans Van Beuningen in Rotterdam. A specialist in Dutch design, she is the author of works including *Dutch Art Nouveau and Art Deco Ceramics 1880–1940: The Boijmans Van Beuningen Collection* (2001), *Jaap Gidding: Art Deco in Nederland* (2006), and *Dutch Design: A History* (2009).

Cet ouvrage a bénéficié du soutien de La Mission Culturelle et Universitaire Française aux Etats-Unis.

This work received support from La Mission Culturelle et Universitaire Française aux Etats-Unis through its publishing assistance program.

Acknowledgments

We are most grateful to all the designers, manufacturers, gallerists, collectors, scholars, and curators who have lent their aid, testimony, or documentation to this book.

In particular, we wish to thank Alain Baillon, Jean-Louis Barrault, Jacques Barsac, Laurence Bartoletti, Stephan Baylay, Bernard Bayle, Daniel Bernard, Marc Berthier, Marianne Brabant, Rachel Brishoual, Michel Buffet, Guillaume de Casson, Carol Chabert, Edgar de Compigny, Guillemette Delaporte, Chloé Demey, Dominique Dorwling-Carter, Thomas Fritsch, Agnès de Gorter, Jean-Bernard Hebey, Françoise Jollant Kneebone, Matthias and Philippe Jousse, Chantal Lachkar, Francis Lacloche, Karine Lacquemant, François Laffanour, Thibault Lannuzel, Claire Leymonnerie, Sylvain Marcoux, Job Meilhuizen, Jean-Philippe Mercier, Sophie Motsch, Pernette Perriand, Pierre Perrigault, Isabelle de Ponfilly, Catherine Prouvé, Danielle Quarante, Dirk-Jan Rol, Pierre Romanet, Geneviève Rudolf, Thibault Seguin, Tim Vermeulen, Marieke Wiegel, and Myriam Zuber Cupissol.

We owe a debt of gratitude to Béatrice Salmon, who has supported this project from its outset.

Finally, our warmest thanks go to Marc Pillet for his encouragement, time, and scholarly advice throughout the development of this work.

Slipcase and Cover Images
Slipcase front (clockwise from top left): See plates 248, 283, 669, 475, 267, 391, 661, 308, 698, 71, 350, 470, and 345.
Slipcase back, bottom row (left to right): See plates 583, 332, and 465.
Slipcase back, second row from bottom (left to right): See plates 76, 92, and 485.
Slipcase back, images around the text (clockwise from top right): See plates 182, 691, 516, 650, 176, 709, 429, 111, and 599.
Front cover: Verner Panton, *Panton* chair, 1967. See plate 152.
Back cover: Philippe Starck, *Juicy Salif* citrus squeezer, 1988. See plate 553.

Translations from the French by Elizabeth Heard ("Introduction," "Old Continent and New World," "France"), Linda Gardiner ("The Landscape of Design Today"), Molly Stevens ("Scandinavia"), and Jane Marie Todd ("Italy")

Translation from the Dutch by Tamr Translations, London ("Belgium and the Netherlands")

For the original edition
Design: Lot 49/Mateo Baronnet
Copy editor: Suzanne Madon
Caption editor: Cécile Niesseron
Photo research: Marion Jouan
Production: Ingrid Boeringer
Prepress: Nord Compo, Villeneuve-d'Ascq

For the English-language edition
Editor: David Fabricant
Copy editor: Ashley Benning
Production manager: Louise Kurtz
Slipcase design: Misha Beletsky

First published in the United States of America in 2014 by Abbeville Press, 137 Varick Street, New York, NY 10013

First published in France in 2013 by Citadelles & Mazenod, 8, rue Gaston de Saint-Paul, 75116 Paris

 The text of this book was set in FF Tisa. Printed in China.

First edition
10 9 8 7 6 5 4 3 2 1

Library of Congress Cataloging-in-Publication Data
Art du design. English
The art of things : product design since 1945 / edited by Dominique Forest ; with Jeremy Aynsley, Ásdís Ólafsdóttir, Anty Pansera, Constance Rubini, Penny Sparke, and Mienke Simon Thomas. — First edition.
pages cm
"First published in France in 2013 by Citadelles & Mazenod, 8, rue Gaston de Saint-Paul, 75116 Paris."
Includes bibliographical references and index.
ISBN 978-0-7892-1208-5 (hardback)
1. Design—History—20th century. 2. Design—History—21st century. 3. Industrial design—History—20th century. 4. Industrial design—History—21st century. I. Forest, Dominique, editor. II. Title.
NK1390.A7813 2014
745.209'045--dc23
2014019717

For bulk and premium sales and for text adoption procedures, write to Customer Service Manager, Abbeville Press, 137 Varick Street, New York, NY 10013, or call 1-800-ARTBOOK.

Visit Abbeville Press online at www.abbeville.com.